Data-Driven Global Optimization Methods and Applications

This book presents recent advances in data-driven global optimization methods, combining theoretical foundations with real-world applications to address complex engineering optimization challenges.

This book begins with an overview of the state-of-the-art, key technologies and standard benchmark problems in the field. It then delves into several innovative approaches: space reduction-based, hybrid surrogate model-based and multi-surrogate model-based global optimization, followed by surrogate-assisted constrained global optimization, discrete global optimization and high-dimensional global optimization. These methods represent a variety of optimization techniques that excel in both optimization capability and efficiency, making them ideal choices for complex engineering optimization problems. Through benchmark test problems and real-world engineering applications, this book illustrates the practical implementation of these methods, linking established theories with cutting-edge research in industrial and engineering optimization.

This is a professional book and an academic reference, which will provide valuable insights for researchers, students, engineers and practitioners in a variety of fields, including optimization methods and algorithms, engineering design and manufacturing and artificial intelligence and machine learning.

Huachao Dong is an Associate Professor at the School of Marine Science and Technology at Northwestern Polytechnical University, China. His research includes underwater vehicle design, digital design, multidisciplinary optimization, digital twins for underwater vehicles and data-driven global optimization, with over 50 peer-reviewed papers and one book published.

Peng Wang is a Professor at the School of Marine Science and Technology at Northwestern Polytechnical University, China. His research focuses on surrogate-based design optimization, multidisciplinary design optimization, multicriteria decision-making and the design of underwater vehicles, with over 150 peer-reviewed papers and six books published.

Jinglu Li is an Assistant Researcher at Harbin Engineering University, China. His research includes underwater vehicle design, multidisciplinary optimization, digital twins and data-driven global optimization, and he has published over 20 peer-reviewed papers.

Data-Driven Global Optimization Methods and Applications

Huachao Dong, Peng Wang, and Jinglu Li

CRC Press
Taylor & Francis Group
Boca Raton London New York

CRC Press is an imprint of the
Taylor & Francis Group, an **informa** business

SCIENCE PRESS

This project is supported by the National Natural Science Foundation of China (Grant No. 52175251).

The cover image is created by Huachao Dong.

First edition published 2026
by CRC Press
2385 NW Executive Center Drive, Suite 320, Boca Raton FL 33431

and by CRC Press
4 Park Square, Milton Park, Abingdon, Oxon, OX14 4RN

CRC Press is an imprint of Taylor & Francis Group, LLC

ISBN: 978-1-041-06575-3 (hbk)
ISBN: 978-1-041-06606-4 (pbk)
ISBN: 978-1-003-63626-7 (ebk)

DOI: 10.1201/9781003636267

Typeset in Minion
by codeMantra

Contents

Figures

Tables

Preface

THE RAPID ADVANCEMENT OF computer technology and the growing demand for high-precision industrial design have established simulation as a cornerstone of modern engineering practices. In the fields of mechanical and electronic systems, many complex and computationally expensive black-box models are widely used. These models are characterized by their reliance on known input-output relationships without revealing internal operations. Expensive black-box models, such as automotive crash simulations, aerodynamic calculations for aircraft design, underwater vehicle shape optimization and structural stability analysis, often require substantial computational resources. Each simulation run may range from several minutes to several hours, and optimizing the design parameters for these models can result in prohibitively high computational costs.

To overcome these challenges, data-driven optimization techniques have emerged as a promising solution. By leveraging data and computational intelligence, these methods significantly enhance efficiency and accuracy in optimization processes, offering transformative potential for complex engineering design tasks. These techniques can be broadly categorized into offline and online data-driven optimization. Offline optimization involves generating a large dataset at the outset, constructing surrogate models with satisfactory accuracy, and keeping these models static throughout the optimization process. While this approach is straightforward and easy to implement, especially for system optimization, it has notable limitations: it lacks adaptability, heavily relies on initial sample points and often exhibits poor local approximation accuracy near the optimum, making it less suitable for global optimization tasks. In contrast, online optimization dynamically updates the database and surrogate models during the iterative process. This adaptability enhances prediction accuracy near the optimum, enabling precise solutions while significantly reducing computational costs. Online methods are particularly well-suited for scenarios demanding high accuracy and efficiency in global optimization.

Despite their advantages, existing data-driven optimization methods often rely on single-point sampling strategies, which lead to a high number of iterations and hinder parallel computation. Future developments in optimization should focus on enabling parallel execution of expensive simulations during iterative processes, highlighting the critical role of multi-point sampling strategies. Furthermore, single surrogate models, while effective for specific problem types, may exhibit significant prediction errors when applied to others. For example, polynomial response surface models are well-suited for approximating polynomial-type problems but struggle with trigonometric function-based problems. This underscores the need for hybrid surrogate modeling techniques or multi-source prediction optimization strategies that combine the strengths of different models to improve overall performance and robustness. In addition, given the inherent error tolerances in real-world manufacturing processes, solutions derived from discrete optimization often better align with actual production requirements. As such, advancing global optimization techniques tailored for discrete data-driven problems is a pressing research priority.

Given the current state of development and the challenges in this field, the authors and their research team have undertaken extensive studies in related areas. This book consolidates and presents the data-driven global optimization methods developed by the team over recent years. The content is organized into the following chapters:

- **Chapter 1** introduces the development status of advanced data-driven optimization methods.
- **Chapter 2** provides background knowledge on data-driven optimization techniques.
- **Chapter 3** presents commonly used test functions for validating data-driven optimization methods.
- **Chapter 4** introduces a multi-start space reduction method based on Kriging models.
- **Chapter 5** describes a global optimization method combining Kriging and polynomial response surface models.

- **Chapter 6** presents a hybrid global optimization method combining radial basis function and Kriging models.
- **Chapter 7** introduces a score-based multi-surrogate global optimization method.
- **Chapter 8** describes a surrogate-based constrained global optimization algorithm using space reduction.
- **Chapter 9** presents a Kriging-assisted teaching-learning-based constrained optimization method.
- **Chapter 10** describes a Kriging-assisted discrete global optimization method.
- **Chapter 11** introduces a surrogate-assisted gray wolf optimization for high-dimensional, computationally expensive black-box problems.

This research has been supported by several grants, including the National Natural Science Foundation of China (Grant No. 52175251) and the Postdoctoral Fellowship Program of CPSF under Grant Number GZC20242194. The authors express their sincere gratitude for this support. Special thanks are extended to the Autonomous Underwater Vehicle Team of Northwestern Polytechnical University, current doctoral students Xiao-Yao Han, Wenxin Wang, Weibin Ma, Yunyi Zhang and Wenyi Long as well as master's candidates Jing Pan and Jingxue Shen for their assistance in preparing this book.

Data-driven global optimization methods represent a relatively new and rapidly evolving research field. The techniques introduced in this book reflect cutting-edge developments from the past 5 years, delivering high optimization efficiency and robust performance. This book is designed as a reference for researchers and engineers involved in the design of complex electromechanical systems. To support comprehension and practical application, this book includes numerous mathematical examples and engineering case studies, making it a valuable resource for both theoretical exploration and real-world problem-solving.

Given the authors' limited expertise, errors and omissions may inevitably occur in this book. The authors welcome feedback and constructive criticism from readers to improve future editions and enhance the quality of the work.

CHAPTER 1

Introduction

1.1 OVERVIEW

The rapid advancement of computer technology and the increasing demand for high-precision industrial products have made simulation-based computation an indispensable tool in modern engineering design. In the field of mechanical and electronic engineering, there are numerous complex and costly black-box models (Miller et al., 2011; Steer et al., 2002). A black-box model is defined as a model where the input-output relationship is known, but the internal computational mechanisms remain unknown (Bunge, 1963). Costly black-box models refer to those models in which a set of inputs produces a set of outputs at the expense of significant computational resources, such as in automotive crash simulations, aerodynamic calculations for aircraft shapes, underwater vehicle design and structural stability analysis (Liebeck, 2004; Qin et al., 2004). Each simulation can take anywhere from several minutes to several hours. When designers seek a feasible set of design parameters within the design space for costly black-box models, the computational cost is typically very high. To address this issue, data-driven optimization (DDO) techniques have emerged. Since DDO typically involves the use of surrogate models, this approach is also referred to as surrogate-based optimization (SBO) in the fields of mechanical design and aerospace engineering. Figure 1.1 illustrates the simulation system of the blend-wing-body underwater glider, showcasing the computational time involved.

DOI: 10.1201/9781003636267-1

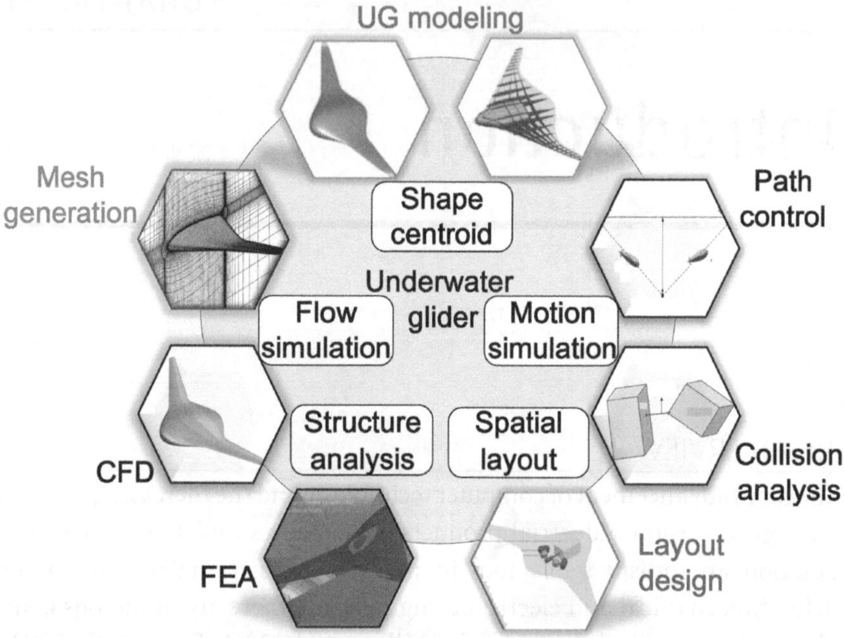

FIGURE 1.1 Simulation system of a blend-wing-body underwater glider.

Over the past two decades, computer-aided design and engineering (CAD/CAE) have experienced rapid development. Complex computational models and time-consuming simulations are frequently used to model system behavior and improve design quality. It has been reported that a single automotive crash simulation conducted by Ford can take between 36 and 160 hours (Antoine & Kroo, 2005; Gu, 2001; Gur et al., 2010; Zhang et al., 2006). Consider a two-dimensional optimization problem where 50 iterations are required, with each iteration involving one crash simulation. The total computational time would then range from 75 days to 11 months. Consequently, traditional optimization solvers become infeasible when applied to complex and time-consuming black-box models. Reducing the number of evaluations of the complex black-box model is crucial for minimizing computational costs. Traditional global optimization methods, like genetic algorithms (GA), explore the design space randomly and update the population. After hundreds or thousands of evaluations of the objective and constraint functions, an optimal solution can be found. However, the heavy

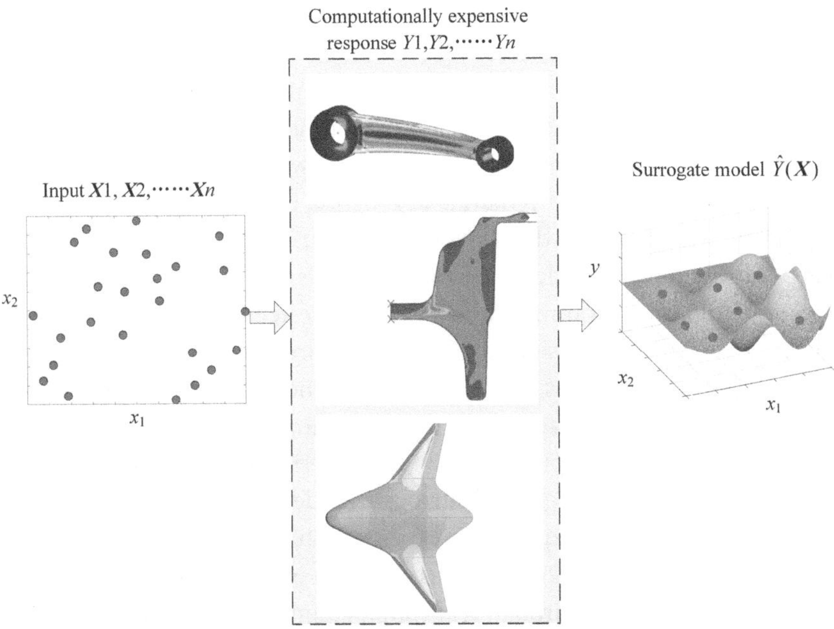

FIGURE 1.2 Application process of DACE in engineering design.

reliance on objective function analysis in methods like GA makes them unsuitable for handling computationally expensive simulation-based optimization problems.

In 1989, Sacks et al. (1989) introduced the concept of design and analysis of computer experiments (DACE). Figure 1.2 illustrates the application process of DACE in engineering design. Typically, multiple sets of computer experiments require repeated execution of computational codes, and each execution is time-consuming, which is referred to as the "expensive simulation" problem. A set of inputs undergoes expensive simulation to produce a set of outputs, which serve as responses and can form the objective or constraint functions in an optimization problem. As optimization progresses, the computational cost increases significantly with each iteration. To reduce this computational burden, the input and output values obtained from the simulation experiments are used to construct a "cheaper" surrogate model (also known as an approximation model), which replaces the original complex system and predicts the output for unknown inputs. To this day, many researchers continue to explore optimization based on surrogate models.

Wang and Shan (2006) pointed out that computationally intensive design problems are becoming increasingly common in industry, with computational loads typically arising from expensive simulation analyses or complex simulation procedures aimed at approximating real physical test results. Simpson et al. (2008) noted that over the past two decades, surrogate model techniques have achieved remarkable progress in the field of experimental design analysis. Based on the performance of surrogate models, future efforts should focus on multi-fidelity surrogate models and the feasibility of using surrogate models in commercial software. Forrester and Keane (2009) highlighted that aerospace design calculations require long runtimes and expensive computer simulations, thereby driving the need for efficient applications of surrogate models in aerospace design optimization. Younis and Dong (2010a) stated that computationally intensive simulation analyses support modern engineering design, and surrogate models can effectively reduce the number of evaluations required for expensive objective and constraint simulations. Tabatabaei et al. (2015) emphasized that obtaining objective and constraint function values through real computational experiments incurs high computational costs, such as in thermodynamic analysis, structural analysis, fluid dynamics analysis, or complex simulations involving differential equations. The basic idea to address this time-consuming issue is to build a computationally inexpensive surrogate model to replace the real experiment. Bartz-Beielstein and Zaefferer (2017) noted that SBO plays an increasingly important role in today's modeling, simulation and optimization processes. Additionally, surrogate model optimization techniques can effectively solve complex optimization problems with discrete design domains in the real world. Liu et al. (2018) pointed out that surrogate models, as a widely adopted technique, can reduce the number of time-consuming simulation calculations and adaptive surrogate model techniques, which learn from existing data and models, have gained considerable attention from researchers.

As shown in Figure 1.3, traditional optimization methods often directly link complex black-box analysis models to optimization solvers for iterative calculations. General optimization algorithms typically require numerous iterations to achieve an optimal result, and if the analysis model is an expensive black-box model, the computational burden increases significantly. For example, if a GA calls the complex black-box model 1,000 times to obtain an optimal solution, with each iteration taking 1 minute to compute the output, the total computational time would be 1,000 minutes. The substantial increase in computational load necessitates a reduction in the number of evaluations of the analysis model (Younis & Dong, 2010b).

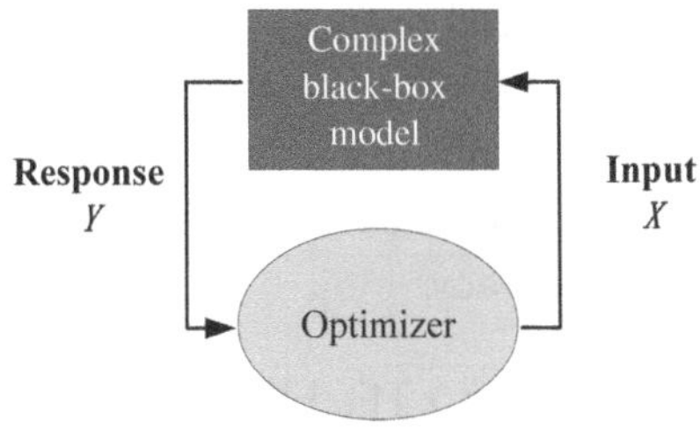

FIGURE 1.3 Direct integration of optimizer with black-box model for optimization.

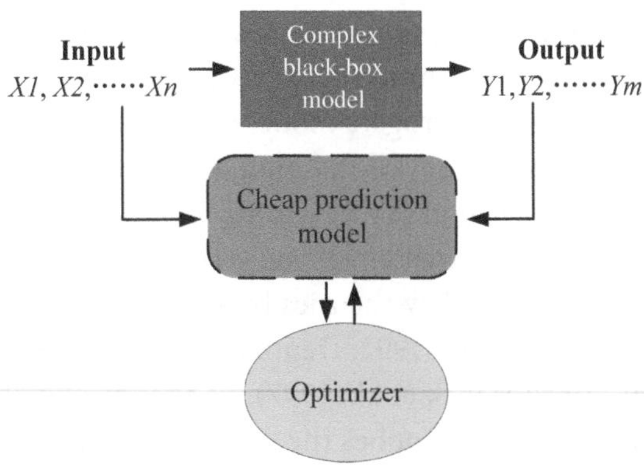

FIGURE 1.4 Optimization of complex black-box model using surrogate model and optimizer.

As shown in Figure 1.4, by systematically conducting multiple experimental analyses, multiple sets of corresponding input-output pairs can be obtained. By combining these input-output pairs, a "cheap predictive model," or surrogate model, can be constructed. Classical optimization algorithms can then be directly applied to the surrogate model to iteratively obtain an optimal solution. However, the so-called optimal solution is a predicted estimate of the "optimal solution," and its accuracy depends on the experimental analysis method and the number of tests conducted. Achieving a balance between reducing computational costs and obtaining satisfactory results requires intelligent strategies, which will be discussed in detail in the following sections.

In summary, SBO is an optimization strategy based on surrogate models. Figure 1.4 simply illustrates the general relationships between complex black-box analysis models, surrogate models, optimization solvers and their respective inputs and outputs. However, to obtain an optimal solution to practical problems, the surrogate model needs to be updated iteratively to improve its predictive accuracy adaptively. Additionally, algorithms need to intelligently select the best predictive results to balance computational cost and accuracy.

1.2 APPLICATION OF DDO TECHNIQUES IN SIMULATION SYSTEMS

With the development of simulation technologies and the increasing complexity of modern product designs, simulation analysis has been frequently applied in system design and optimization, providing precise analysis but also resulting in high computational costs. Consequently, DDO techniques have become a key solution for optimizing time-consuming simulation systems. Common surrogate model methods used in DDO include polynomial response surfaces (PRS), Kriging, radial basis functions (RBF) and support vector regression (SVR) (Haftka et al., 2016).

NASA funded early research on response surface methods, which led to the development of several key theories based on response surface methodology (RSM) (Cox & John, 1992; Dennis & Torczon, 1997; Giunta et al., 1997; Otto et al., 1997; Wujek et al., 1997). RSM typically utilizes polynomials as basis functions and applies the least squares method to construct a predictive model (Box & Wilson, 2018). Virginia Tech developed a variable complexity response surface modeling (VCRSM) approach (Giunta et al., 1996), which uses information of varying fidelity to reduce the design space, supplementing expensive samples only in the regions most likely to contain the optimal solution, thus reducing the computational cost. The University of Notre Dame developed a concurrent subspace optimization (CSSO) method (Renaud & Gabriele, 1991; Renaud & Gabriele, 1994; Wujek et al., 1996) and applied it to multidisciplinary design optimization (MDO) to coordinate the optimization of various subspaces. Haftka et al. (1998) and Hardy (1971) also conducted extensive research on RSM in mechanical and aerospace engineering.

In the past decade, most researchers have shifted their focus from PRS methods to a variety of surrogate model techniques, including RBF (Dyn et al., 1986), Kriging (Cressie, 1988), SVR (Smola & lkopf, 2004) and artificial neural networks (ANNs) (Paliwal & Kumar, 2009). Numerous scholars both domestically and internationally have proposed optimization methods based on these surrogate models and applied them to engineering design fields. In aerospace engineering, SBO has been used for designing high-speed civil transport aircraft (Booker et al., 1998), wing shape optimization (Rai & Madavan, 2000), diffuser shape optimization (Madsen et al., 2000) and supersonic turbines (Papila et al., 2002). Iuliano and Pérez (2016) proposed an SVR-based SBO method to optimize aerodynamic shapes. This method combines evolutionary algorithms (EA) with an intelligent estimation search with sequential learning (IES-SL) sampling strategy to efficiently explore the design space. The surrogate model constructed by SVR replaces computational fluid dynamics (CFD) to calculate the objective function values, ultimately achieving the globally optimal aerodynamic shape while reducing computational costs. Iuliano and Pérez (2016) introduced a surrogate model method that implements proper orthogonal decomposition (POD) of aerodynamic flow fields and reconstructs aerodynamic flow fields at unknown design points using RBF. Additionally, to achieve global optimization, this method was coupled with EA and two sampling strategies based on goal enhancement and prediction error reduction were proposed. As a result, only 100 CFD calls were needed to obtain the global optimal solution. Ulaganathan and Asproulis (2013) argued that a key challenge in the development of aerospace systems lies in understanding system behavior. While high-precision computations provide valuable insights for high-specification designs and enhanced understanding of system responses, their high computational cost limits their application across the entire system. They suggested a surrogate-based analysis (SBA) method based on Kriging and Hammersley sequence sampling for accurate aerodynamic predictions, which was combined with a GA for global optimization on the surrogate model. This approach achieved satisfactory aerodynamic efficiency while significantly reducing computational costs. Glaz et al. (2008) compared the prediction accuracy of Kriging, RBF and RSM surrogate models in helicopter vibration problems. They did not focus on how to search the design space to capture the global optimum, but rather on the adaptability of the surrogate model methods to vibration reduction problems. They ultimately found that Kriging provided the best average accuracy for this problem.

Based on the surrogate model methods, SBA techniques have gradually been applied in engineering design. Today, due to their powerful predictive capabilities, SBA has expanded into fields such as structural design, aerodynamic shape design, multidisciplinary optimization design and electronic system simulation design. Leading research institutions, including Virginia Tech, the University of Notre Dame, Rensselaer Polytechnic

Institute, Old Dominion University and NASA Langley Research Center, have been at the forefront of developing SBA to address optimization design problems in engineering (Balabanov & Venter, 2004; Schmit & Farshi, 1974; Stanford et al., 2013; Sun et al., 2011; Yamazaki, 2012; Yamazaki & Mavriplis, 2013). Take finite element analysis (FEA) as an example, which is commonly used for structural simulation design. Directly coupling FEA with general optimization solvers to find an optimal solution can lead to high computational costs. An earlier approach involved constructing an approximate empirical formula using first-order sensitivity analysis (Sacks et al., 1989), with the optimization process sequentially executed on this formula. Pedersen (1981) employed sequential linear programming (SLP) to solve structural optimization problems; Fleury and Braibant (1986) proposed the convex linearization method (CONLIN); and Svanberg (1987) introduced the method of moving asymptotes (MMA). These methods extracted the response and first-order sensitivity information from the current design point, and therefore, they are collectively referred to as single-point approximation methods. Later, Haftka et al. (1987) and Fadel et al. (1990) developed a two-point approximation method using both the current and previous points' values and derivative information. Rasmussen (1990) further proposed an accumulated approximation technique that utilizes the values and gradients at the current point while also incorporating all previously obtained points' values and derivatives. Finally, Toropov (1989) summarized the concept of multi-point approximations (MA), where regression analysis is used to predict the response at the current point in each iteration. By leveraging information from previous solutions, optimization is carried out within a locally valid sub-region to reduce the number of FEA evaluations.

Besides, DDO methods also have significant potential in system optimization design, particularly in reducing the number of calls to time-consuming simulation units. For example, Mohammad Zadeh and Sadat Shirazi (2017) employed a two-layer multidisciplinary optimization method to design a complex satellite system, replacing time-consuming simulation units with quadratic response surface (QRS) models that meet accuracy requirements, thus reducing the number of calls. Similarly, Wang et al. (2017b) proposed a novel system optimization method for lithium-ion battery thermal management system design, where surrogate models replace costly responses such as temperature and pressure variations, greatly improving computational efficiency. Wang et al. (2017b) introduced an improved collaborative optimization algorithm for automotive structural design. By constructing QRS models, they effectively reduced the computational load caused by FEA, yielding satisfactory results. Although these methods reduce the computational costs of optimizing time-consuming simulation systems to varying degrees, they all employ offline DDO techniques. Specifically, a surrogate model is constructed using a large number of samples that meet accuracy requirements for optimization, but the surrogate model is not updated during the optimization process. While this approach is simple to implement and easy to apply in system optimization, it lacks adaptability, heavily relies on initial sample points and does not provide high local approximation accuracy at optimal locations, making it unsuitable for global optimization.

Online DDO methods, on the other hand, involve using a sampling strategy during the iteration process to collect samples and automatically update the surrogate model. This dynamic process typically improves the prediction accuracy near the optimal location, allowing for precise optimal solutions with fewer computational costs. Recent studies have increasingly applied online DDO methods in system optimization workflows. For instance, Ollar et al. (2017) optimized the overall design of a wing anti-collision system by constructing Kriging models for two time-consuming analysis units—linear static and explicit dynamics. The entire optimization process was carried out using a local trust region method, with Kriging models continuously updated during iterations, ultimately determining the optimal solution with fewer computational costs. Pires et al. (2013) employed an RBF-based EA to minimize the total cost of a complex thermal system, constructing an RBF model for the time-consuming objective. In each iteration, the EA searches for the optimal sample predicted by the RBF model and iteratively updates it until a satisfactory solution is found. Yao et al. (2012) proposed a new method combining multidisciplinary feasibility and collaborative subspace optimization strategies. This method approximates time-consuming state variables, objectives and constraints using surrogate models and updates the surrogate model by supplementing the dataset with predicted optimal solutions obtained during each optimization step, facilitating rapid identification of the real optimal target. While these online DDO methods can focus samples in regions predicted to be optimal, they struggle with handling large-scale, highly nonlinear simulation systems. To achieve global optimization, more intelligent sampling strategies are required to adaptively balance the "exploitation of surrogate models" and "effective exploration of the design space" (Liu et al., 2018).

1.3 DEVELOPMENT OF DATA-DRIVEN GLOBAL OPTIMIZATION TECHNIQUES

A significant amount of research has been conducted by scholars on data-driven global optimization (DDGO). Jones et al. (1998) first proposed the efficient global optimization (EGO) algorithm, which constructs an expected improvement function using Kriging and updates the sample points by maximizing this function. Regis and Shoemaker (2007) introduced a stochastic response surface method that simultaneously considers space-filling and the prediction of optimal values to select candidate points for supplementation. In recent years, domestic scholars have also carried out extensive research on DDGO. For example, Long et al. (2015) used a space intelligence exploration strategy to accelerate the convergence speed of adaptive response surface optimization, which was validated through various test functions and wing plate structural design. Jie et al. (2015) proposed a multi-surrogate global optimization algorithm that constructed a new model combining Kriging and RBF, adjusting internal parameters adaptively to balance global and local exploration. Gu et al. (2012) developed a hybrid adaptive optimization method using three surrogate models, which divided candidate points into several subsets and selected a different number of samples for updating the surrogate model based on the importance of each subset, applied to an automotive crash example.

Haftka et al. (2016) from the University of Florida pointed out that improving the multi-point sampling capability (parallelism) is crucial for DDGO. Collecting multiple sample points in each iteration and performing simulation analyses in parallel can significantly shorten the design cycle. Both domestic and international teams have since researched multi-point sampling techniques for DDGO and published new methods. For instance, the Shoemaker team at Cornell University (Krityakierne et al., 2016) employed a non-dominated sorting method to find supplemental sample points for single-objective optimization problems; Zhan et al. (2017) from Huazhong University of Science and Technology captured multiple extreme points of the expected improvement function as supplemental sample sets; Li et al. (2016) from Dalian University of Technology proposed a new domain decomposition technique to enhance multi-point sampling capabilities based on the EGO algorithm.

Most existing DDO methods use single-point sampling strategies, such as the classic expected improvement (EI) or minimize prediction (MP). These sampling strategies often lead to numerous iterations during optimization, which is not conducive to parallel computation. As Professor Haftka mentioned, the future development of optimization should involve parallel execution of expensive simulations during iterations, making the development of multi-point sampling strategies particularly important. Additionally, a single surrogate model may perform well for certain problems but produce large prediction errors for others. For example, PRS models can provide accurate approximations for polynomial-type problems but may struggle with precise expressions for problems involving trigonometric functions. Therefore, developing hybrid surrogate model optimization methods or multi-source prediction optimization techniques can lead to more robust results. Furthermore, considering the error precision in real-world structural manufacturing processes, the optimal solution obtained from discrete optimization is often more consistent with actual production conditions. Thus, developing discrete DDGO techniques is also of significant importance.

1.4 CHAPTER SUMMARY

This chapter provides an overview of advanced DDO methods, highlighting the historical development of DDO techniques and their application in practical simulation systems. It demonstrates the significant advantages of DDO approaches in addressing computationally expensive black-box problems. These methods effectively learn from and mine historical data, construct surrogate models, predict potentially beneficial samples, accelerate the exploration of design space and greatly reduce the number of calls to time-consuming simulation models, thus holding significant implications for simulation-based product design and optimization.

REFERENCES

Antoine, N. E., & Kroo, I. M. (2005). Framework for Aircraft Conceptual Design and Environmental Performance Studies. *AIAA Journal*, *43*(10), 2100–2109.

Balabanov, V., & Venter, G. (2004). Multi-fidelity optimization with high-fidelity analysis and low-fidelity gradients. *10th AIAA/ISSMO Multidisciplinary Analysis and Optimization Conference*, Albany, New York, USA, August 30 to September 01, 2004.

Bartz-Beielstein, T., & Zaefferer, M. (2017). Model-based methods for continuous and discrete global optimization. *Applied Soft Computing*, *55*, 154–167.

Booker, A. J., Dennis Jr, J., Frank, P. D., Serafini, D. B., & Torczon, V. (1998). Optimization using surrogate objectives on a helicopter test example. *Computational Methods for Optimal Design and Control: Proceedings of the AFOSR Workshop on Optimal Design and Control*, Arlington, Virginia, USA, September 30 to October 3, 1997.

Box, G. E. P., & Wilson, K. B. (2018). On the Experimental Attainment of Optimum Conditions. *Journal of the Royal Statistical Society: Series B (Methodological)*, *13*(1), 1–38. https://doi.org/10.1111/j.2517-6161.1951.tb00067.x

Bunge, M. (1963). A General Black Box Theory. *Philosophy of Science*, *30*(4), 346–358.

Cox, D. D., & John, S. (1992). A statistical method for global optimization. *[Proceedings] 1992 IEEE International Conference on Systems, Man, and Cybernetics*, Chicago, Illinois, USA, October 18–21, 1992.

Cressie, N. (1988). Spatial Prediction and Ordinary Kriging. *Mathematical Geology*, *20*, 405–421.

Dennis, J., & Torczon, V. (1997). Managing Approximation Models in Optimization. *Multidisciplinary Design Optimization: State-of-the-Art*, *5*, 330–347.

Dyn, N., Levin, D., & Rippa, S. (1986). Numerical Procedures for Surface Fitting of Scattered Data by Radial Functions. *SIAM Journal on Scientific and Statistical Computing*, *7*(2), 639–659. https://doi.org/10.1137/0907043

Fadel, G. M., Riley, M. F., & Barthelemy, J. M. (1990). Two Point Exponential Approximation Method for Structural Optimization. *Structural Optimization*, *2*(2), 117–124. https://doi.org/10.1007/BF01745459

Fleury, C., & Braibant, V. (1986). Structural Optimization: A New Dual Method Using Mixed Variables. *International Journal for Numerical Methods in Engineering*, *23*(3), 409–428. https://doi.org/10.1002/nme.1620230307

Forrester, A. I. J., & Keane, A. J. (2009). Recent Advances in Surrogate-Based Optimization. *Progress in Aerospace Sciences*, *45*(1), 50–79. https://doi.org/10.1016/j.paerosci.2008.11.001

Giunta, A. A., Balabanov, V., Haim, D., Grossman, B. M., Mason, W. H., Watson, L. T., & Haftka, R. T. (1996). Wing design for a high-speed civil transport using a design of experiments methodology. *6th Symposium on Multidisciplinary Analysis and Optimization*, Bellevue, Washington, USA, September 4–6, 1996.

Giunta, A. A., Balabanov, V., Kaufman, M., Burgee, S., Grossman, B., Haftka, R., Mason, W., & Watson, L. (1997). Variable-Complexity Response Surface Design of an HSCT Configuration. In N. M. Alexandrov & M. Y. Hussaini (Eds.), *Multidisciplinary design optimization* (pp. 53–69). Springer.

Glaz, B., Friedmann, P. P., & Liu, L. (2008). Surrogate Based Optimization of Helicopter Rotor Blades for Vibration Reduction in Forward Flight. *Structural and Multidisciplinary Optimization*, *35*, 341–363.

Gu, J., Li, G. Y., & Dong, Z. (2012). Hybrid and Adaptive Meta-Model-Based Global Optimization. *Engineering Optimization*, *44*(1), 87–104. https://doi.org/10.1080/0305215x.2011.564768

Gu, L. (2001). A comparison of polynomial based regression models in vehicle safety analysis. *International Design Engineering Technical Conferences and Computers and Information in Engineering Conference*, Anaheim, California, USA, August 17–20, 2025.

Gur, O., Bhatia, M., Schetz, J. A., Mason, W. H., Kapania, R. K., & Mavris, D. N. (2010). Design Optimization of a Truss-Braced-Wing Transonic Transport Aircraft. *Journal of Aircraft*, *47*(6), 1907–1917. https://doi.org/10.2514/1.47546

Haftka, R. T., Nachlas, J. A., Watson, L. T., Rizzo, T., & Desai, R. (1987). Two-Point Constraint Approximation in Structural Optimization. *Computer Methods in Applied Mechanics and Engineering*, *60*(3), 289–301. https://doi.org/10.1016/0045-7825(87)90136-8

Haftka, R. T., Scott, E. P., & Cruz, J. R. (1998). Optimization and Experiments: A Survey. *Applied Mechanics Reviews*, *51*(7), 435–448.

Haftka, R. T., Villanueva, D., & Chaudhuri, A. (2016). Parallel Surrogate-Assisted Global Optimization with Expensive Functions – A Survey. *Structural and Multidisciplinary Optimization*, *54*(1), 3–13. https://doi.org/10.1007/s00158-016-1432-3

Hardy, R. L. (1971). Multiquadric Equations of Topography and Other Irregular Surfaces. *Journal of Geophysical Research*, *76*(8), 1905–1915.

Iuliano, E., & Pérez, E. A. (2016). *Application of surrogate-based global optimization to aerodynamic design* (Vol. 22). Springer.

Jie, H., Wu, Y., & Ding, J. (2015). An Adaptive Metamodel-Based Global Optimization Algorithm for Black-Box Type Problems. *Engineering Optimization*, *47*(11), 1459–1480.

Jones, D. R., Schonlau, M., & Welch, W. J. (1998). Efficient Global Optimization of Expensive Black-Box Functions. *Journal of Global Optimization*, *13*(4), 455–492. https://doi.org/10.1023/A:1008306431147

Krityakierne, T., Akhtar, T., & Shoemaker, C. A. (2016). SOP: Parallel Surrogate Global Optimization with Pareto Center Selection for Computationally Expensive Single Objective Problems. *Journal of Global Optimization*, *66*, 417–437.

Li, Z., Ruan, S., Gu, J., Wang, X., & Shen, C. (2016). Investigation on Parallel Algorithms in Efficient Global Optimization Based on Multiple Points Infill Criterion and Domain Decomposition. *Structural and Multidisciplinary Optimization*, *54*, 747–773.

Liebeck, R. H. (2004). Design of the Blended Wing Body Subsonic Transport. *Journal of Aircraft*, *41*(1), 10–25. https://doi.org/10.2514/1.9084

Liu, H., Ong, Y.-S., & Cai, J. (2018). A Survey of Adaptive Sampling for Global Metamodeling in Support of Simulation-Based Complex Engineering Design. *Structural and Multidisciplinary Optimization*, *57*, 393–416.

Long, T., Wu, D., Guo, X., Wang, G. G., & Liu, L. (2015). Efficient Adaptive Response Surface Method Using Intelligent Space Exploration Strategy. *Structural and Multidisciplinary Optimization*, *51*(6), 1335–1362. https://doi.org/10.1007/s00158-014-1219-3

Madsen, J. I., Shyy, W., & Haftka, R. T. (2000). Response Surface Techniques for Diffuser Shape Optimization. *AIAA Journal*, *38*(9), 1512–1518. https://doi.org/10.2514/2.1160

Miller, W., Smith, C. W., & Evans, K. E. (2011). Honeycomb Cores with Enhanced Buckling Strength. *Composite Structures*, *93*(3), 1072–1077. https://doi.org/10.1016/j.compstruct.2010.09.021

Mohammad Zadeh, P., & Sadat Shirazi, M. (2017). Multidisciplinary Design Optimization Architecture to Concurrent Design of Satellite Systems. *Proceedings of the Institution of Mechanical Engineers, Part G: Journal of Aerospace Engineering*, *231*(10), 1898–1916.

Ollar, J., Jones, R., & Toropov, V. (2017). Sub-space metamodel-based multidisciplinary optimization of an aircraft wing subjected to bird strike. *58th AIAA/ASCE/AHS/ASC Structures, Structural Dynamics, and Materials Conference*, Grapevine, Texas, USA, January 9–13, 2017.

Otto, J., Paraschivoiu, M., Yesilyurt, S., & Patera, A. T. (1997). Bayesian-Validated Computer-Simulation Surrogates for Optimization and Design: Error Estimates and Applications. *Mathematics and Computers in Simulation*, *44* (4), 347–367.

Paliwal, M., & Kumar, U. A. (2009). Neural Networks and Statistical Techniques: A Review of Applications. *Expert Systems with Applications*, *36*(1), 2–17. https://doi.org/10.1016/j.eswa.2007.10.005

Papila, N., Shyy, W., Griffin, L., & Dorney, D. J. (2002). Shape Optimization of Supersonic Turbines Using Global Approximation Methods. *Journal of Propulsion and Power*, *18*(3), 509–518. https://doi.org/10.2514/2.5991

Pedersen, P. (1981). The integrated approach of FEM-SLP for solving problems of optimal design. In E. J. Haug & J. Gea (Eds.), *Optimization of distributed parameter structures* (Vol. 1, pp. 757–780). Sijthoff and Noordhoff.

Pires, T. S., Cruz, M. E., & Colaço, M. J. (2013). Response Surface Method Applied to the Thermoeconomic Optimization of a Complex Cogeneration System Modeled in a Process Simulator. *Energy*, *52*, 44–54.

Qin, N., Vavalle, A., Le Moigne, A., Laban, M., Hackett, K., & Weinerfelt, P. (2004). Aerodynamic Considerations of Blended Wing Body Aircraft. *Progress in Aerospace Sciences*, *40*(6), 321–343. https://doi.org/10.1016/j.paerosci.2004.08.001

Rai, M. M., & Madavan, N. K. (2000). Aerodynamic Design Using Neural Networks. *AIAA Journal*, *38*(1), 173–182.

Rasmussen, J. (1990). Accumulated approximation: a new method for structural optimization by iterative improvement. *NASA. Langley Research Center, The Third Air Force (NASA Symposium on Recent Advances in Multidisciplinary Analysis and Optimization*, San Francisco, California, USA, September 24–26, 1990.

Regis, R. G., & Shoemaker, C. A. (2007). A Stochastic Radial Basis Function Method for the Global Optimization of Expensive Functions. *Informs Journal on Computing*, *19*(4), 497–509.

Renaud, J., & Gabriele, G. (1991). Sequential global approximation in nonhierarchic system decomposition and optimization. *International Design Engineering Technical Conferences and Computers and Information in Engineering Conference*, Anaheim, California, USA, August 17–20, 2025.

Renaud, J. E., & Gabriele, G. A. (1994). Approximation in Nonhierarchic System Optimization. *AIAA Journal*, *32*(1), 198–205. https://doi.org/10.2514/3.11967

Sacks, J., Welch, W. J., Mitchell, T. J., & Wynn, H. P. (1989). Design and Analysis of Computer Experiments. *Statistical Science*, *4*(4), 409–423.

Schmit, L. A., & Farshi, B. (1974). Some Approximation Concepts for Structural Synthesis. *AIAA Journal*, *12*(5), 692–699. https://doi.org/10.2514/3.49321

Simpson, T., Toropov, V., Balabanov, V., & Viana, F. (2008). Design and analysis of computer experiments in multidisciplinary design optimization: a review of how far we have come-or not. *12th AIAA/ISSMO Multidisciplinary Analysis and Optimization Conference*, Victoria, British Columbia, Canada, September 10–12, 2008.

Smola, A. J., & lkopf, B. S. (2004). A Tutorial on Support Vector Regression. *Statistics and Computing, 14*(3), 199–222.

Stanford, B., Beran, P., & Kobayashi, M. (2013). Simultaneous Topology Optimization of Membrane Wings and Their Compliant Flapping Mechanisms. *AIAA Journal, 51*(6), 1431–1441. https://doi.org/10.2514/1.J052118

Steer, M. B., Bandler, J. W., & Snowden, C. M. (2002). Computer-Aided Design of RF and Microwave Circuits and Systems. *IEEE Transactions on Microwave Theory and Techniques, 50*(3), 996–1005. https://doi.org/10.1109/22.989983

Sun, G., Li, G., Zhou, S., Xu, W., Yang, X., & Li, Q. (2011). Multi-Fidelity Optimization for Sheet Metal Forming Process. *Structural and Multidisciplinary Optimization, 44*(1), 111–124. https://doi.org/10.1007/s00158-010-0596-5

Svanberg, K. (1987). The Method of Moving Asymptotes—A New Method for Structural Optimization. *International Journal for Numerical Methods in Engineering, 24*(2), 359–373. https://doi.org/10.1002/nme.1620240207

Tabatabaei, M., Hakanen, J., Hartikainen, M., Miettinen, K., & Sindhya, K. (2015). A Survey on Handling Computationally Expensive Multiobjective Optimization Problems Using Surrogates: Non-nature Inspired Methods. *Structural and Multidisciplinary Optimization, 52*, 1–25.

Toropov, V. V. (1989). Simulation Approach to Structural Optimization. *Structural Optimization, 1*, 37–46.

Ulaganathan, S., & Asproulis, N. (2013). Surrogate Models for Aerodynamic Shape Optimisation. In S. Koziel & L. Leifsson (Eds.), *Surrogate-based modeling and optimization: Applications in engineering* (pp. 285–312). Springer. https://doi.org/10.1007/978-1-4614-7551-4_12

Wang, G. G., & Shan, S. (2006). Review of Metamodeling Techniques in Support of Engineering Design Optimization. *Journal of Mechanical Design, 129*(4), 370–380. https://doi.org/10.1115/1.2429697

Wang, W., Gao, F., Cheng, Y., & Lin, C. (2017a). Multidisciplinary Design Optimization for Front Structure of an Electric Car Body-in-White Based on Improved Collaborative Optimization Method. *International Journal of Automotive Technology, 18*, 1007–1015.

Wang, X., Li, M., Liu, Y., Sun, W., Song, X., & Zhang, J. (2017b). Surrogate Based Multidisciplinary Design Optimization of Lithium-Ion Battery Thermal Management System in Electric Vehicles. *Structural and Multidisciplinary Optimization, 56*, 1555–1570.

Wujek, B., Renaud, J. E., & Batill, S. (1997). A Concurrent Engineering Approach for Multidisciplinary Design in a Distributed Computing Environment. In N. M. Alexandrov & M. Yousuff Hussaini (Eds.), *Multidisciplinary design optimization: state of the art* (pp. 189–208). Society for Industrial and Applied Mathematics.

Wujek, B. A., Renaud, J. E., Batill, S. M., & Brockman, J. B. (1996). Concurrent Subspace Optimization Using Design Variable Sharing in a Distributed Computing Environment. *Concurrent Engineering, 4*(4), 361–377.

Yamazaki, W. (2012). Efficient robust design optimization by variable fidelity kriging model. *53rd AIAA/ASME/ASCE/AHS/ASC Structures, Structural Dynamics and Materials Conference 20th AIAA/ASME/AHS Adaptive Structures Conference 14th AIAA*, Honolulu, Hawaii, USA, April 23–26, 2012.

Yamazaki, W., & Mavriplis, D. J. (2013). Derivative-Enhanced Variable Fidelity Surrogate Modeling for Aerodynamic Functions. *AIAA Journal*, *51*(1), 126–137. https://doi.org/10.2514/1.J051633

Yao, W., Chen, X., Ouyang, Q., & Van Tooren, M. (2012). A Surrogate Based Multistage-Multilevel Optimization Procedure for Multidisciplinary Design Optimization. *Structural and Multidisciplinary Optimization*, *45*, 559–574.

Younis, A., & Dong, Z. (2010a). Metamodelling and Search Using Space Exploration and Unimodal Region Elimination for Design Optimization. *Engineering Optimization*, *42*(6), 517–533.

Younis, A., & Dong, Z. (2010b). Trends, Features, and Tests of Common and Recently Introduced Global Optimization Methods. *Engineering Optimization*, *42*(8), 691–718. https://doi.org/10.1080/03052150903386674

Zhan, D., Qian, J., & Cheng, Y. (2017). Balancing Global and Local Search in Parallel Efficient Global Optimization Algorithms. *Journal of Global Optimization*, *67*, 873–892.

Zhang, Y., Zhu, P., Chen, G. L., & Lin, Z. Q. (2006). Study on Structural Lightweight Design of Automotive Front Side Rail Based on Response Surface Method. *Journal of Mechanical Design*, *129*(5), 553–557. https://doi.org/10.1115/1.2712223

CHAPTER 2

Data-Driven Optimization Framework

2.1 SAMPLING METHODS

Data-driven optimization (DDO) begins with the use of experimental design methods (design of experiment, DOE) to perform initial data sampling. DOE is a mathematical and statistical approach for planning and analyzing experiments (Myers et al., 2016), primarily aimed at obtaining ideal experimental results with a minimal number of experiments, shorter experimental duration and lower costs.

2.1.1 Traditional Design of Experiment Methods

Traditional DOE methods include full factorial design, fractional factorial design, central composite design (CCD) (Chen, 1995) and Box–Behnken design (BBD) (Box & Behnken, 1960). Full factorial design considers all possible combinations of design factors and levels. Here, factors refer to design parameters or variables, while levels represent specific values assigned to a given factor within the design space. The main advantage of full factorial design is its ability to provide comprehensive information, allowing for a robust estimation of both the main effects of design variables on the response and the interaction effects between variables. However, the primary drawback is the substantial increase in the number of required experiments, which results in higher labor and resource consumption. The goal of fractional factorial design is to select a subset of valuable information from the full factorial design, making

DOI: 10.1201/9781003636267-2

the experiment more efficient. A fractional factorial design can be viewed as a subset of a full factorial experiment.

BBD was proposed by George in 1960 and is primarily applied in PRS design. BBD is a standalone second-order design that does not include embedded fractional factorial designs. It selects the midpoint of the boundaries of the design space as well as the center point of the entire design, typically choosing three design levels for each dimension. BBD is particularly useful for problems where design variables have a nonlinear relationship with the response values. Similarly, CCD is also applied to nonlinear problems and is mainly used in PRS design. However, CCD typically requires the inclusion of axial points. Figures 2.1 and 2.2 illustrate the sampling methods of BBD and CCD in three-dimensional space, showing that both DOE methods provide good coverage of the entire design space.

The GS method is similar to the previously described full factorial design. GS divides each dimension of the design space into several equal parts, and all grid points obtained by intersecting the divisions across

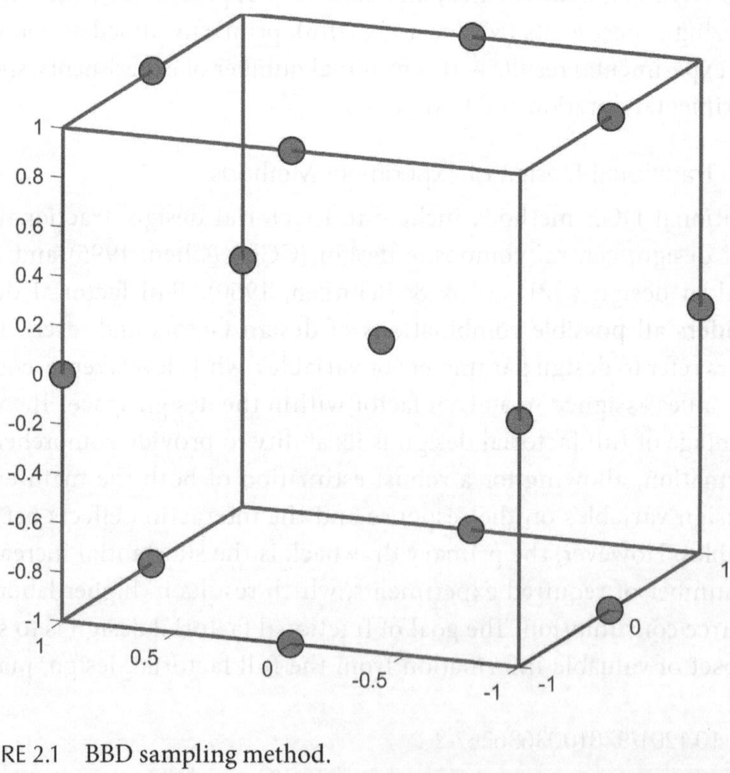

FIGURE 2.1 BBD sampling method.

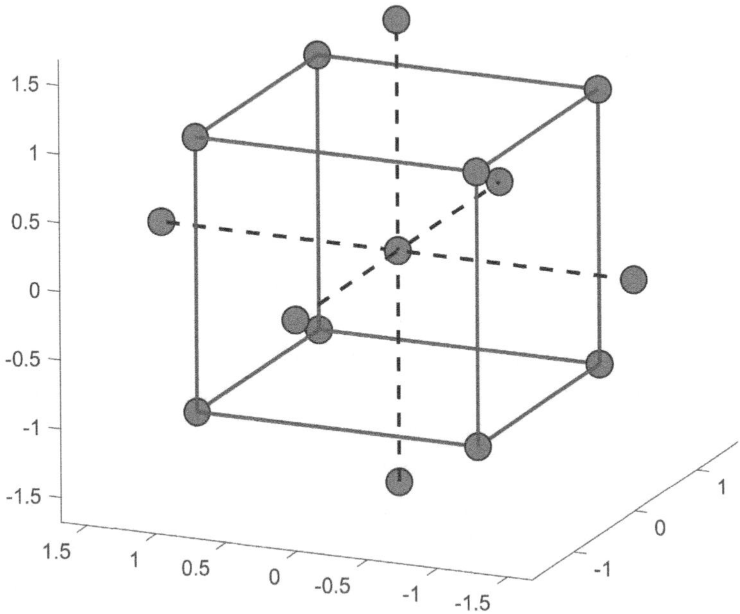

FIGURE 2.2 CCD sampling method.

dimensions are considered as design points. It is important to note that at least two nodes are selected for each dimension. Equation (2.1) provides the relationship between the number of design points m and the dimensionality n, where $q(i)$ denotes the number of design nodes in the i-th dimension.

$$m = \prod_{i=1}^{n} q(i) \tag{2.1}$$

There are various experimental design methods, and choosing an appropriate one typically involves considering the following factors: (1) the cost of a single experiment, (2) the size of the design space, and (3) the type of surrogate model the designer needs to construct.

If the experimental cost is high, it is preferable to choose a DOE strategy that generates fewer sample points. If the experimental cost is relatively low, increasing the sample size can be considered, and even full factorial design or GS may be viable options. If the design space is large (i.e., the design dimensionality is high), DOE methods that correlate the number of sample points with the number of dimensions should not be used. Different surrogate modeling techniques are suited to different DOE strategies. For instance, PRSs are often combined with CCD sampling methods to construct approximation models. In summary, experimental design is the first step in the DDO framework, and its choice should be based on the overall design process.

2.1.2 Latin Hypercube Sampling

LHS is a widely used statistical sampling method (Iman, 2008). Figure 2.3 illustrates 25 sample points from an LHS process, while Figure 2.4 shows 25 sample points from a GS for comparison. In the statistical sampling process, each row and column of the grid can contain only one sample point. LHS refers to a square matrix in which no two elements in the same row or column are identical. Figure 2.5 provides a visual representation of the Latin hypercube and LHS.

Figure 2.5a shows one possible arrangement of the four letters 'LHSD' in a Latin hypercube. As seen in figure, each row and column contains a unique permutation of the letters 'LHSD,' ensuring that each letter occupies a distinct row and column in the matrix. Figure 2.5b–d shows the three random outcomes of LHS with four points.

By combining Figures 2.3 and 2.5, it is evident that LHS is random but effectively covers the entire design space. For continuous design problems,

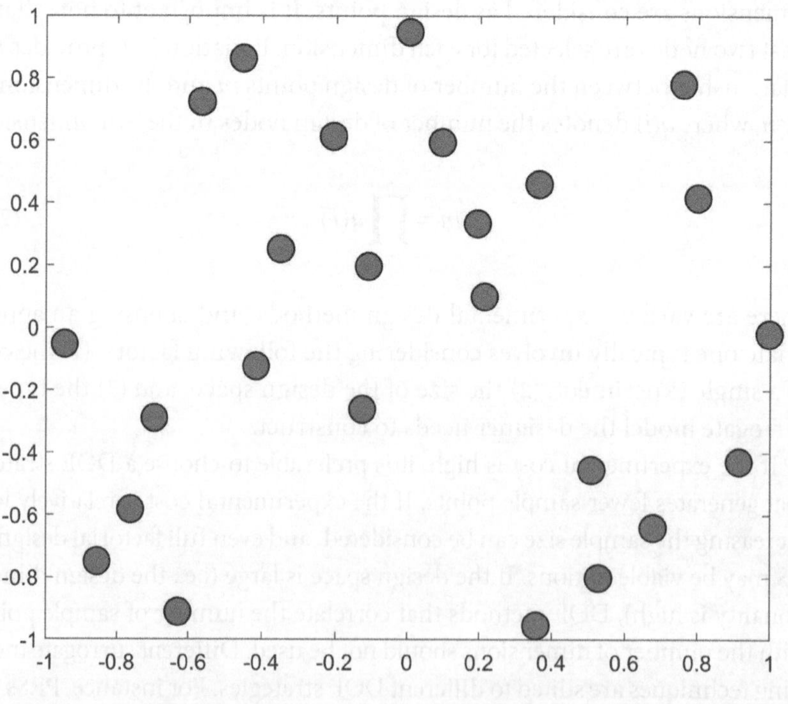

FIGURE 2.3 LHS (25 samples).

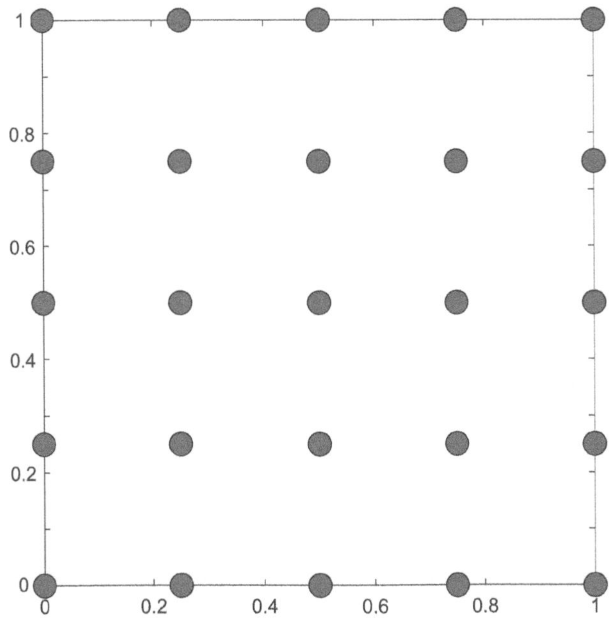

FIGURE 2.4 Grid sampling (25 samples).

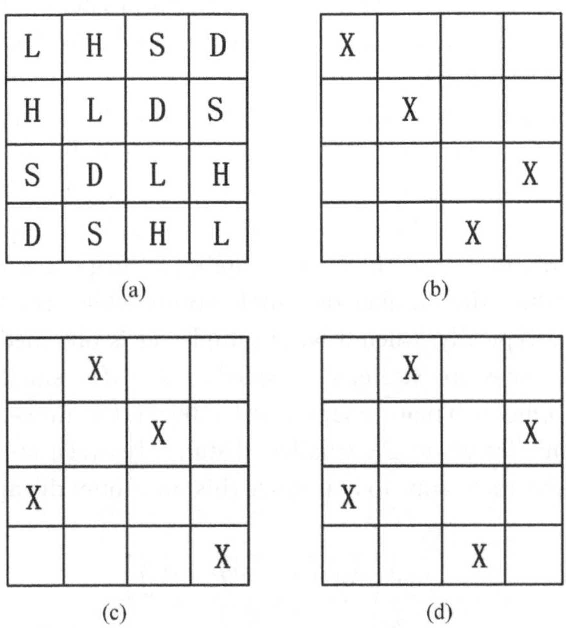

FIGURE 2.5 Explanation of Latin hypercube and Latin hypercube sampling. (a) Permutation without repetition. (b) Random situation 1. (c) Random situation 2. (d) Random situation 3.

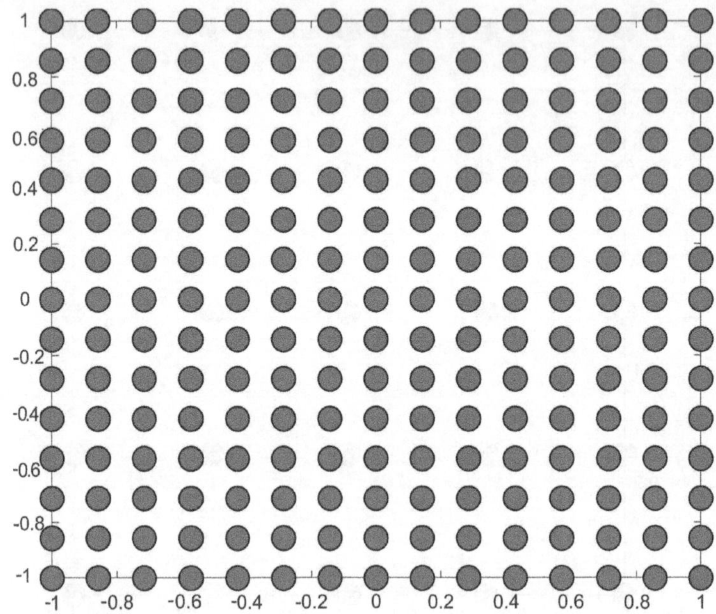

FIGURE 2.6 Grid sampling (225 samples obtained).

LHS divides each dimension of the space into m equal parts (considering a two-dimensional space), and the design points are randomly placed within $m \times m$ grid areas.

As mentioned earlier, GS evenly covers the design space, but this comes at the cost of a significant increase in the number of experiments. Figure 2.6 shows 225 sample points obtained through GS, with 15 design levels for each dimension. Executing all 225 sample points can lead to a costly computation. A key focus of sampling strategy research is how to effectively reduce the number of sample points while retaining valuable information. Typically, when a large sample set is obtained in a practical problem, a selection strategy is needed to identify a smaller, more efficient subset. One such mature sampling strategy is the 'max–min' strategy, where the 'min' refers to the smallest distance between any two sample points, and the 'max' aims to maximize this minimum distance.

$$\max\left(\min_{i \neq j}\left(dis_{ij} = \left\|P_i - P_j\right\|\right)\right) \tag{2.2}$$

Equation (2.2) provides the calculation formula for the max–min strategy. Figures 2.7–2.9 show the optimal results selected by the max–min criterion for 100, 1,000 and 10,000 iterations, respectively.

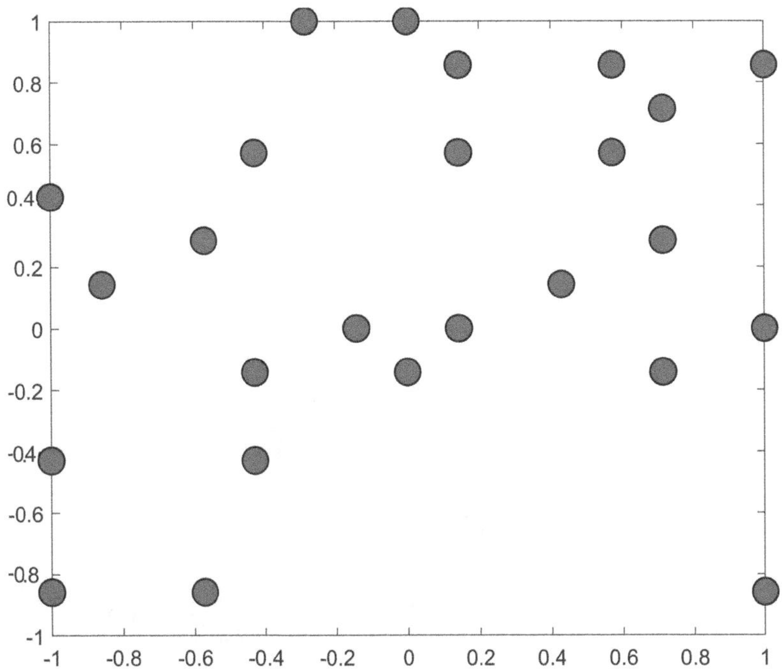

FIGURE 2.7 After 100 iterations using max–min criterion.

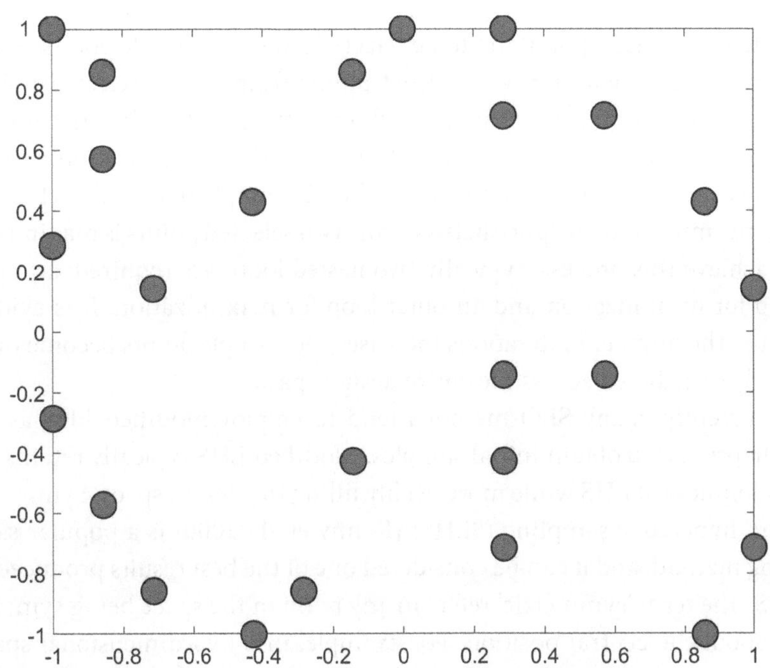

FIGURE 2.8 After 1,000 iterations using max–min criterion.

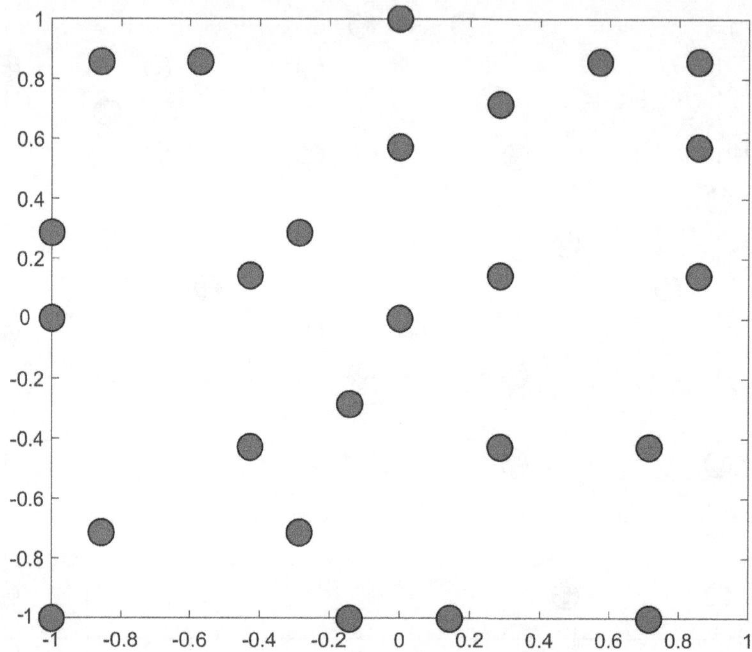

FIGURE 2.9 After 10^4 iterations using max–min criterion.

Suppose that 25 points are to be selected from 225 sample points, where P_i and P_j represent any two distinct points from the 25 selected points. The minimization process involves finding the smallest distance among all pairwise combinations of the 25 points. The maximization process involves selecting the 25 points from the 225 sample points in such a way that the minimum distance between any two selected points is maximized. To achieve this process, typically two nested loops are required: an inner loop for minimization and an outer loop for maximization. It is evident that as the number of iterations increases, the sample points become more evenly distributed across the entire design space.

Currently, many SBO methods tend to employ modified LHS as the DOE process to obtain initial samples. Modified LHS typically retains the randomness of LHS while more evenly filling the design space. Symmetric Latin hypercube sampling (SLHS) (Kenny et al., 2000) is a popular sampling method, and it can be considered one of the best results produced by LHS. The term 'symmetric' refers to any point in the space being symmetric about the central position. For example, in a two-dimensional space, suppose six design samples are needed. The first dimension is divided into six equal parts, with levels 1, 2, 3, 4, 5 and 6 assigned in order. The second dimension randomly generates a sequence from 1 to 3, here given as 3, 1, 2. The remaining numbers of three are calculated as $(6+1-2)$, $(6+1-1)$ and $(6+1-3)$. Additionally, there is a 50% chance that any of the first three numbers in this sequence will be swapped with their corresponding counterparts in the last three, and in this case, the second and fifth numbers are exchanged. The final sequence for the second dimension becomes three, $(6+1-1)$, 2, $(6+1-2)$, 1, $(6+1-3)$. Figure 2.10a shows the final result of SLHS in a two-dimensional space for six points. When an odd number of points is required, the central point is selected, and the remaining points are symmetrically distributed about the center. Figure 2.10b illustrates the situation for seven sample points.

Similarly, the optimal Latin hypercube sampling (OLHS) algorithm has been widely adopted for optimization purposes, such as genetic algorithm-optimal Latin hypercube sampling (GA-OLHS) and enhanced stochastic evolutionary algorithm-optimal Latin hypercube sampling (ESEA-OLHS) (Jin et al., 2005). To ensure the sample points uniformly fill the design space, OLHS typically utilizes a global optimization solver to determine an optimal criterion, such as the aforementioned max–min criterion, entropy principle or centered discrepancy criterion L_2.

Shannon (1948) quantified information content using entropy, where a lower entropy value indicates more precise information. Minimizing the 'posterior entropy' is equivalent to finding a set of experimental design

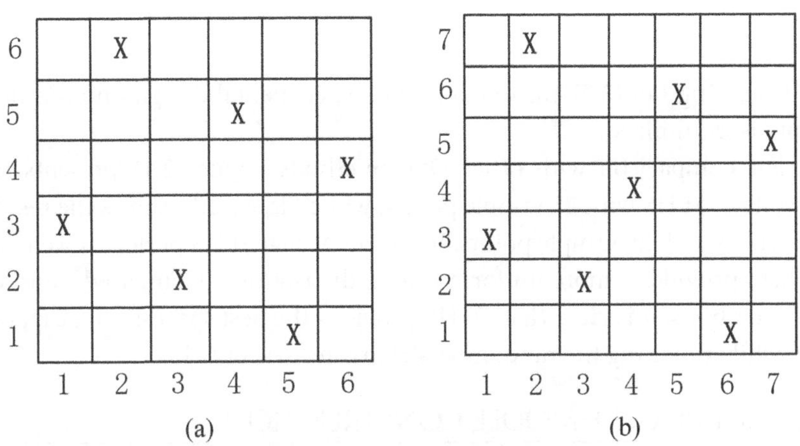

FIGURE 2.10 Symmetric Latin hypercube sampling points for even and odd cases. (a) Even case. (b) Odd case.

points with the least amount of information. Koehler and Owen (1996) further demonstrated that the entropy principle criterion is equivalent to the following minimization expression:

$$-\log_{10}|\boldsymbol{R}| \tag{2.3}$$

where $\boldsymbol{R}$ is the correlation matrix with elements defined in Eq. (2.4).

$$R_{ij} = \exp\left(\sum_{k=1}^{m} \theta_k |x_{ik} - x_{jk}|^t\right), \quad 1 \le i, \quad j \le n; \quad 1 \le t \le 2 \tag{2.4}$$

where $\theta_k \, (k = 1,\ldots,m)$ is the correlation coefficient.

The centered discrepancy criterion L_2 is a method for measuring the difference between the empirical cumulative distribution function and the uniform cumulative distribution function of an experimental design. In other words, L_2 is used to express the non-uniformity of an experimental design. Hickernell (1998) proposed three formulas for L_2, among which the centered L_2 formula is the most expressive.

$$CL_2(\mathbf{X}) = \left(\frac{13}{12}\right)^2 - \frac{2}{n}\sum_{i=1}^{n}\prod_{k=1}^{m}\left(1 + \frac{1}{2}|x_{ik} - 0.5| - \frac{1}{2}|x_{ik} - 0.5|^2\right)$$
$$+ \frac{1}{n^2}\sum_{i=1}^{n}\sum_{j=1}^{n}\prod_{k=1}^{m}\left(1 + \frac{1}{2}|x_{ik} - 0.5| - \frac{1}{2}|x_{jk} - 0.5| - \frac{1}{2}|x_{ik} - x_{jk}|\right) \tag{2.5}$$

Minimizing Eq. (2.5) ensures that the experimental design's non-uniformity is minimized.

For comparison with other DOE methods, Figure 2.11 presents the results of SLHS with 25 sample points, while Figure 2.12 shows the results of OLHS with 25 sample points. It is evident that SLHS performs well, but OLHS provides a more uniform spatial distribution. Compared to previous methods, it is clear that OLHS provides the best space-filling capability while retaining the randomness characteristic of LHS.

2.2 SURROGATE MODEL CONSTRUCTION

Common surrogate models include PRS, RBF, Kriging, and SVR. All of these methods generally incorporate interpolation and regression concepts. RBF and Kriging are commonly used interpolation methods, PRS

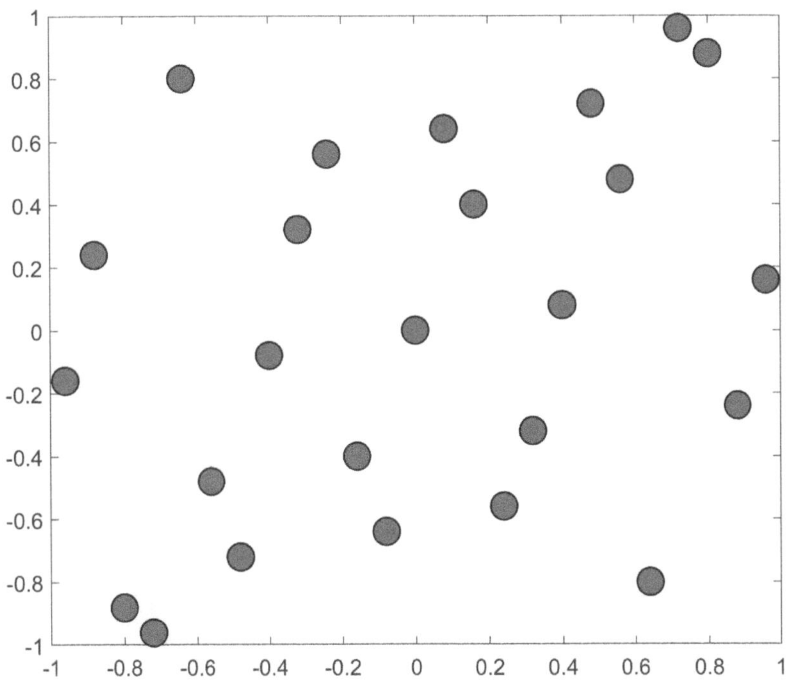

FIGURE 2.11 SLHS with 25 samples.

uses polynomial least squares regression, and SVR is a regression analysis method derived from machine learning for classification.

2.2.1 Polynomial Response Surface

PRS has been widely and effectively applied in numerous engineering designs, as it can accurately represent convex function problems. The approximate expression of PRS is obtained through least squares. The first-order and second-order polynomial functions of PRS are shown in Eqs. (2.6) and (2.7).

$$\hat{y}(\boldsymbol{x}) = \beta_0 + \sum_{i=1}^{n} \beta_i x_i \tag{2.6}$$

$$\hat{y}(\boldsymbol{x}) = \beta_0 + \sum_{i=1}^{n} \beta_i x_i + \sum_{i=1}^{n} \beta_{ii} x_i^2 + \sum_{i} \sum_{j} \beta_{ij} x_i x_j \tag{2.7}$$

$$N_{sampling} > \frac{1}{2} d^2 + \frac{3}{2} d + 1 \tag{2.8}$$

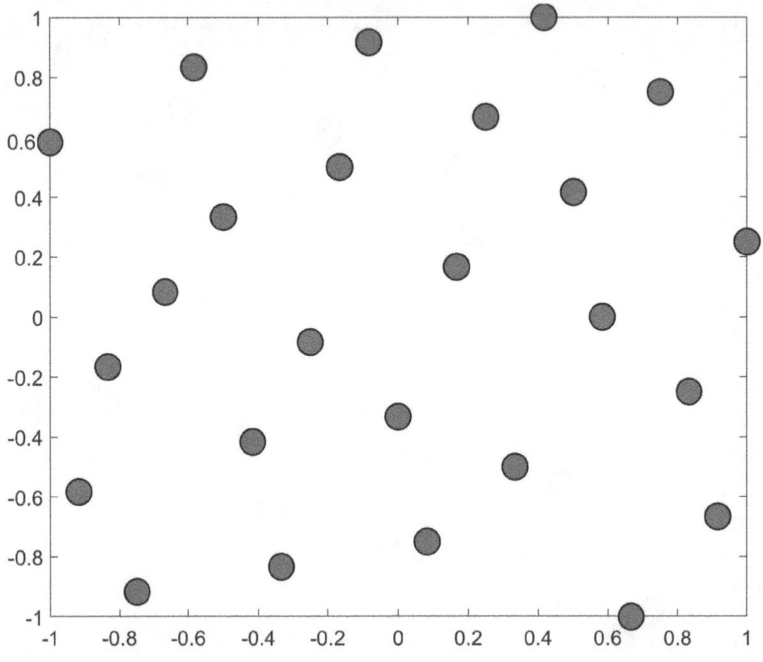

FIGURE 2.12 OLHS with 25 samples.

where n represents the number of design variables. β_i denotes the coefficients of univariate polynomials. β_{ii} represents the coefficients of the quadratic terms. β_{ij} indicates the coefficients of the hinge terms between two variables. $y(\boldsymbol{x})$ is the true function, while $\hat{y}(\boldsymbol{x})$ is its approximate expression. $N_{sampling}$ denotes the number of samples. Generally, if $N_{sampling}$ does not satisfy the condition in Eq. (2.8), the PRS model will exhibit significant prediction errors.

Given the sample points and corresponding response values, the polynomial parameters can be determined based on Eq. (2.9):

$$\beta = [X'X]^{-1} X'y \tag{2.9}$$

where X represents the design matrix of the sample points. y contains the response values for all the sample points. PRS is relatively easy to construct, and its continuous and smooth nature aids in the rapid convergence of optimization problems with noise. However, due to its simplicity, it is often difficult for PRS to accurately predict and express nonlinear problems. PRS has a wide range of applications, including robust optimization, multidisciplinary optimization, adaptive strategies for global optimization and manufacturing analysis.

2.2.2 Radial Basis Function

RBF was initially proposed by Hardy as an interpolation strategy. Then Dyn made the RBF method more practical, smoothing the data while retaining the interpolation function. RBF expresses the overall approximation function as a weighted sum of a series of basis functions, where the basis functions are derived from the Euclidean distance between known sample points or between known sample points and the points to be tested.

Given a set of sample points $\boldsymbol{X}=\left\{x^{(1)},x^{(2)},\ldots,x^{(n)}\right\}^T$ and the corresponding real response values $\boldsymbol{y}=\left\{y^{(1)},y^{(2)},\ldots,y^{(n)}\right\}^T$, the approximate expression is given by Eq. (2.10):

$$\hat{y}(x)=w^T\psi=\sum_{i=1}^{n_c} w_i\psi\left(\left\|x-c^{(i)}\right\|\right) \tag{2.10}$$

where $c(i)$ represents the center of the i-th basis dunction, $\psi(\bullet)$ is the basis function, x means a unobserved point; w_i denotes the weight coefficients. There are various forms of basis functions commonly used in RBF interpolation, each defined by different mathematical expressions. Some of the most widely used forms include

Linear function:

$$\psi(r)=r \tag{2.11}$$

Cubic function:

$$\psi(r)=r^3 \tag{2.12}$$

Thin-plate splines:

$$\psi(r)=r^2\ln r \tag{2.13}$$

Gaussian function:

$$\psi(r)=e^{-r^2/2\sigma^2} \tag{2.14}$$

Multiquadric function:

$$\psi(r)=\left(r^2+\sigma^2\right)^{1/2} \tag{2.15}$$

Inverse multiquadric function:

$$\psi(r)=\left(r^2+\sigma^2\right)^{-1/2} \tag{2.16}$$

The weight coefficients w in Eq. (2.10) can be obtained through the interpolation conditions.

$$\hat{y}\left(\boldsymbol{x}^{(j)}\right)=\sum_{i=1}^{n_c} w_i\psi\left(\left\|\boldsymbol{x}^{(j)}-\boldsymbol{c}^{(i)}\right\|\right)=y^{(j)},\quad j=1,\ldots,n. \tag{2.17}$$

In Eqs. (2.10) and (2.17), $n_c=n$ and $x(i)=c(i)$, then the Kram matrix can be represented as:

$$\psi_{ij}=\psi\left(\left\|x^{(i)}-x^{(j)}\right\|\right) \tag{2.18}$$

Thus, the weight coefficient matrix can be easily obtained:

$$\boldsymbol{w}=\boldsymbol{\psi}^{-1}\boldsymbol{y} \tag{2.19}$$

By observing Eqs. (2.10)–(2.19), it can be observed that RBF is highly similar to artificial neural networks. In fact, RBF is essentially a simple single-layer neural network.

2.2.3 Kriging

In statistics, specifically in geostatistics, Kriging (also known as Gaussian process regression) is an interpolation strategy that is fundamentally different from piecewise-polynomial spline methods. This strategy is modeled through Gaussian process interpolation and is influenced by the prior covariance. Under suitable prior assumptions, Kriging provides the best linear unbiased prediction for the interpolated values, which has led to its widespread application in statistical sciences. Another important and rapidly developing application is in engineering, where deterministic computer simulation outputs are used as the interpolation targets. In this context, Kriging is employed as a surrogate model tool to address black-box problems. In many engineering design problems, a single simulation analysis can take several hours or even days. Therefore, the Kriging interpolation method can quickly predict the response to inputs, significantly reducing the number of costly simulation runs.

The Kriging surrogate model has been widely applied due to its exceptional ability to solve nonlinear problems. To construct a Kriging model for a function $f(\boldsymbol{x})$, where $\boldsymbol{x}$ is an n-dimensional vector, the function $F(\boldsymbol{x})$ is defined to represent the deterministic response of $f(\boldsymbol{x})$, as expressed in the following formula:

$$F(\boldsymbol{x})=\mu+Z(\boldsymbol{x}) \tag{2.20}$$

where μ is defined as a constant. $Z(\boldsymbol{x})$ is a stochastic process with the following statistical behavior:

$$E[Z(\boldsymbol{x})]=0$$

$$\mathrm{Cov}[Z(\boldsymbol{x}),Z(\boldsymbol{x}')]=\sigma^2 R(\Theta,\boldsymbol{x},\boldsymbol{x}') \tag{2.21}$$

$$R(\Theta,\boldsymbol{x},\boldsymbol{x}')=\prod_{j=1}^{n} R_j(\theta_j,x_j-x_j')$$

where σ^2 represents the process variance of the response value. $R(\Theta,\boldsymbol{x},\boldsymbol{x}')$ is the correlation model between any two points $\boldsymbol{x}$ and $\boldsymbol{x}'$. $\Theta=\{\theta_1,\theta_2,\ldots,\theta_n\}$ is the parameter of the correlation model, namely the correlation parameters. In this book, the Gaussian correlation function is used for modeling.

$$R(\theta_j,x_j,x_j')=\exp\left(-\theta_j\left|x_j-x_j'\right|^2\right) \tag{2.22}$$

Next, assume there are N sample points $\boldsymbol{x}^{(1)},\boldsymbol{x}^{(2)},\ldots,\boldsymbol{x}^{(N)}$, and the corresponding response values for the function $f(\boldsymbol{x})$ are calculated. According to Eq. (2.20), the Kriging model is represented as:

$$f\left(\boldsymbol{x}^{(i)}\right)=F\left(\boldsymbol{x}^{(i)}\right)=\mu+Z\left(\boldsymbol{x}^{(i)}\right) \tag{2.23}$$

In the Kriging model, the three parameters μ, σ^2, Θ are obtained through maximum likelihood estimation (MLE):

$$\hat{\mu}=\frac{\mathbf{1}^T\boldsymbol{R}^{-1}\boldsymbol{f}}{\mathbf{1}^T\boldsymbol{R}^{-1}\mathbf{1}}$$

$$\hat{\sigma}^2=\frac{(\boldsymbol{f}-\mathbf{1}\hat{\mu})^{\mathrm{T}}\boldsymbol{R}^{-1}(\boldsymbol{f}-\mathbf{1}\hat{\mu})}{N} \tag{2.24}$$

$$\ln(\Theta)=-\frac{N}{2}\ln(2\pi)-\frac{N}{2}\ln\hat{\sigma}^2-\frac{1}{2}\ln|\boldsymbol{R}|$$

where $\boldsymbol{f}=[f(\boldsymbol{x}^{(1)}),f(\boldsymbol{x}^{(2)}),\ldots,f(\boldsymbol{x}^{(N)})]^T$. $\boldsymbol{R}$ is a covariance matrix of size $N\times N$, where the element in the i-th row and j-th column is $R(\Theta,\boldsymbol{x}^{(i)},\boldsymbol{x}^{(j)})$.

Finally, the mean square error (MSE) is minimized.

$$\hat{s}^2(\boldsymbol{x})=\mathrm{Var}\left[\hat{f}(\boldsymbol{x})-F(\boldsymbol{x})\right] \tag{2.25}$$

Meanwhile, the following non-Bayesian constraint needs to be satisfied:

$$E\left[\hat{f}(\boldsymbol{x})\right]=E[F(\boldsymbol{x})] \tag{2.26}$$

The predicted function, $\hat{f}(\boldsymbol{x})$, obtained through the best linear unbiased estimation, is expressed as:

$$\hat{f}(\boldsymbol{x})=\hat{\mu}+\boldsymbol{r}^T(\boldsymbol{x})\boldsymbol{R}^{-1}(\boldsymbol{f}-\mathbf{1}\hat{\mu}) \tag{2.27}$$

where $\boldsymbol{r}(\boldsymbol{x})$ is a N-dimensional vector. The i-th element of $\boldsymbol{r}(\boldsymbol{x})$ is $R(\Theta,\boldsymbol{x},\boldsymbol{x}^{(i)})$. $\boldsymbol{x}$ is any sample point for which prediction is required. The final form of the estimated MSE is

$$\hat{s}^2(\boldsymbol{x})=\hat{\sigma}^2\left[1-\boldsymbol{r}^T(\boldsymbol{x})\boldsymbol{R}^{-1}\boldsymbol{r}(\boldsymbol{x})+\frac{(1-\mathbf{1}^{\mathrm{T}}\boldsymbol{R}^{-1}\boldsymbol{r}(\boldsymbol{x}))^2}{\mathbf{1}^{\mathrm{T}}\boldsymbol{R}^{-1}\mathbf{1}}\right] \tag{2.28}$$

Figure 2.13 illustrates the prediction diagram of Kriging in a one-dimensional example. The circles represent the known samples, the curve indicates the predicted function values, and the surrounding area represents the prediction uncertainty. From the figure, it can be observed that the uncertainty is close to 0 at the known sample points, and the uncertainty increases as the distance from the known sample points grows.

2.3 DYNAMIC SAMPLING TECHNIQUES

In most cases, a surrogate model is constructed using known sample data. However, if optimization is performed solely on the surrogate model to obtain the optimal solution, this optimal solution may not correspond to the true global optimum. This is because the surrogate model is constructed based on available information, and while it has predictive capabilities, it is not always perfectly accurate. To improve the accuracy of the surrogate model, one approach is to increase the initial sample size—by adding more samples during the experimental design phase—so that the model utilizes

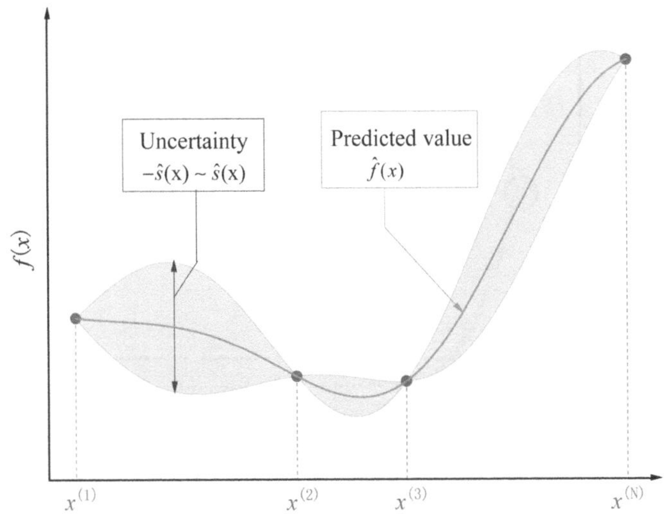

FIGURE 2.13 Illustration of Kriging prediction on a 1D example.

more real data. However, this typically results in a significant increase in computational cost. Another approach is to construct a rough surrogate model that captures the general trend of the original model. Such a model typically requires fewer samples. Then, promising regions of the surrogate model are identified to select new sample points. The previous samples are stored in a database, and new samples are chosen iteratively to update the database, with the surrogate model being updated accordingly. This iterative process improves the accuracy of the model at certain preferred locations. This second approach avoids large-scale blind sampling in the early stages and adopts a strategy of optimizing while incrementally adding new samples, thus saving significant computational costs.

2.3.1 Minimizing the Predictor

Minimizing the predictor (MP), namely a constructed surrogate, to obtain a new sample is a commonly used updating strategy (Hastie et al., 2004), as illustrated in Figure 2.14. Suppose the surrogate model is sufficiently accurate, and a robust optimization solver is used to find the minimum of this surrogate model. After many iterations, the global optimum will be reached. At this point, high-precision computational simulations are performed at the predicted optimal solution, and the high-precision response obtained will often differ from the response predicted by the surrogate model. This set of high-precision results is then added to the

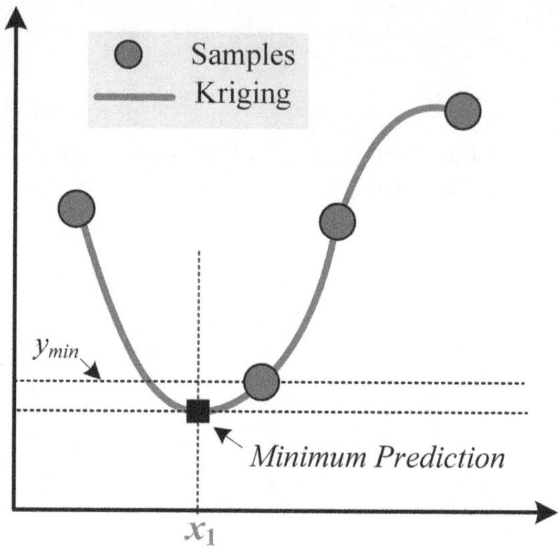

FIGURE 2.14 MP strategy.

original database, and the surrogate model is reconstructed. This process is repeated, gradually reducing the deviation between the predicted and true values, ultimately obtaining the true minimum. MP strategy is a relatively simple and intuitive sample update strategy, where the predicted optimal solution or a nearby solution is used as the update sample. However, a drawback of this approach is that the optimization process may become trapped in local optimum regions and fail to escape. After constructing the surrogate models for the objective function and constraint functions, the following optimization problem is solved, where n represents the number of constraint functions.

$$\begin{aligned} &Minimize\ \hat{y}(X) \\ &st\ \ \hat{g}_i \leq 0, \quad i = 1,2,\ldots,n \end{aligned} \tag{2.29}$$

When the objective is a smooth and continuous function, the MP sampling method will at least find a local optimal solution of the surrogate model. However, the convergence rate depends on the properties of the function.

2.3.2 Maximum Improvement Probability Criterion

Maximum improvement probability criterion (MIPC) aims to find the next sample point x that maximizes the probability of improving the current best observed value, $y_{\min}$. Let $Y \sim N[\hat{y}(x), s^2(x)]$ be a random variable following a normal distribution, and the improvement degree over $y_{\min}$ is denoted as $I = y_{\min} - Y(x)$. Therefore, the probability that the predicted objective value is better than the current best observed value is given by:

$$P\left[Y < y_{\min}\right] = \Phi\left(\frac{y_{\min} - \hat{y}(x)}{s(x)}\right) \tag{2.30}$$

$$P\left[I(x)\right] = \frac{1}{\hat{s}\sqrt{2\pi}} \int_{-\infty}^{0} e^{-\left[I - \hat{y}(x)\right]^2 / (2s^2)} dI \tag{2.31}$$

Figure 2.15 provides a graphical interpretation of Eq. (2.30), along with a Gaussian normal distribution in the vertical direction, where the mean is $\hat{y}(x)$ and the variance is $s^2(x)$. This Gaussian distribution represents the uncertainty of the predicted result $\hat{y}(x)$. The area below the dashed line indicates the probability of improvement over the current best value, and the enclosed area represents the improvement probability.

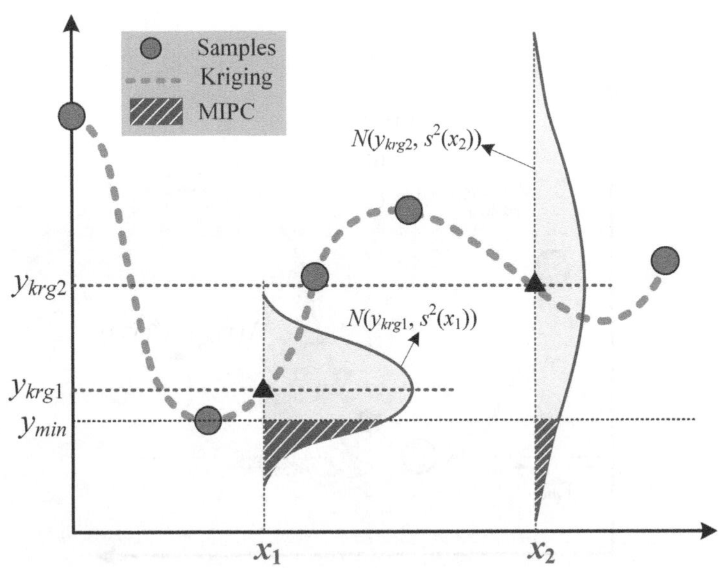

FIGURE 2.15 MIPC strategy.

2.3.3 Maximum Improvement Expectation Criterion

Maximum improvement expectation criterion (MIEC) refers to the expected improvement at an unobserved point x. Let $Y \sim N[\hat{y}(x), s^2(x)]$ be a random variable following a normal distribution, $\hat{y}$ is the predicted value from the surrogate model, s^2 is the estimated MSE. Given $\hat{y}$ and s^2, not only can the probability of improvement be calculated, but the expected improvement can also be estimated. The expected improvement calculation is shown in Eq. (2.32).

$$E[I(x)] = \begin{cases} \left(y_{\min} - \hat{y}(x)\right)\Phi\left(\dfrac{y_{\min} - \hat{y}(x)}{s^2(x)}\right) + s\phi\left(\dfrac{y_{\min} - \hat{y}(x)}{s^2(x)}\right), & s > 0 \\ 0 & s = 0 \end{cases} \tag{2.32}$$

where $\Phi(\cdot)$ and $\phi(\cdot)$ represent the cumulative distribution function and the probability density function of the standard normal distribution, respectively.

In Figure 2.16, the expected improvement can be intuitively understood as the area below the current optimal value, which represents the average

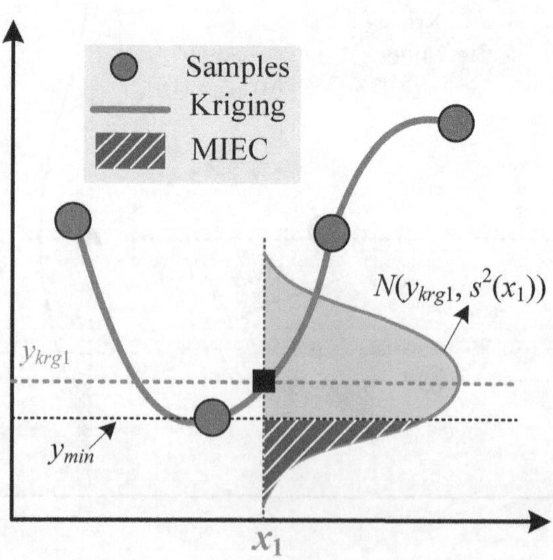

FIGURE 2.16 MIEC strategy.

value of the integral of the probability density under the Gaussian distribution function. When $\hat{s}^2(x)=0$, $P[I(x)]=E[I(x)]=0$.

Another classic update method is the trust-region method (TR). Alexandrov et al. (1998) have rigorously proven that the TR method can converge to a local optimum, regardless of the starting point, under the condition that the gradient information of the real model at the interpolation points is available. TR can also match the gradient of the objective function using the first-order scaling method suggested by Haftka (1991) or the second-order scaling method proposed by Eldred et al. (2004). In general, both TR and MP belong to strategies that exploit the design space through the use of surrogate models, often referred to as exploitation-based infill criteria. While MP can easily miss the true global optimum when dealing with highly nonlinear problems, TR guarantees the search for a local optimum from any starting point, although it does not ensure finding the global optimum.

To determine the global optimum, a new element, namely space exploration, needs to be introduced. Pure design space exploration can essentially be viewed as filling gaps between known design points with new samples. The simplest approach is a sequential space sampling plan, such as Sobol sequences or LP arrays, although these methods perform poorly when the number of samples is small (Sobol, 1979; Statnikov & Matusov, 2012). New sample points can also be determined by the max–min criterion. If the residual estimate of the surrogate model is available, selecting the location with the largest residual to add a new sample is also a viable strategy. However, pure space exploration can sometimes be time-consuming, as designers are typically less concerned with the overall accuracy of the surrogate model and more focused on the precision at the global optimum location.

2.4 CHAPTER SUMMARY

This chapter provides an overview of the DDO process, detailing the initial sampling techniques, surrogate modeling methods and dynamic sampling strategies employed in DDO. The initial sampling methods, as the foundation of DDO, determine the distribution of the initial samples. Surrogate modeling, as the key component of the process, ensures the accuracy of the model predictions. Dynamic sampling strategies, as the core of DDO, guarantee a thorough search of the design space.

REFERENCES

Alexandrov, N. M., Dennis Jr, J. E., Lewis, R. M., & Torczon, V. (1998). A Trust-Region Framework for Managing the Use of Approximation Models in Optimization. *Structural Optimization*, *15*(1), 16–23.

Box, G. E., & Behnken, D. W. (1960). Some New Three Level Designs for the Study of Quantitative Variables. *Technometrics*, *2*(4), 455–475.

Chen, W. (1995). *A robust concept exploration method for configuring complex systems*. Georgia Institute of Technology.

Eldred, M., Giunta, A., & Collis, S. (2004). Second-order corrections for surrogate-based optimization with model hierarchies. *10th AIAA/ISSMO Multidisciplinary Analysis and Optimization Conference*, Albany, New York, August 30 to September 1, 2004.

Haftka, R. T. (1991). Combing Global and Local Approximations. *AIAA Journal*, *29*(9), 1523–1525.

Hastie, T., Tibshirani, R., & Friedman, J. (2004). The Elements of Statistical Learning. 2001. *Journal of the Royal Statistical Society*, *167*(1), 192–192.

Hickernell, F. (1998). A Generalized Discrepancy and Quadrature Error Bound. *Mathematics of Computation*, *67*(221), 299–322.

Iman, R. L. (2008). Latin hypercube sampling. In E. L. Melnick & B. S. Everitt (Eds.), *Encyclopedia of quantitative risk analysis and assessment*. John Wiley & Sons, Ltd. https://doi.org/10.1002/9780470061596.risk0299

Jin, R., Chen, W., & Sudjianto, A. (2005). An efficient Algorithm for Constructing Optimal Design of Computer Experiments. *Journal of Statistical Planning and Inference*, *134*(1), 268–287.

Kenny, Q. Y., Li, W., & Sudjianto, A. (2000). Algorithmic Construction of Optimal Symmetric Latin Hypercube Designs. *Journal of Statistical Planning and Inference*, *90*(1), 145–159.

Koehler, J., & Owen, A. (1996). Computer Experiments. In S. Ghosh & C. R. Rao (Eds.), *Handbook of statistics* (pp. 261–308). Elsevier Science New York.

Myers, R. H., Montgomery, D. C., & Anderson-Cook, C. M. (2016). *Response surface methodology: Process and product optimization using designed experiments*. John Wiley & Sons.

Shannon, C. E. (1948). A Mathematical Theory of Communication. *The Bell System Technical Journal*, *27*(3), 379–423.

Sobol, I. M. (1979). On the Systematic Search in a Hypercube. *SIAM Journal on Numerical Analysis*, *16*(5), 790–793. https://doi.org/10.1137/0716058

Statnikov, R. B., & Matusov, J. B. (2012). *Multicriteria optimization and engineering*. Springer Science & Business Media.

CHAPTER 3

Benchmark Functions for Data-Driven Optimization Methods

3.1 INTRODUCTION

In recent years, various optimization algorithms have rapidly developed, addressing optimization problems that are challenging for traditional numerical optimization methods. Benchmark function testing is one of the most commonly used methods by researchers to assess the performance and robustness of optimization algorithms. In the subsequent chapters of this book, a wide range of benchmark functions is employed to validate the accuracy and efficiency of various optimization methods. This chapter provides a comprehensive summary and classification of these functions. Specifically, it introduces single-objective optimization test functions (Jamil & Yang, 2013; Surjanovic & Bingham, 2013), constrained and unconstrained optimization test functions (Adorio & Diliman, 2005; Akbari & Kazerooni, 2020; Jamil & Yang, 2013; Liang et al., 2006; Liu et al., 2021; Liu et al., 2017; Mezura-Montes & Cetina-Domínguez, 2012; Surjanovic & Bingham, 2013), discrete optimization test functions (Dong et al., 2020; Li et al., 2013; Müller et al., 2013; Müller et al., 2014; Pichitlamken et al., 2006) and high-dimensional optimization test functions (Adorio & Diliman, 2005; Jamil & Yang, 2013; Surjanovic & Bingham, 2013).

DOI: 10.1201/9781003636267-3

Besides, functions that possess multiple local optima are referred to as multimodal functions. These functions are used to test an algorithm's ability to escape local minima. If the exploration process of an algorithm is poorly designed, it will fail to effectively search for the global optimum, causing the algorithm to become trapped in local minima. For many algorithms, escaping from multimodal functions with numerous local minima represents a major challenge. Another difficulty is the search process for plate-shaped functions, as the minimal variation in the function makes it difficult for the algorithm to gather useful information to guide the search process.

For any new optimization algorithm, it is essential to compare it with other existing algorithms using a wide range of test functions to validate its performance. If the problems are overly simplified and lack diversity, the effectiveness of the algorithm in comparison to other methods may not be accurately evaluated. Therefore, to assess the quality of an algorithm, it is necessary to identify the specific problems on which it performs better. This helps describe the types of problems the algorithm is suited for. The results can be considered reliable only when the number of benchmark functions is sufficiently large and the types of problems covered are diverse, such as unimodal, multimodal, discrete, and high-dimensional problems. Without loss of generality, this book focuses on minimization problems, as maximization problems can be transformed into minimization problems by changing the sign of the objective function. The mathematical definitions of the test functions used in this book are provided below.

3.2 UNCONSTRAINED OPTIMIZATION PROBLEMS

3.2.1 Unconstrained Low-Dimensional Problems

3.2.1.1 Generalized Polynomial Function

The generalized polynomial function, as shown in Figure 3.1, is defined by the following mathematical expression:

$$f(x)=(1.5-x_1(1-x_2))^2+\left(2.25-x_1(1-x_2^2)\right)^2+\left(2.625-x_1(1-x_2^3)\right)^2$$
$$n=2 \quad -2\le x_1\le 2, \quad -2\le x_2\le 2 \tag{3.1}$$

Design objective: Single objective

Function characteristics: Continuous, unimodal

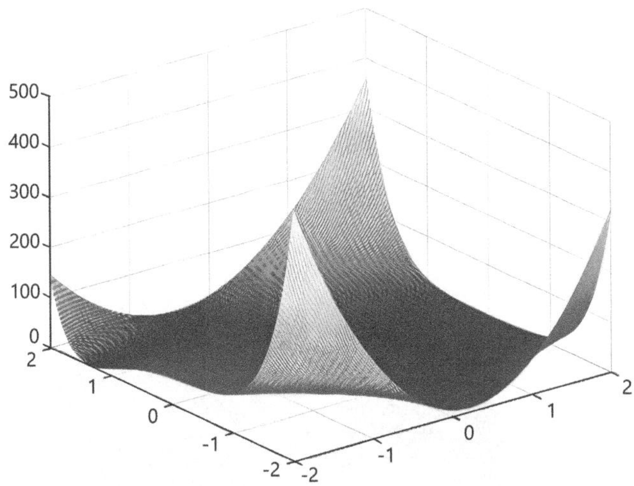

FIGURE 3.1 Generalized polynomial function.

Dimensions: 2-Dimensional

Optimal value: 0.523

3.2.1.2 Zakharov Function

The Zakharov function is shown in Figure 3.2, and its mathematical expression is given by Eq. (3.2).

$$f(\boldsymbol{x}) = \sum_{i=1}^{2} x_i^2 + \left(\frac{1}{2} \sum_{i=1}^{2} i x_i \right)^2 + \left(\frac{1}{2} \sum_{i=1}^{2} i x_i \right)^4 \tag{3.2}$$

$$n = 2 \quad -5 \le x_1 \le 10, \quad -5 \le x_2 \le 10$$

Design objective: Single objective

Function characteristics: Continuous, unimodal

Dimensions: 2-Dimensional

Optimal value: 0

3.2.1.3 Beale Function

The Beale function is shown in Figure 3.3, and its mathematical expression is given by Eq. (3.3).

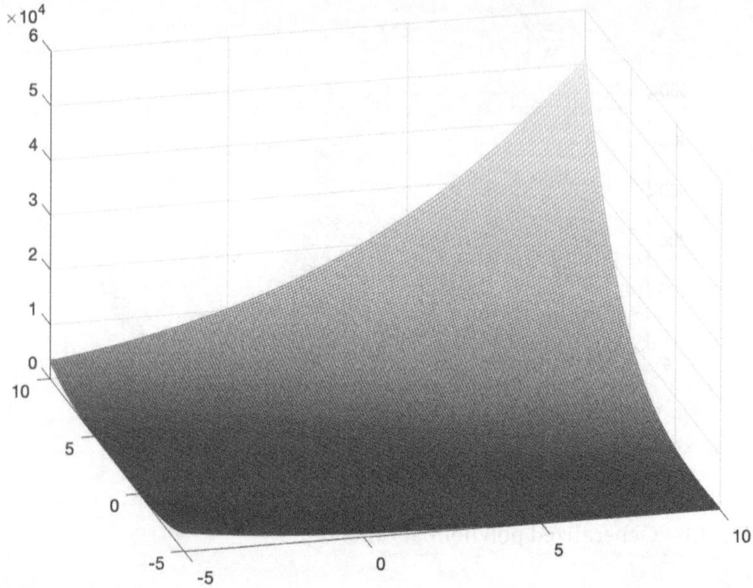

FIGURE 3.2 Zakharov function.

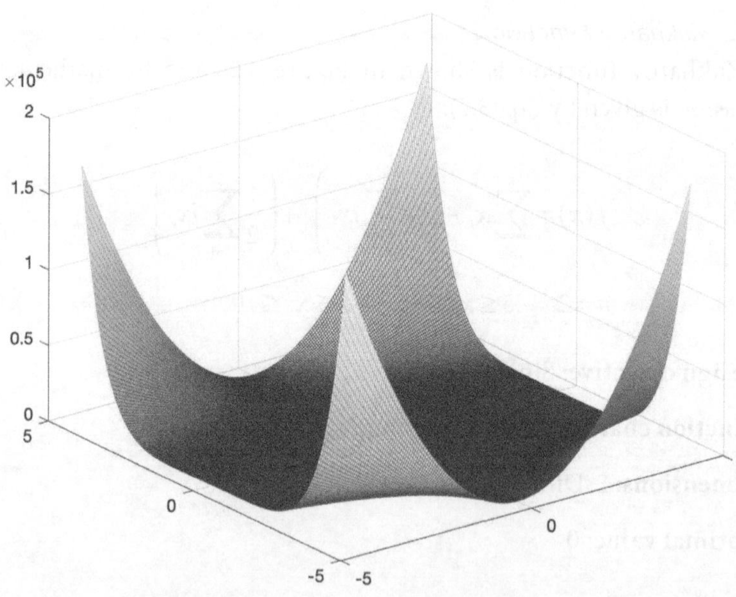

FIGURE 3.3 Beale function.

$$f(\boldsymbol{x}) = (1.5 - x_1 + x_1x_2)^2 + (2.25 - x_1 + x_1x_2^2)^2 + (2.625 - x_1 + x_1x_2^3)^2$$
$$n = 2 \quad -4.5 \le x_1 \le 4.5, \quad -4.5 \le x_2 \le 4.5 \tag{3.3}$$

Design objective: Single objective

Function characteristics: Continuous

Dimensions: 2-Dimensional

Optimal value: 0

3.2.1.4 Six-Hump Camel-Back Function

The six-hump camel-back function is shown in Figure 3.4, and its mathematical expression is given by Eq. (3.4).

$$f(\boldsymbol{x}) = 4x_1^2 - 2.1x_1^4 + \frac{1}{3}x_1^6 + x_1x_2 - 4x_2^2 + 4x_2^4$$
$$n = 2 \quad -2 \le x_i \le 2, \quad i = 1,2 \tag{3.4}$$

Design objective: Single objective

Function characteristics: Continuous

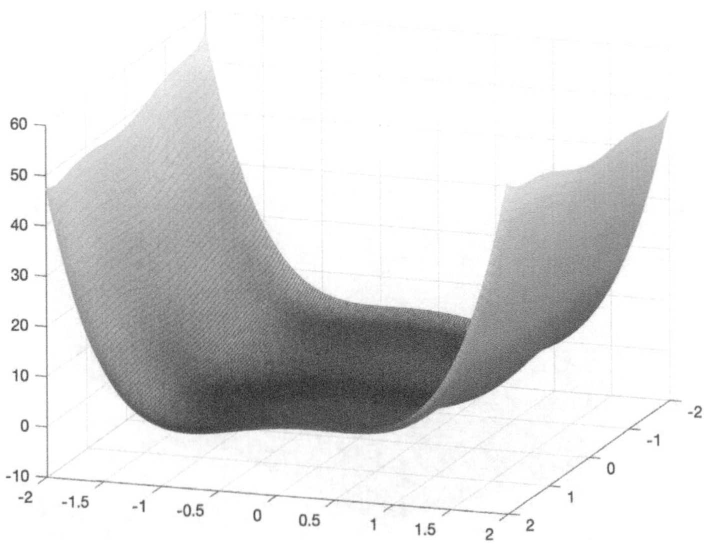

FIGURE 3.4 Six-hump camel-back function.

Dimensions: 2-Dimensional

Optimal value: –1.0320

3.2.1.5 Branin Function

The mathematical expression of the Branin function is given as follows:

$$f(x)=(x_2-5.1(x_1/2\pi)^2+5x_1/\pi-6)^2+10(1-1/8\pi)\cos(x_1)+10$$
$$n=2 \quad -5\le x_i\le 10,\quad i=1 \tag{3.5}$$
$$0\le x_i\le 15,\quad i=2$$

Design objective: Single objective

Function characteristics: Continuous

Dimensions: 2-Dimensional

Optimal value: 0.397

3.2.1.6 Leon Function

The Leon function is shown in Figure 3.5, and its mathematical expression is given by Eq. (3.6).

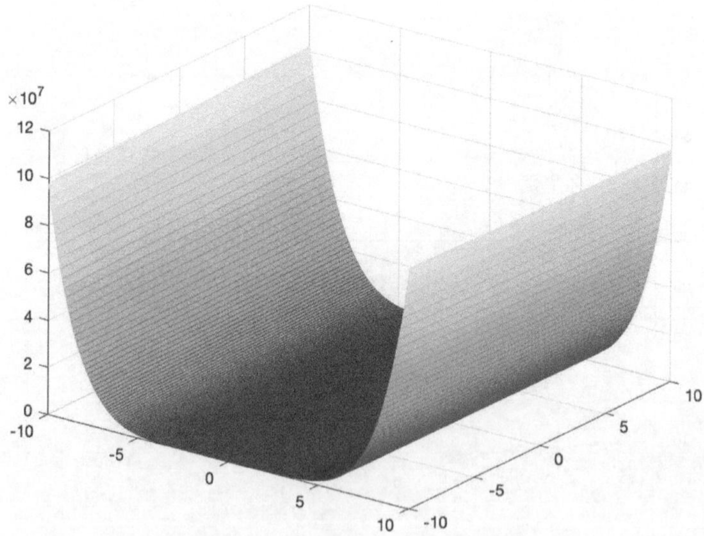

FIGURE 3.5 Leon function.

$$f(\boldsymbol{x}) = 100(x_2 - x_1^3)^2 + (x_1 - 1)^2$$
$$n = 2 \quad -10 \le x_i \le 10, \quad i = 1,2 \tag{3.6}$$

Design objective: Single objective

Function characteristics: Continuous, unimodal

Dimensions: 2-Dimensional

Optimal value: 0

3.2.1.7 Griewank Function

The mathematical expression is given as follows:

$$f(\boldsymbol{x}) = \sum_{i=1}^{2} \frac{x_i^2}{200} + \prod_{i=1}^{2} \cos\left(x_1/\sqrt{i}\right) + 1$$
$$n = 2 \quad -100 \le x_i \le 100, \quad i = 1,2 \tag{3.7}$$

Design objective: Single objective

Function characteristics: Continuous

Dimensions: 2-Dimensional

Optimal value: 0

3.2.1.8 Ackley Function

The Ackley function is shown in Figure 3.6, and its mathematical expression is given by Eq. (3.8).

$$f(\boldsymbol{x}) = 20 + e - 20e^{-\frac{1}{5}\sqrt{\frac{1}{n}\sum_{i=1}^{n} x_i^2}} - e^{-\frac{1}{n}\sum_{i=1}^{n}\cos(2\pi x_i)} \tag{3.8}$$
$$n = 2 \quad -30 \le x_i \le 30, \quad i = 1,\ldots,n$$

Design objective: Single objective

Function characteristics: Continuous, multimodal

Dimensions: 2-Dimensional

Optimal value: 0

Description: The Ackley function features an almost flat outer region with a large hole at its center. This function possesses numerous local minima.

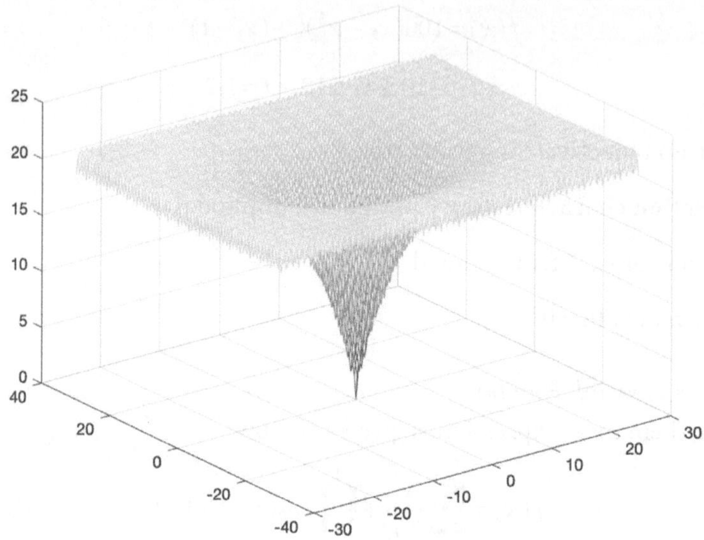

FIGURE 3.6 Ackley function.

3.2.1.9 Griewank Function

The Griewank function (GW/GW2/GW10) is shown in Figure 3.7, and its mathematical expression is given by Eq. (3.9).

$$f(\boldsymbol{x}) = \sum_{i=1}^{n} \frac{x_i^2}{4{,}000} - \prod_{i=1}^{n} \cos\left(\frac{x_i}{\sqrt{i}}\right) + 1$$

$$n = 2 \quad -10 \le x_i \le 10, \quad i = 1,\ldots,n \tag{3.9}$$

$$n = 10 \quad -600 \le x_i \le 600, \quad i = 1,\ldots,n$$

Design objective: Single objective

Function characteristics: Continuous, multimodal

Dimensions: n-Dimensional (the figure displays its two-dimensional version, and this book uses both two-dimensional and ten-dimensional versions)

Optimal value: 0

Description: The GW function possesses several local minima. Although there is only one global optimum, the nearby peaks are extremely close, posing a significant challenge to the algorithm's ability to escape from local minima.

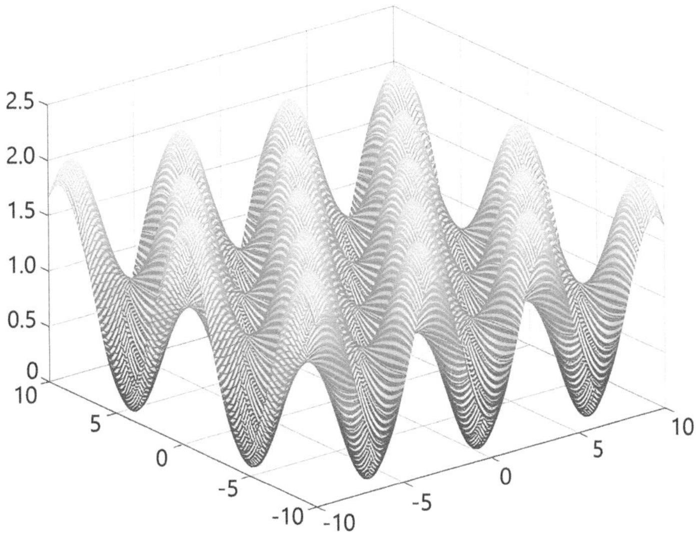

FIGURE 3.7 Griewank function.

3.2.1.10 Peaks Function

The Peaks function is shown in Figure 3.8, and its mathematical expression is given by Eq. (3.10).

$$f(\boldsymbol{x}) = 3(1-x_1)^2 e^{-x_1^2-(x_2+1)^2} - 10\left(\frac{x_1}{5} - x_1^3 - x_2^5\right)e^{-x_1^2-x_2^2} - \frac{1}{3}e^{-(x_1+1)^2-x_2^2}$$

$$n=2 \quad -3 \le x_1 \le 3, \quad -4 \le x_2 \le 4 \tag{3.10}$$

Design objective: Single objective

Function characteristics: Continuous, multimodal

Dimensions: 2-Dimensional

Optimal value: −6.551

3.2.1.11 Styblinski–Tang Function

The graph of the Styblinski–Tang function (ST/ST5) is shown in Figure 3.9, and its mathematical expression is given in Eq. (3.11).

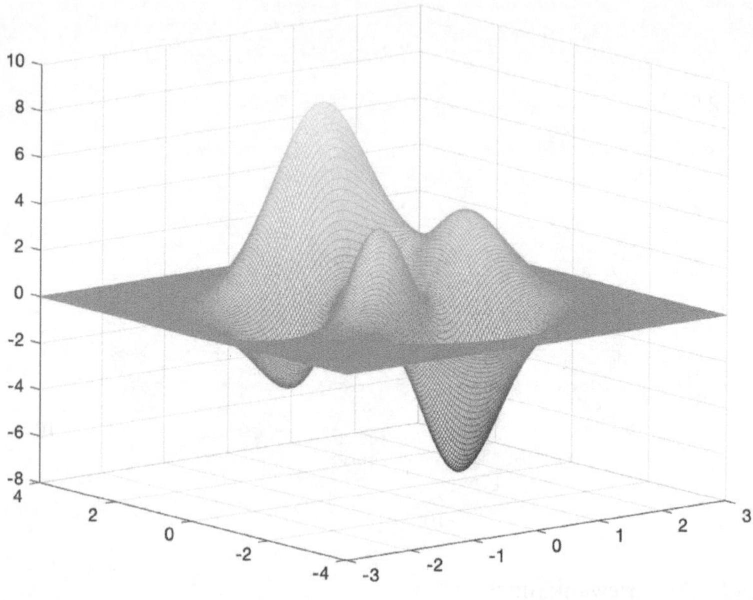

FIGURE 3.8 Peaks function.

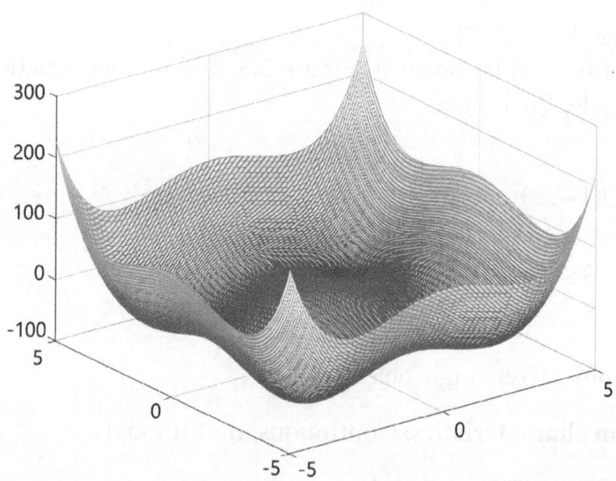

FIGURE 3.9 Styblinski–Tang function.

$$f(\boldsymbol{x}) = \frac{1}{2}\sum_{1}^{n}\left(x_i^4 - 16x_i^2 + 5x_i\right)$$
$$n = 2 \quad -5 \le x_i \le 5, \quad i = 1,\ldots,n \tag{3.11}$$

Design objective: Single objective

Function characteristics: Continuous, multimodal

Dimensions: n-Dimensional (the figure displays its two-dimensional version, and this book uses both two-dimensional and five-dimensional versions)

Optimal value: –78.332 (two-dimensional); –195.831 (five-dimensional)

3.2.1.12 Alpine Function

The graph of the Alpine function is shown in Figure 3.10, and its mathematical expression is given in Eq. (3.12).

$$f(\boldsymbol{x}) = \prod_{i=1}^{2} \sqrt{x_i}\,\sin(x_i) \tag{3.12}$$

$$n=2 \quad 0 \le x_1 \le 10, \quad 0 \le x_2 \le 10$$

Design objective: Single objective

Function characteristics: Continuous, multimodal

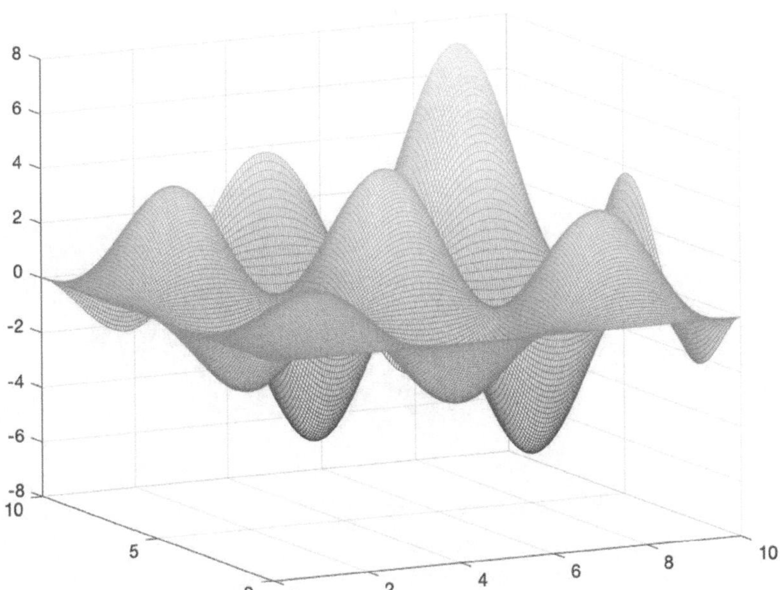

FIGURE 3.10 Alpine function.

Dimensions: 2-Dimensional

Optimal value: –6.130

3.2.1.13 F1 Function

The graph of the F1 function is shown in Figure 3.11, and its mathematical expression is given in Eq. (3.13).

$$\begin{aligned} &f(\boldsymbol{x}) = x_1^2 + x_2^2 - \cos(18x_1) - (18x_2) \\ &n = 2 \quad -1 \leq x_1 \leq 1, \quad -1 \leq x_2 \leq 1 \end{aligned} \tag{3.13}$$

Design objective: Single objective

Function characteristics: Continuous, multimodal

Dimensions: 2-Dimensional

Optimal value: –2

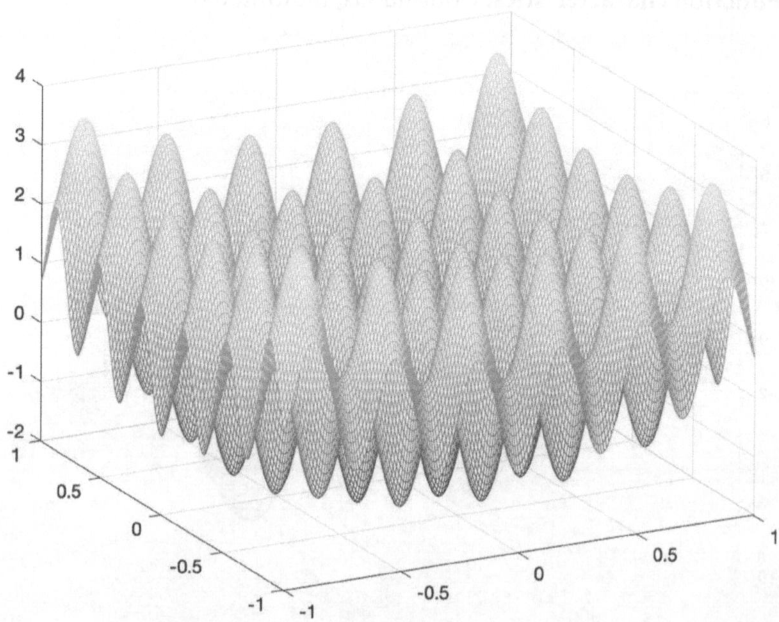

FIGURE 3.11 F1 function.

3.2.1.14 Himmelblau Function

The graph of the Himmelblau function is shown in Figure 3.12, and its mathematical expression is given in Eq. (3.14).

$$f(\boldsymbol{x}) = 3(1-x_1)^2 e^{-x_1^2-(x_2+1)^2} - 10\left(\frac{x_1}{5} - x_1^3 - x_2^5\right)e^{-x_1^2-x_2^2} - \frac{1}{3}e^{-(x_1+1)^2-x_2^2}$$

$$n=2 \quad -3 \leq x_1 \leq 3, \quad -4 \leq x_2 \leq 4 \tag{3.14}$$

Design objective: Single objective

Function characteristics: Continuous, multimodal

Dimensions: 2-Dimensional

Optimal value: 0

Description: The function has four extrema, all of which are global optimal points.

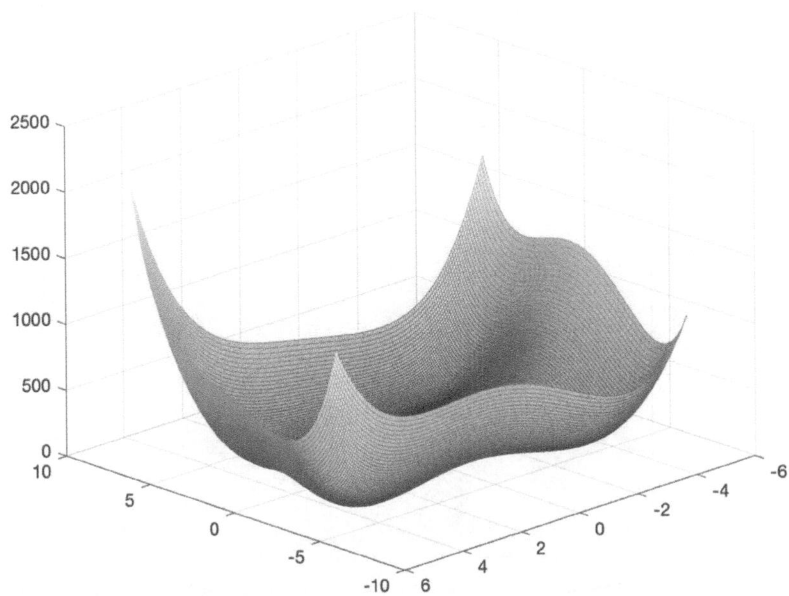

FIGURE 3.12 Himmelblau function.

3.2.1.15 Shubert Function

The graph of the Shubert function is shown in Figure 3.13, and its mathematical expression is given in Eq. (3.15).

$$f(\boldsymbol{x})=\left(\sum_{i=1}^{5} i\cos((i+1)x_1+i)\right)\left(\sum_{i=1}^{5} i\cos((i+1)x_2+i)\right) \tag{3.15}$$

$$n=2 \quad -2\le x_1\le 2, \quad -2\le x_2\le 2$$

Design objective: Single objective

Function characteristics: Continuous, multimodal

Dimensions: 2-Dimensional

Optimal value: –186.7309

3.2.1.16 Banana Function

The graph of the Banana function (BA/Rosenbrock) is shown in Figure 3.14, and its mathematical expression is given in Eq. (3.16).

$$f(\boldsymbol{x})=\sum_{i=1}^{n-1}\left[100(x_{i+1}-x_i^2)^2+(x_i-1)^2\right] \tag{3.16}$$

$$n=2 \quad -2\le x_i\le 2, \quad i=1,\ldots,n$$

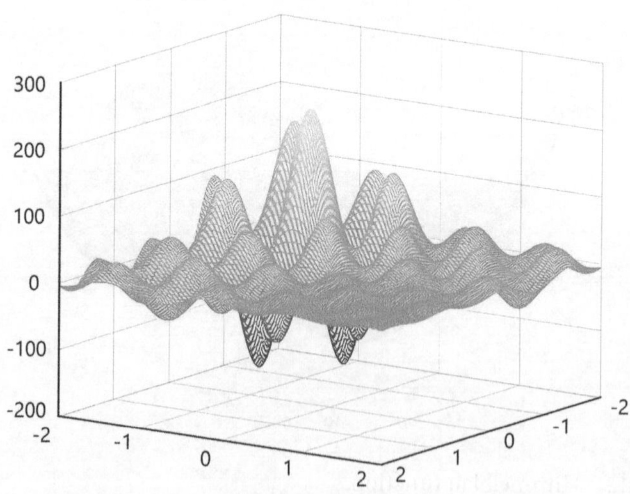

FIGURE 3.13 Shubert function.

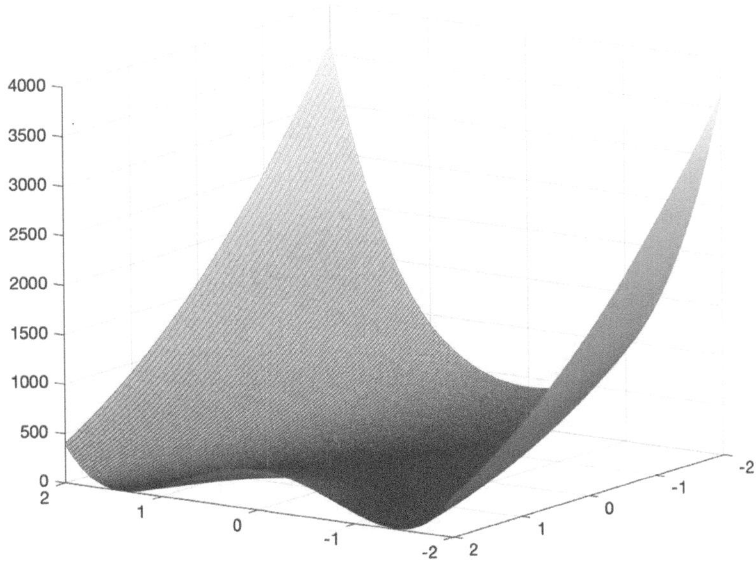

FIGURE 3.14 Banana function.

Design objective: Single objective

Function characteristics: Continuous, multimodal

Dimensions: n-Dimensional (the figure displays its two-dimensional version, and this book uses its two-dimensional version)

Optimal value: 0

3.2.1.17 Sasena Function

The graph of the Sasena function is shown in Figure 3.15, and its mathematical expression is given in Eq. (3.17).

$$f(\boldsymbol{x}) = 2 + 0.01(x_2 - x_1^2)^2 + (1 - x_1)^2 + 2(2 - x_2)^2 + 7\sin(0.5x_1)\sin(0.7x_1x_2)$$
$$n = 2 \quad 0 \le x_1 \le 5, \quad 0 \le x_2 \le 5 \tag{3.17}$$

Design objective: Single objective

Function characteristics: Continuous

Dimensions: 2-Dimensional

Optimal value: −1.457

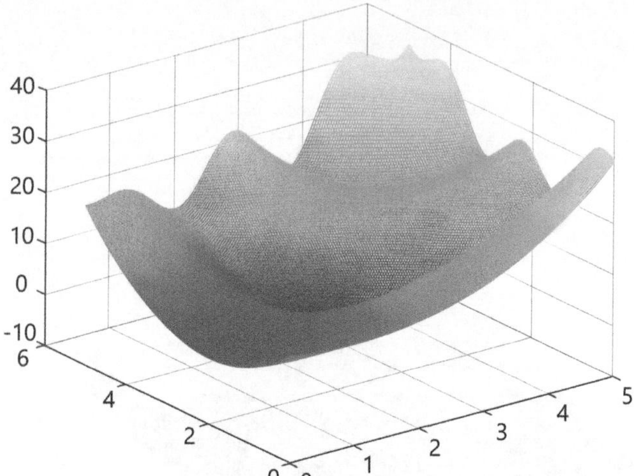

FIGURE 3.15 Sasena function.

3.2.1.18 Goldstein–Price Function

The Goldstein–Price function is shown in Figure 3.16, and its mathematical expression is given in Eq. (3.18).

$$\begin{aligned} f(\boldsymbol{x}) = & \left[1+(x_1+x_2+1)^2(19-14x_1+3x_1^2-14x_2+6x_1x_2+3x_2^2)\right] \\ & \times\left[30+(2x_1-3x_2)^2(18-32x_1+12x_1^2+48x_2-36x_1x_2+27x_2^2\right] \\ & n=2 \quad -2\le x_1\le 2, \quad -2\le x_2\le 2 \end{aligned} \tag{3.18}$$

Design objective: Single objective

Function characteristics: Continuous, multimodal

Dimensions: 2-Dimensional

Optimal value: 3

3.2.1.19 Rastrigin Function

The Rastrigin function is shown in Figure 3.17, and its mathematical expression is given in Eq. (3.19).

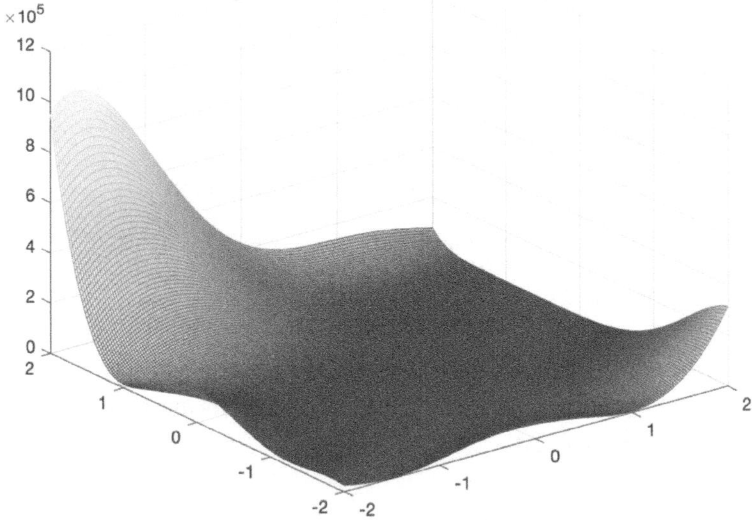

FIGURE 3.16 Goldstein–Price function.

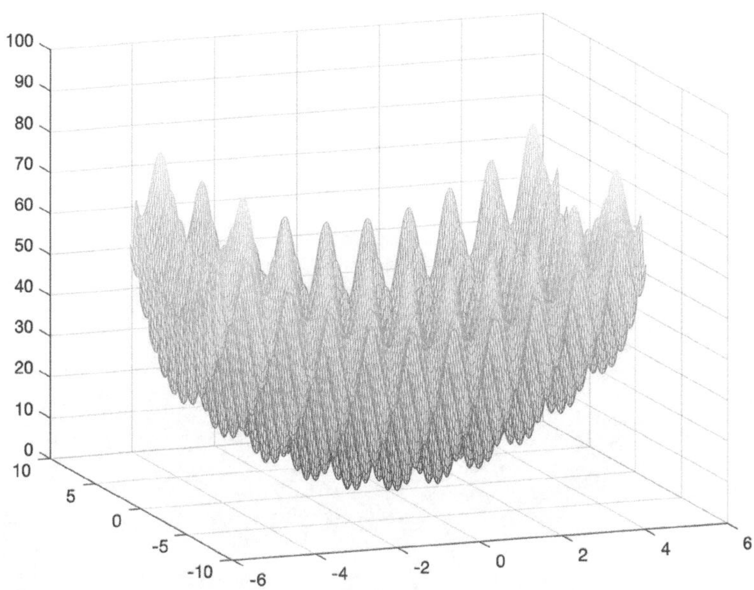

FIGURE 3.17 Rastrigin function.

$$f(\boldsymbol{x})=20+\sum_{i=1}^{2}\left(x_i^2-10\cos(2\pi x_i)\right)$$
$$n=2 \quad -5.12\le x_1\le 5.12, \quad -5.12\le x_2\le 5.12 \tag{3.19}$$

Design objective: Single objective

Function characteristics: Continuous, multimodal

Dimensions: 2-Dimensional

Optimal value: 0

3.2.1.20 Alpine1 Function

The Alpine1 function is shown in Figure 3.18, and its mathematical expression is given in Eq. (3.20).

$$f(\boldsymbol{x})=\sum_{i=1}^{2}\left|x_i\sin(x_i)+0.1x_i\right|$$
$$n=2 \quad -10\le x_1\le 10, \quad -10\le x_2\le 10 \tag{3.20}$$

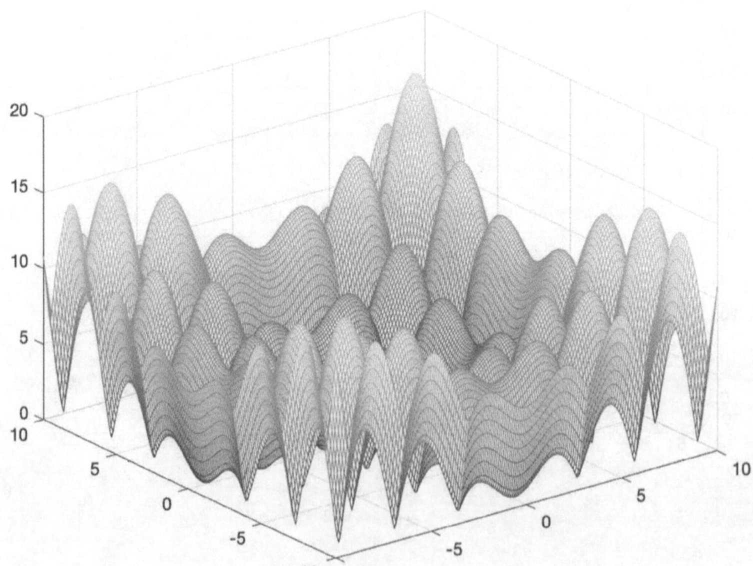

FIGURE 3.18 Alpine1 function.

Design objective: Single objective

Function characteristics: Continuous, multimodal

Dimensions: 2-Dimensional

Optimal value: 0

3.2.1.21 Alpine2 Function

The Alpine2 function is shown in Figure 3.19, and its mathematical expression is given in Eq. (3.21).

$$f(\boldsymbol{x}) = \prod_{i=1}^{2} \sqrt{x_i} \sin(x_i) \tag{3.21}$$

$$n = 2 \quad 0 \le x_1 \le 10, \quad 0 \le x_2 \le 10$$

Design objective: Single objective

Function characteristics: Continuous, multimodal

Dimensions: 2-Dimensional

Optimal value: −6.13

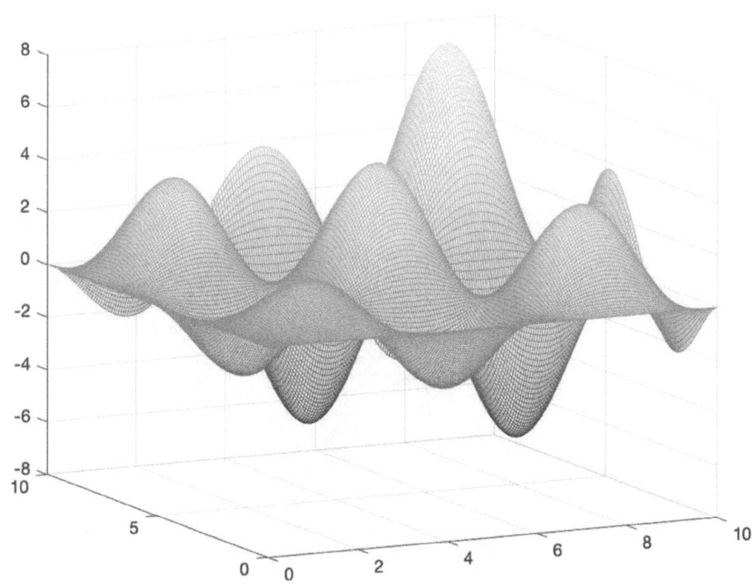

FIGURE 3.19 Alpine2 function.

3.2.1.22 Bird Function

The Bird function is shown in Figure 3.20, and its mathematical expression is given in Eq. (3.22).

$$f(\boldsymbol{x}) = \sin(x_1)e^{(1-\cos(x_2))^2} + \cos(x_2)e^{(1-\sin(x_1))^2} + (x_1 - x_2)^2$$
$$n=2 \quad -2\pi \le x_1 \le 2\pi, \quad -2\pi \le x_2 \le 2\pi \tag{3.22}$$

Design objective: Single objective

Function characteristics: Continuous, multimodal

Dimensions: 2-Dimensional

Optimal value: –106.76

3.2.1.23 Easom Function

The Easom function is shown in Figure 3.21, and its mathematical expression is given in Eq. (3.23).

$$f(\boldsymbol{x}) = -\cos(x_1)\cos(x_2)e^{(-(x_1-\pi)^2-(x_2-\pi)^2)}$$
$$n=2 \quad -10 \le x_1 \le 10, \quad -10 \le x_2 \le 10 \tag{3.23}$$

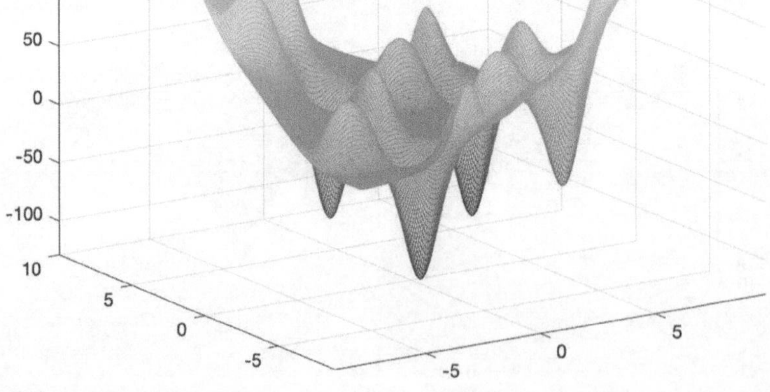

FIGURE 3.20 Bird function.

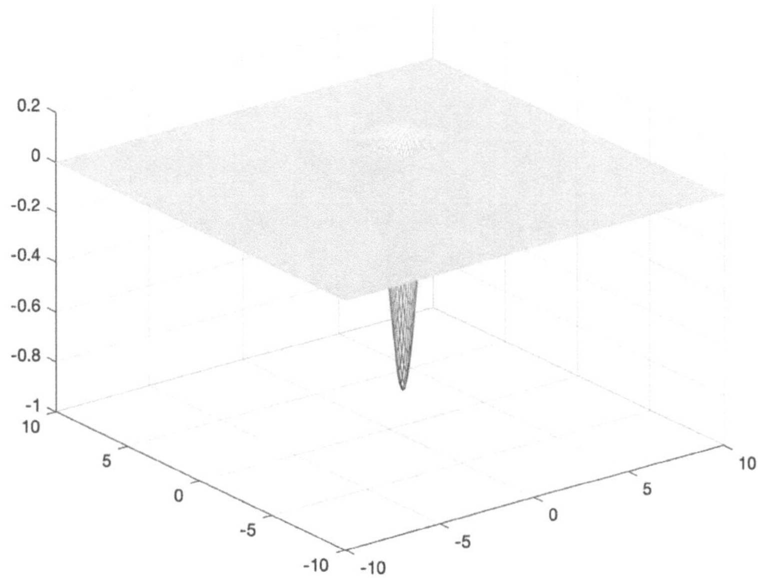

FIGURE 3.21 Easom function.

Design objective: Single objective

Function characteristics: Continuous, unimodal

Dimensions: 2-Dimensional

Optimal value: –1

3.2.1.24 Schaffer2 Function

The Schaffer2 function is shown in Figure 3.22, and its mathematical expression is given in Eq. (3.24).

$$f(\boldsymbol{x}) = 0.5 + \frac{\sin^2(x_1^2 - x_2^2) - 0.5}{\left[1 + 0.001(x_1^2 + x_2^2)\right]^2} \tag{3.24}$$

$$n = 2 \quad -2 \le x_1 \le 2, \quad -2 \le x_2 \le 2$$

Design objective: Single objective

Function characteristics: Continuous, multimodal

Dimensions: 2-Dimensional

Optimal value: 0

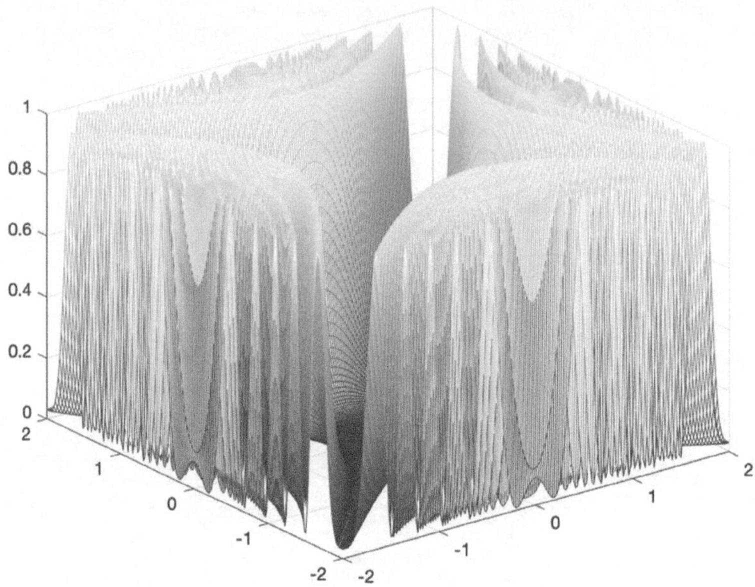

FIGURE 3.22 Schaffer2 function.

3.2.1.25 Levy Function

The Levy function is shown in Figure 3.23, and its mathematical expression is given in Eq. (3.25).

$$f(\boldsymbol{x}) = \sin^2(\pi y_1) + \sum_{i=1}^{n-1}\left[(y_i - 1)^2(1 + 10\sin^2(\pi y_i + 1))\right]$$

$$+(y_n - 1)^2(1 + 10\sin^2(2\pi y_n)) \quad y_i = 1 + \frac{x_i - 1}{4} \tag{3.25}$$

$$n = 4 \quad -10 \le x_i \le 10, \quad i = 1,\ldots,n$$

Design objective: Single objective

Function characteristics: Continuous, multimodal

Dimensions: n-Dimensional (the figure displays its two-dimensional version, and this book uses its four-dimensional version)

Optimal value: 0

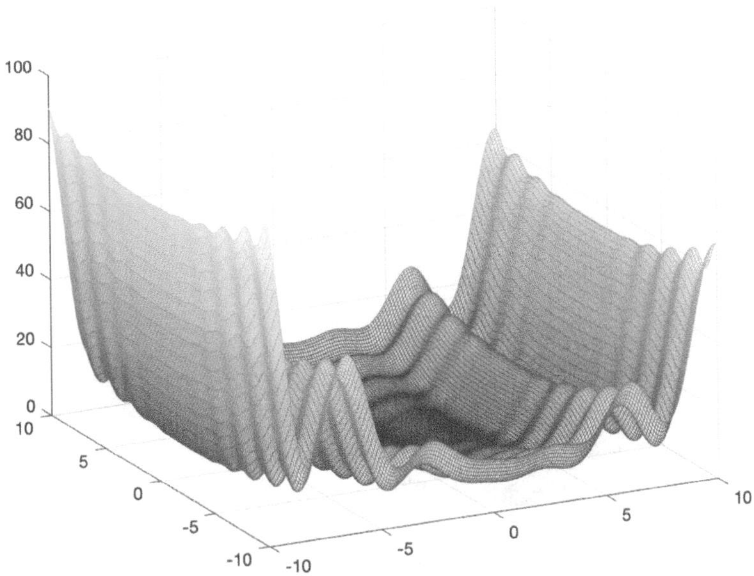

FIGURE 3.23 Levy function.

3.2.1.26 Dixon–Price Function

The Dixon–Price function is shown in Figure 3.24, and its mathematical expression is given in Eq. (3.26).

$$f(\boldsymbol{x}) = (x_1 - 1)^2 + \sum_{i=2}^{n} i\left(2x_i^2 - x_{i-1}\right)^2 \tag{3.26}$$

$$n = 4 \quad -10 \le x_i \le 10, \quad i = 1,\dots,n$$

Design objective: Single objective

Function characteristics: Continuous, unimodal

Dimensions: n-Dimensional (the figure displays its two-dimensional version, and this book uses its four-dimensional version)

Optimal value: 0

3.2.1.27 Shekel Function

The mathematical expression is given in Eq. (3.27).

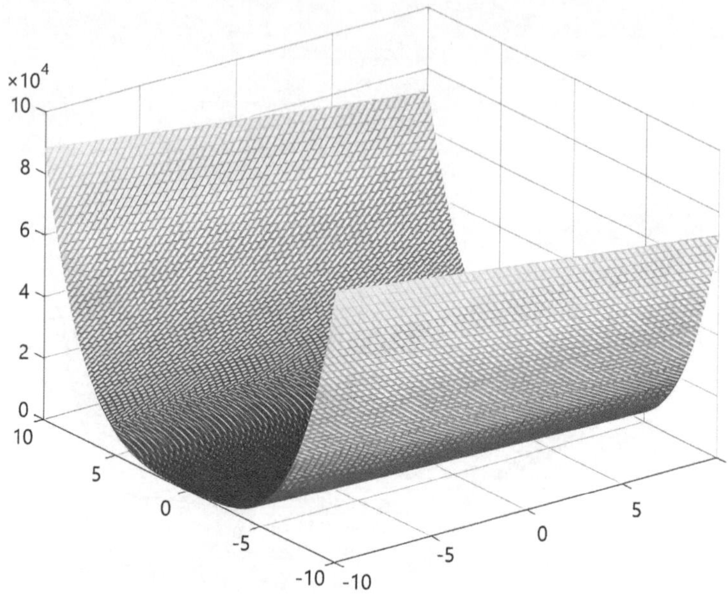

FIGURE 3.24 Dixon–Price function.

$$f(\boldsymbol{x}) = -\sum_{i=1}^{10}\left(c_i + \sum_{j=1}^{4}(x_j - a_{ji})^2\right)^{-1}$$

$$a = \begin{bmatrix} 4\,1\,8\,6\,3\,2\,5\,8\,6\,7 \\ 4\,1\,8\,6\,7\,9\,5\,1\,2\,3.6 \\ 4\,1\,8\,6\,3\,2\,3\,8\,6\,7 \\ 4\,1\,8\,6\,7\,9\,3\,1\,2\,3.6 \end{bmatrix}$$

$$c = [0.1\;0.2\;0.2\;0.4\;0.4\;0.6\;0.3\;0.7\;0.5\;0.5] \tag{3.27}$$

$$0 \le x_i \le 10, \quad i = 1,2,3,4$$

Design objective: Single objective

Function characteristics: Continuous, multimodal

Dimensions: 4-Dimensional

Optimal value: –10.1532

3.2.1.28 Hartman6 Function

The mathematical expression is given in Eq. (3.28).

$$f(\boldsymbol{x}) = -\sum_{i=1}^{4} a_i \exp\left[-\sum_{j=1}^{6} B_{ij}(x_j - Q_{ij})^2\right]$$

$$a = [1, 1.2, 3, 3.2]^T, \quad B = \begin{bmatrix} 10 & 3 & 17 & 3.5 & 1.7 & 8 \\ 0.05 & 10 & 17 & 0.1 & 8 & 14 \\ 3 & 3.5 & 1.7 & 10 & 17 & 8 \\ 17 & 8 & 0.05 & 10 & 0.1 & 14 \end{bmatrix} \quad (3.28)$$

$$Q = 10^{-4} \begin{bmatrix} 1{,}312 & 1{,}696 & 5{,}569 & 124 & 8{,}283 & 5{,}886 \\ 2{,}329 & 4{,}135 & 8{,}307 & 3{,}736 & 1{,}004 & 9{,}991 \\ 2{,}348 & 1{,}451 & 3{,}522 & 2{,}883 & 3{,}047 & 6{,}650 \\ 4{,}047 & 8{,}828 & 8{,}732 & 5{,}743 & 1{,}091 & 381 \end{bmatrix}$$

$$n = 6 \quad 0 \le x_i \le 1, \quad i = 1, \ldots, n$$

Design objective: Single objective

Function characteristics: Continuous

Dimensions: 6-Dimensional

Optimal value: –3.322

3.2.2 Unconstrained High-Dimensional Problems

3.2.2.1 Schwefel3 Function

The Schwefel3 function is shown in Figure 3.25, and its mathematical expression is given in Eq. (3.29).

$$f(\boldsymbol{x}) = \sum_{i=2}^{n} (x_i - 1)^2 + (x_1 - x_i^2)^2 \quad (3.29)$$

$$n = 8 \quad 0 \le x \le 5, \quad i = 1, \ldots, n$$

Design objective: Single objective

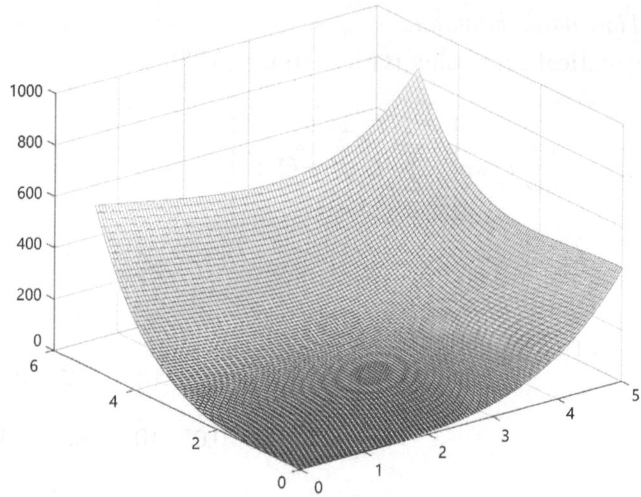

FIGURE 3.25 Schwefel3 function.

Function characteristics: Continuous, unimodal

Dimensions: *n*-Dimensional (this book uses its eight-dimensional version)

Optimal value: 0

3.2.2.2 Convex Function

The mathematical expression is given in Eq. (3.30).

$$\min f(\boldsymbol{x}) = 3.1x_1^2 + 7.6x_2^2 + 6.9x_3^2 + 0.004x_4^2 + 19x_5^2 + 3x_6^2 + x_7^2 + 4x_8^2$$
$$s.t.\ \ x_i \in \{-10, 9, \ldots, 9, 10\},\quad i = 1, \ldots, 8 \tag{3.30}$$

Design objective: Single objective

Function characteristics: Discrete

Dimensions: 8-Dimensional

3.2.2.3 Nvs09 Function

The mathematical expression is given in Eq. (3.31).

$$\min f(\boldsymbol{x}) = 3.1x_1^2 + 7.6x_2^2 + 6.9x_3^2 + 0.004x_4^2 + 19x_5^2 + 3x_6^2 + x_7^2 + 4x_8^2$$
$$s.t.\ \ x_i \in \{-10, 9, \ldots, 9, 10\},\quad i = 1, \ldots, 8 \tag{3.31}$$

Design objective: Single objective

Function characteristics: Discrete

Dimensions: 10-Dimensional

3.2.2.4 AlteredNvs09 Function

The mathematical expression is given in Eq. (3.32).

$$\min f(\boldsymbol{x}) = \sum_{i=1}^{10} \left(\log(x_i - 2)^2 + \log(10 - x_i)^2 \right) - \left(\prod_{i=1}^{10} x_i \right)^{0.2} \tag{3.32}$$

$$s.t.\ \ x_i \in \{3,4,\ldots,99\},\ \ i = 1,\ldots,10$$

Design objective: Single objective

Function characteristics: Discrete

Dimensions: 10-Dimensional

3.2.2.5 Paviani Function

The mathematical expression is given in Eq. (3.33).

$$f(\boldsymbol{x}) = \sum_{i=1}^{n} \left(\log(x_i - 2)^2 + \log(10 - x_i)^2 \right) - \left(\prod_{i=1}^{n} x_i \right)^{0.2} \tag{3.33}$$

$$n = 10\ \ 2.1 \le x_i \le 9.9,\ \ i = 1,\ldots,n$$

Design objective: Single objective

Function characteristics: Discrete

Dimensions: 10-Dimensional

Optimal value: −45.8

3.2.2.6 Trid Function

The Trid function (Trid/Trid6/Trid10) is shown in Figure 3.26, and its mathematical expression is given in Eq. (3.34).

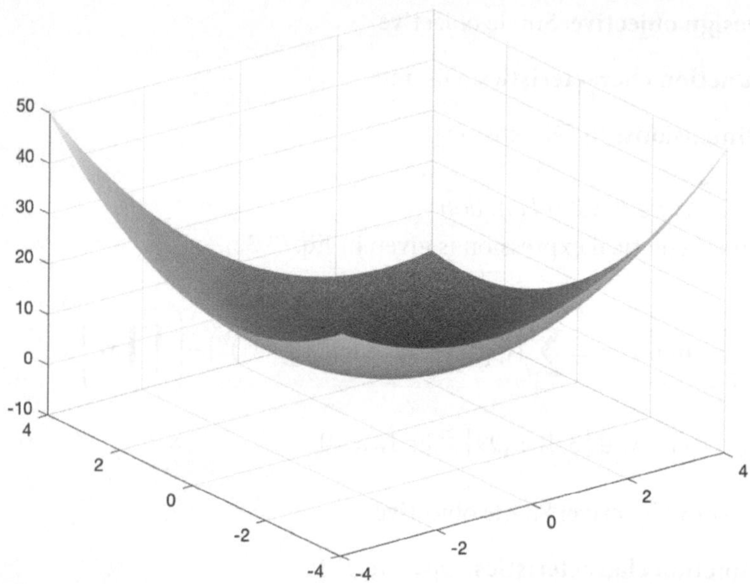

FIGURE 3.26 Trid function.

$$f(\boldsymbol{x})=\sum_{i=1}^{n}(x_i-1)^2-\sum_{i=2}^{n}x_i x_{i-1}$$

$$n=10 \quad -n^2 \le x_i \le n^2,\quad i=1,\ldots,n \tag{3.34}$$

$$n=6 \quad -n^2 \le x_i \le n^2,\quad i=1,\ldots,n$$

Design objective: Single objective

Function characteristics: Continuous, unimodal

Dimensions: n-Dimensional (this book uses its six- and ten-dimensional versions)

Optimal value: –50 (six-dimensional); –210 (ten-dimensional)

3.2.2.7 Rastrigin01 Function

The mathematical expression is given in Eq. (3.35).

$$\min f(\boldsymbol{x})=\sum_{i=1}^{n}x_i^2-\cos(2\pi x_i)$$

$$s.t.\quad x_i \in \{-1,0,1,2,3\},\quad i=1,\ldots,12 \tag{3.35}$$

Design objective: Single objective

Function characteristics: Discrete

Dimensions: 12-Dimensional

3.2.2.8 Rastrigin02 Function

The mathematical expression is given in Eq. (3.36).

$$\min f(\boldsymbol{x}) = \sum_{i=1}^{n} x_i^2 - \cos(2\pi x_i)$$
$$s.t. \quad x_i \in \{-10,-9,\ldots,29,30\}, \quad i=1,\ldots,12 \tag{3.36}$$

Design objective: Single objective

Function characteristics: Discrete

Dimensions: 12-Dimensional

3.2.2.9 Sum Squares Function

The sum squares function is shown in Figure 3.27, and its mathematical expression is given in Eq. (3.37).

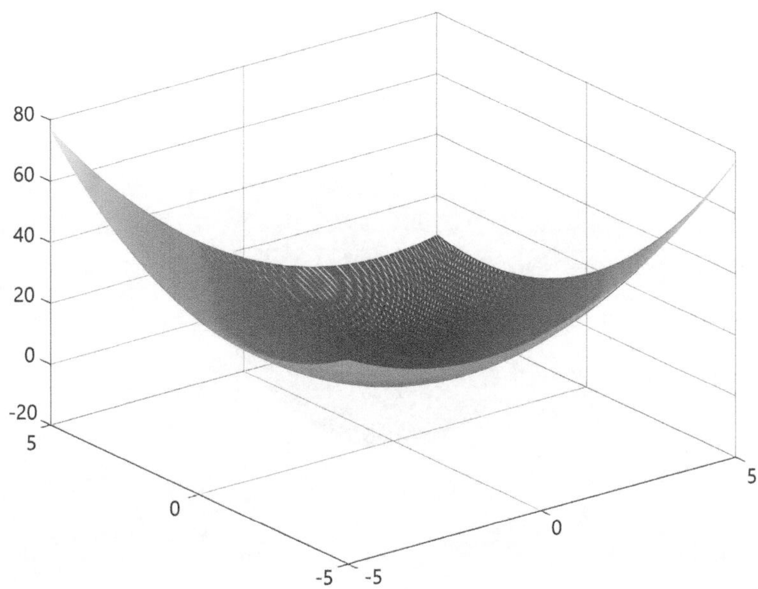

FIGURE 3.27 Sum squares function.

$$f(\boldsymbol{x}) = \sum_{i=1}^{n} i x_i^2$$

$$n = 15 \quad -5 \le x_i \le 5, \quad i = 1,\ldots,n \tag{3.37}$$

$$n = 20 \quad -10 \le x_i \le 10, \quad i = 1,\ldots,n$$

Design objective: Single objective

Function characteristics: Continuous, unimodal

Dimensions: n-Dimensional (this book uses its 15- and 20-dimensional versions)

Optimal value: 0

3.2.2.10 Sphere Function

The sphere function is shown in Figure 3.28, and its mathematical expression is given in Eq. (3.38).

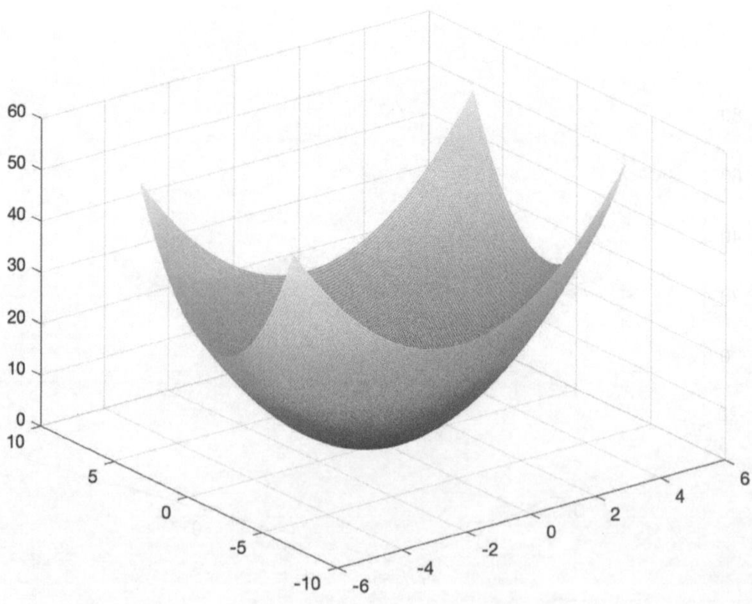

FIGURE 3.28 Sphere function.

$$
\begin{aligned}
& f(\boldsymbol{x})=\sum_{i=1}^{n} x_i^2 \\
& n=20 \quad -5.12 \le x_i \le 5.12, \quad i=1,\ldots,n \\
& n=15 \quad -5.12 \le x_i \le 5.12, \quad i=1,\ldots,n \\
& n=10 \quad -5.12 \le x_i \le 5.12, \quad i=1,\ldots,n
\end{aligned} \tag{3.38}
$$

Design objective: Single objective

Function characteristics: Continuous, unimodal

Dimensions: n-Dimensional (this book uses its 10-, 15- and 20-dimensional versions)

Optimal value: 0

3.2.2.11 F16 Function

The mathematical expression is given in Eq. (3.39).

$$
\begin{aligned}
& f(\boldsymbol{x})=\sum_{i=1}^{16}\sum_{j=1}^{16} a_{ij}(x_i^2+x_i+1)(x_j^2+x_j+1) \\
& a_{ij(row1-8)}=\begin{bmatrix}
1\,0\,0\,1\,0\,0\,1\,1\,0\,0\,0\,0\,0\,0\,0\,1 \\
0\,1\,1\,0\,0\,0\,1\,0\,0\,1\,0\,0\,0\,0\,0\,0 \\
0\,0\,1\,0\,0\,0\,1\,0\,1\,1\,0\,0\,0\,1\,0\,0 \\
0\,0\,0\,1\,0\,0\,1\,0\,0\,0\,1\,0\,0\,0\,1\,0 \\
0\,0\,0\,0\,1\,1\,0\,0\,0\,1\,0\,1\,0\,0\,0\,1 \\
0\,0\,0\,0\,0\,1\,0\,1\,0\,0\,0\,0\,0\,0\,1\,0 \\
0\,0\,0\,0\,0\,0\,1\,0\,0\,0\,1\,0\,1\,0\,0\,0 \\
0\,0\,0\,0\,0\,0\,0\,1\,0\,1\,0\,0\,0\,0\,1\,0
\end{bmatrix}
\quad a_{ij(row9-16)}=\begin{bmatrix}
0\,0\,0\,0\,0\,0\,0\,0\,1\,0\,0\,1\,0\,0\,0\,1 \\
0\,0\,0\,0\,0\,0\,0\,0\,0\,1\,0\,0\,0\,1\,0\,0 \\
0\,0\,0\,0\,0\,0\,0\,0\,0\,0\,1\,0\,1\,0\,0\,0 \\
0\,0\,0\,0\,0\,0\,0\,0\,0\,0\,0\,1\,0\,1\,0\,0 \\
0\,0\,0\,0\,0\,0\,0\,0\,0\,0\,0\,0\,1\,1\,0\,0 \\
0\,0\,0\,0\,0\,0\,0\,0\,0\,0\,0\,0\,0\,1\,0\,0 \\
0\,0\,0\,0\,0\,0\,0\,0\,0\,0\,0\,0\,0\,0\,1\,0 \\
0\,0\,0\,0\,0\,0\,0\,0\,0\,0\,0\,0\,0\,0\,0\,1
\end{bmatrix} \\
& n=16 \quad -1 \le x_i \le 1, \quad i=1,\ldots,n
\end{aligned} \tag{3.39}
$$

Design objective: Single objective

Function characteristics: Continuous

Dimensions: 16-Dimensional

Optimal value: 25.875

3.3 CONSTRAINED OPTIMIZATION PROBLEMS

3.3.1 Constrained Low-Dimensional Problems

3.3.1.1 g06

The mathematical expression of g06 is given in Eq. (3.40).

$$
\begin{aligned}
& f(\boldsymbol{x})=(x_1-10)^3+(x_2-20)^3 \\
& subject: \\
& g_1(\boldsymbol{x})=-(x_1-5)^2-(x_2-5)^2+100\leq 0 \\
& g_2(\boldsymbol{x})=(x_1-5)^2+(x_2-5)^2-82.81\leq 0 \\
& n=2 \quad 13\leq x_i\leq 100(i=1) \\
& \qquad\quad 0\leq x_i\leq 100(i=2)
\end{aligned}
\tag{3.40}
$$

Design objective: Single objective

Function characteristics: Continuous or discrete

Dimensions: 2-Dimensional

Optimal value: –6,961.8138 (when the functions are continuous)

Active constraints: g_1, g_2

Description: In the discrete case, the range of values for design variables is $x_1 \in \{13, 14, \ldots, 100\}$, $x_2 \in \{0, 1, 2, \ldots, 100\}$.

3.3.1.2 g08

The mathematical expression of g08 (G8) is given in Eq. (3.41).

$$
\begin{aligned}
& f(\boldsymbol{x})=-\frac{\sin^3(2\pi x_1)\sin(2\pi x_2)}{x_1^3(x_1+x_2)} \\
& subject: \\
& g_1(\boldsymbol{x})=x_1^2-x_2+1\leq 0 \\
& g_2(\boldsymbol{x})=1-x_1+(x_2-4)^2\leq 0 \\
& n=2 \quad 0\leq x_i\leq 10(i=1,2)
\end{aligned}
\tag{3.41}
$$

Design objective: Single objective

Function characteristics: Continuous

Dimensions: 2-Dimensional

Optimal value: –0.0958

3.3.1.3 g24

The mathematical expression of g24 is given in Eq. (3.42).

$$
\begin{aligned}
& f(\boldsymbol{x}) = -x_1 - x_2 \\
& subject: \\
& g_1(\boldsymbol{x}) = -2x_1^4 + 8x_1^3 - 8x_1^2 + x_2 - 2 \le 0 \\
& g_2(\boldsymbol{x}) = -4x_1^4 + 32x_1^3 - 88x_1^2 + 96x_1 + x_2 - 36 \le 0 \\
& n = 2 \quad 0 \le x_i \le 3 (i = 1) \\
& 0 \le x_i \le 4 (i = 2)
\end{aligned}
\tag{3.42}
$$

Design objective: Single objective

Function characteristics: Continuous

Dimensions: 2-Dimensional

Optimal value: –5.5080

3.3.1.4 Gomez

The mathematical expression of Gomez is given in Eq. (3.43).

$$
\begin{aligned}
& f(\boldsymbol{x}) = \left(4 - 2.1x_1^2 + \frac{1}{3}x_1^4\right)x_1^2 + x_1x_2 + (-4 + 4x_2^2)x_2^2 \\
& subject: \\
& -\sin(4\pi x_1) + 2\sin^2(2\pi x_2) \le 0 \\
& n = 2 \quad -0.5 \le x_i \le 0.5 (i = 1) \\
& \qquad -1 \le x_i \le 0 (i = 2)
\end{aligned}
\tag{3.43}
$$

Design objective: Single objective

Function characteristics: Continuous

Dimensions: 2-Dimensional

Optimal value: –0.9711

3.3.1.5 Sasena

The mathematical expression of Sasena is given in Eq. (3.44).

$$f(\boldsymbol{x})=2+0.01(x_2-x_1^2)^2+(1-x_1)^2+2(2-x_2)^2+7\sin(0.5x_1)\sin(0.7x_1x_2)$$

$$subject:$$

$$-\sin(x_1-x_2-\pi/8)\le 0$$

$$n=2 \quad 0\le x_i\le 5(i=1,2) \tag{3.44}$$

Design objective: Single objective

Function characteristics: Continuous

Dimensions: 2-Dimensional

Optimal value: –1.1743

3.3.1.6 Brianin

The mathematical expression of Brianin is given in Eq. (3.45).

$$f(\boldsymbol{x})=\left(x_2-\frac{5.1}{4\pi^2}x_1^2+\frac{5}{\pi}x_1-6\right)^2+10\left(1-\frac{1}{8\pi}\right)\cos(x_1)+10$$

$$subject:$$

$$a=1;b=\frac{5.1}{4\pi^2};c=-\frac{5}{\pi};d=6;h=10;ff=\frac{1}{8\pi} \tag{3.45}$$

$$a(x_2-bx_1^2-cx_1-d)+h(1-ff)\cos x_1-5+h\le 0$$

$$n=2 \quad -5\le x_i\le 10(i=1)$$

$$0\le x_i\le 15(i=2)$$

Design objective: Single objective

Function characteristics: Continuous

Dimensions: 2-Dimensional

Optimal value: 0.3979

3.3.1.7 g12

The mathematical expression of g12 is given in Eq. (3.46).

$$\begin{aligned}
&f(\boldsymbol{x}) = -(100 - (x_1 - 5)^2 - (x_2 - 5)^2 - (x_3 - 5)^2)/100 \\
&subject: \\
&g_1(\boldsymbol{x}) = (x_1 - p)^2 + (x_2 - q)^2 + (x_3 - r)^2 - 0.0625 \le 0 \\
&n = 3 \quad 0 \le x_i \le 10 (i = 1,2,3); \quad p,q,r = 1,2,\ldots,9
\end{aligned} \tag{3.46}$$

Design objective: Single objective

Function characteristics: Continuous

Dimensions: 3-Dimensional

Optimal value: −1

3.3.1.8 g04

The mathematical expression of g04 (g04/G4/Him) is given in Eq. (3.47).

$$\begin{aligned}
&f(\boldsymbol{x}) = 5.3578547x_3^2 + 0.8356891x_1x_5 + 37.293239x_1 - 40792.141 \\
&subject: \\
&g_1(\boldsymbol{x}) = 85.334407 + 0.0056858x_2x_5 + 0.0006262x_1x_4 - 0.0022053x_3x_5 - 92 \le 0 \\
&g_2(\boldsymbol{x}) = -85.334407 - 0.0056858x_2x_5 - 0.0006262x_1x_4 + 0.0022053x_3x_5 \le 0 \\
&g_3(\boldsymbol{x}) = 80.51249 + 0.0071317x_2x_5 + 0.0029955x_1x_2 + 0.0021813x_3^2 - 110 \le 0 \\
&g_4(\boldsymbol{x}) = -80.51249 - 0.0071317x_2x_5 - 0.0029955x_1x_2 - 0.0021813x_3^2 + 90 \le 0 \\
&g_5(\boldsymbol{x}) = 9.300961 + 0.0047026x_3x_5 + 0.0012547x_1x_3 + 0.0019085x_3x_4 - 25 \le 0 \\
&g_6(\boldsymbol{x}) = -9.300961 - 0.0047026x_3x_5 - 0.0012547x_1x_3 - 0.0019085x_3x_4 + 20 \le 0 \\
&n = 5 \quad 78 \le x_i \le 102 (i = 1) \\
&\qquad\quad 33 \le x_i \le 45 (i = 2) \\
&\qquad\quad 27 \le x_i \le 45 (i = 3,4,5)
\end{aligned} \tag{3.47}$$

Design objective: Single objective

Function characteristics: Continuous or discrete

Dimensions: 5-Dimensional

Optimal value: –30,665.5386 (when the functions are continuous)

Active constraints: g_1, g_6

3.3.1.9 Ex1221

The mathematical expression of Ex1221 is given in Eq. (3.48).

$$\begin{aligned} \min f(\boldsymbol{x}) &= 2x_1 + 3x_2 + 1.5x_3 + 2x_4 - 0.5x_5 \\ s.t. \quad & x_1^2 + x_3 \le 1.25, \quad 1.333x_2 + x_4 \le 3 \\ & x_2^{1.5} + 1.5x_4 \le 3, \quad -x_3 - x_4 + x_5 \le 0 \\ & x_1 + x_3 \le 1.6, \\ & x_1, x_2, x_3 \in \{0,1,\ldots,10\} \quad x_4, x_5 \in \{0,1\} \end{aligned} \tag{3.48}$$

Design objective: Single objective

Function characteristics: Discrete

Dimensions: 5-Dimensional

3.3.1.10 Altered ex1221

The mathematical expression of Altered ex1221 is given in Eq. (3.49).

$$\begin{aligned} \min f(\boldsymbol{x}) &= -2x_1 - 3x_2 - 1.5x_3 - 2x_4 + 0.5x_5 \\ s.t. \quad & x_1 + x_3 \le 1.6, \quad 1.333x_2 + x_4 \le 3 \\ & -x_3 - x_4 + x_5 \le 0 \\ & x_1, x_2, x_3 \in \{0,1,\ldots,10\} \quad x_4, x_5 \in \{0,1\} \end{aligned} \tag{3.49}$$

Design objective: Single objective

Function characteristics: Discrete

Dimensions: 5-Dimensional

3.3.1.11 g16

The mathematical expression of g16 is given in Eq. (3.50).

$$f(\boldsymbol{x}) = 0.000117 y_{14} + 0.1365 + 0.00002358 y_{13} + 0.000001502 y_{16} + 0.0321 y_{12}$$
$$+ 0.004323 y_5 + 0.0001 \frac{c_{15}}{c_{16}} + 37.48 \frac{y_2}{c_{12}} - 0.0000005843 y_{17}$$

subject:

$$g_1(\boldsymbol{x}) = \frac{0.28}{0.72} y_5 - y_4 \le 0,\ g_2(\boldsymbol{x}) = x_3 - 1.5 x_2 \le 0,\ g_3(\boldsymbol{x}) = 3{,}496 \frac{y_2}{c_{12}} - 12 \le 0,$$
$$g_4(\boldsymbol{x}) = 110.6 + y_1 - \frac{62{,}212}{c_{17}} \le 0,\ g_5(\boldsymbol{x}) = 213.1 - y_1 \le 0,\ g_6(\boldsymbol{x}) = y_1 - 405.23 \le 0,$$
$$g_7(\boldsymbol{x}) = 17.505 - y_2 \le 0,\ g_8(\boldsymbol{x}) = y_2 - 1{,}053.6667 \le 0,\ g_9(\boldsymbol{x}) = 11.275 - y_3 \le 0,$$
$$g_{10}(\boldsymbol{x}) = y_3 - 35.03 \le 0, g_{11}(\boldsymbol{x}) = 214.228 - y_4 \le 0,\ g_{12}(\boldsymbol{x}) = y_4 - 665.585 \le 0,$$
$$g_{13}(\boldsymbol{x}) = 7.458 - y_5 \le 0,\ g_{14}(\boldsymbol{x}) = y_5 - 584.463 \le 0,\ g_{15}(\boldsymbol{x}) = 0.961 - y_6 \le 0,$$
$$g_{16}(\boldsymbol{x}) = y_6 - 265.916 \le 0,\ g_{17}(\boldsymbol{x}) = 1.612 - y_7 \le 0,\ g_{18}(\boldsymbol{x}) = y_7 - 7.046 \le 0,$$
$$g_{19}(\boldsymbol{x}) = 0.146 - y_8 \le 0,\ g_{20}(\boldsymbol{x}) = y_8 - 0.222 \le 0,\ g_{21}(\boldsymbol{x}) = 107.99 - y_9 \le 0,$$

$$\begin{aligned}
&g_{22}(\boldsymbol{x}) = y_9 - 273.366 \le 0,\ g_{23}(\boldsymbol{x}) = 922.693 - y_{10} \le 0,\\
&g_{24}(\boldsymbol{x}) = y_{10} - 1{,}286.105 \le 0,\ g_{25}(\boldsymbol{x}) = 926.832 - y_{11} \le 0,\\
&g_{26}(\boldsymbol{x}) = y_{11} - 1{,}444.046 \le 0,\ g_{27}(\boldsymbol{x}) = 18.766 - y_{12} \le 0,\\
&g_{28}(\boldsymbol{x}) = y_{12} - 537.141 \le 0,\ g_{29}(\boldsymbol{x}) = 1{,}072.163 - y_{13} \le 0,\\
&g_{30}(\boldsymbol{x}) = y_{13} - 3{,}247.039 \le 0,\ g_{31}(\boldsymbol{x}) = 8{,}961.448 - y_{14} \le 0,\\
&g_{32}(\boldsymbol{x}) = y_{14} - 26{,}844.086 \le 0,\ g_{33}(\boldsymbol{x}) = 0.063 - y_{15} \le 0,\\
&g_{34}(\boldsymbol{x}) = y_{15} - 0.386 \le 0,\ g_3(\boldsymbol{x}) = 71{,}084.33 - y_{16} \le 0,\\
&g_{36}(\boldsymbol{x}) = y_{16} - 140{,}000 \le 0,\ g_{37}(\boldsymbol{x}) = 2{,}802{,}713 - y_{17} \le 0,\\
&g_{38}(\boldsymbol{x}) = y_{17} - 12{,}146{,}108 \le 0,
\end{aligned} \tag{3.50}$$

Design objective: Single objective

Function characteristics: Continuous

Dimensions: 5-Dimensional

Optimal value: –1.9051

3.3.1.12 g09

The mathematical expression of g09 (G9) is given in Eq. (3.51).

$$
\begin{aligned}
& f(\boldsymbol{x})=(x_1-10)^2+5(x_2-12)^2+x_3^4+3(x_4-11)^2+10x_5^6+7x_6^2 \\
& \qquad +x_7^4-4x_6x_7-10x_6-8x_7 \\
& subject: \\
& g_1(\boldsymbol{x})=-127+2x_1^2+3x_2^4+x_3+4x_4^2+5x_5\le 0 \\
& g_2(\boldsymbol{x})=-282+7x_1+3x_2+10x_3^2+x_4-5x_5\le 0 \\
& g_3(\boldsymbol{x})=-196+23x_1+x_2^2+6x_6^2-8x_7\le 0 \\
& g_4(\boldsymbol{x})=4x_1^2+x_2^2-3x_1x_2+2x_3^2+5x_6-11x_7\le 0 \\
& n=7 \quad -10\le x_i\le 10(i=1,\ldots,7)
\end{aligned}
\tag{3.51}
$$

Design objective: Single objective

Function characteristics: Continuous or discrete

Dimensions: 7-Dimensional

Optimal value: 680.6300 (when the functions are continuous)

Description: In the discrete case, the range of values for design variables is $x_i \in \{-10, -9, \ldots, 9, 10\}$, $i=1, \ldots, 7$.

3.3.2 Constrained High-Dimensional Problems

3.3.2.1 g10

The mathematical expression of g10 is given in Eq. (3.52).

$$
\begin{aligned}
& f(\boldsymbol{x}) = x_1 + x_2 + x_3 \\
& \textit{subject}: \\
& g_1(\boldsymbol{x}) = -1 + 0.0025(x_4 + x_6) \leq 0 \\
& g_2(\boldsymbol{x}) = -1 + 0.0025(x_5 + x_7 - x_4) \leq 0 \\
& g_3(\boldsymbol{x}) = -1 + 0.01(x_8 - x_5) \leq 0 \\
& g_4(\boldsymbol{x}) = -x_1 x_5 + 833.33252 x_4 + 100 x_1 - 83{,}333.333 \leq 0 \\
& g_5(\boldsymbol{x}) = -x_2 x_7 + 1{,}250 x_5 + x_2 x_4 - 1{,}250 x_4 \leq 0 \\
& g_6(\boldsymbol{x}) = -x_3 x_8 + 1{,}250{,}000 + x_3 x_5 - 2{,}500 x_5 \leq 0 \\
& n = 8 \quad 100 \leq x_i \leq 10{,}000 (i = 1) \\
& \qquad\qquad 1{,}000 \leq x_i \leq 10{,}000 (i = 2,3) \\
& \qquad\qquad 10 \leq x_i \leq 1{,}000 (i = 4,\ldots,8)
\end{aligned}
\tag{3.52}
$$

Design objective: Single objective

Function characteristics: Continuous

Dimensions: 8-Dimensional

Optimal value: 7,049.248

3.3.2.2 g18

The mathematical expression of g18 is given in Eq. (3.53).

$$f(\boldsymbol{x}) = -0.5(x_1 x_4 - x_2 x_3 + x_3 x_9 - x_5 x_9 + x_5 x_8 - x_6 x_7)$$

subject:

$$g_1(\boldsymbol{x}) = x_3^2 + x_4^2 - 1 \le 0, \quad g_2(\boldsymbol{x}) = x_9^2 - 1 \le 0, \quad g_3(\boldsymbol{x}) = x_5^2 + x_6^2 - 1 \le 0,$$

$$g_4(\boldsymbol{x}) = x_1^2 + (x_2 - x_9)^2 - 1 \le 0, \quad g_5(\boldsymbol{x}) = (x_1 - x_5)^2 + (x_2 - x_6)^2 - 1 \le 0,$$

$$g_6(\boldsymbol{x}) = (x_1 - x_7)^2 + (x_2 - x_8)^2 - 1 \le 0, \quad g_7(\boldsymbol{x}) = (x_3 - x_5)^2 + (x_4 - x_6)^2 - 1 \le 0,$$

$$g_8(\boldsymbol{x}) = (x_3 - x_7)^2 + (x_4 - x_8)^2 - 1 \le 0, \quad g_9(\boldsymbol{x}) = x_7^2 + (x_8 - x_9)^2 - 1 \le 0,$$

$$g_{10}(\boldsymbol{x}) = x_2 x_3 - x_1 x_4 \le 0, \quad g_{11}(\boldsymbol{x}) = -x_3 x_9 \le 0, \quad g_{12}(\boldsymbol{x}) = x_5 x_9 \le 0,$$

$$g_{13}(\boldsymbol{x}) = x_6 x_7 - x_5 x_8 \le 0.$$

$$n = 9 \quad -10 \le x_i \le 10 (i = 1, \ldots, 8)$$

$$0 \le x_i \le 20 (i = 9) \tag{3.53}$$

Design objective: Single objective

Function characteristics: Continuous

Dimensions: 9-Dimensional

Optimal value: –0.866

3.3.2.3 g07

The mathematical expression of g07 is given in Eq. (3.54).

$$f(\boldsymbol{x}) = x_1^2 + x_2^2 + x_1 x_2 - 14x_1 - 16x_2 + (x_3 - 10)^2 + 4(x_4 - 5)^2 + (x_5 - 3)^2$$

$$+2(x_6 - 1)^2 + 5x_7^2 + 7(x_8 - 11)^2 + 2(x_9 - 10)^2 + (x_{10} - 7)^2 + 45$$

subject:

$$g_1(\boldsymbol{x}) = -105 + 4x_1 + 5x_2 - 3x_7 + 9x_8 \le 0$$

$$g_2(\boldsymbol{x}) = 10x_1 - 8x_2 - 17x_7 + 2x_8 \le 0$$

$$g_3(\boldsymbol{x}) = -8x_1 + 2x_2 + 5x_9 - 2x_{10} - 12 \le 0$$

$$g_4(\boldsymbol{x}) = 3(x_1 - 2)^2 + 4(x_2 - 3)^2 + 2x_3^2 - 7x_4 - 120 \le 0$$

$$g_5(\boldsymbol{x}) = 5x_1 + 8x_2 + (x_3 - 6)^2 - 2x_4 - 40 \le 0$$

$$g_6(\boldsymbol{x}) = x_1^2 + 2(x_2 - 2)^2 - 2x_1x_2 + 14x_5 - 6x_6 \le 0$$

$$g_7(\boldsymbol{x}) = 0.5(x_1 - 8)^2 + 2(x_2 - 4)^2 + 3x_5^2 - x_6 - 30 \le 0 \tag{3.54}$$

$$g_8(\boldsymbol{x}) = -3x_1 + 6x_2 + 12(x_9 - 8)^2 - 7x_{10} \le 0$$

$$n = 10 \quad -10 \le x_i \le 10 (i = 1,\ldots,n)$$

Design objective: Single objective

Function characteristics: Continuous

Dimensions: 10-Dimensional

Optimal value: 24.3062

3.3.2.4 g01

The mathematical expression of g01 (G1/G1m) is given in Eq. (3.55).

$$f(\boldsymbol{x}) = 5\sum_{i=1}^{4} x_i - 5\sum_{i=1}^{4} x_i^2 - \sum_{i=5}^{13} x_i$$

$$subject:$$

$$g_1(\boldsymbol{x}) = 2x_1 + 2x_2 + x_{10} + x_{11} - 10 \le 0$$

$$g_2(\boldsymbol{x}) = 2x_1 + 2x_3 + x_{10} + x_{12} - 10 \le 0$$

$$g_3(\boldsymbol{x}) = 2x_2 + 2x_3 + x_{11} + x_{12} - 10 \le 0$$

$$g_4(\boldsymbol{x}) = -8x_1 + x_{10} \le 0$$

$$g_5(\boldsymbol{x}) = -8x_2 + x_{11} \le 0 \tag{3.55}$$

$$g_6(\boldsymbol{x}) = -8x_3 + x_{12} \le 0$$

$$g_7(\boldsymbol{x}) = -2x_4 - x_5 + x_{10} \le 0$$

$$g_8(\boldsymbol{x}) = -2x_6 - x_7 + x_{11} \le 0$$

$$g_9(\boldsymbol{x}) = -2x_8 - x_9 + x_{12} \le 0$$

$$n = 13 \quad 0 \le x_i \le 1 (i = 1,\ldots,9)$$

$$0 \le x_i \le 100 (i = 10,11,12)$$

$$0 \le x_i \le 1 (i = 13)$$

Design objective: Single objective

Function characteristics: Continuous or discrete

Dimensions: 13-Dimensional

Optimal value: –15 (when the functions are continuous)

Description: In the discrete case, the range of values for design variables is $x_i \in \{0, 1\}, i=1, \ldots, 9, 13$, and $x_i \in \{0, 1, \ldots, 100\}, i=10, 11, 12$. When $x_i \in \{0, 1, \ldots, 100\}, i=1, \ldots, 10, 13$, the corresponding problem is denoted as G1m.

3.3.2.5 g19

The mathematical expression of g19 is given in Eq. (3.56), and the parameters of g19 are given in Table 3.1.

$$
\begin{aligned}
& f(\boldsymbol{x}) = \sum_{j=1}^{5}\sum_{i=1}^{5} c_{ij}x_{(10+i)}x_{(10+j)} + 2\sum_{j=1}^{5} d_j x_{(10+j)}^3 - \sum_{i=1}^{10} b_i x_i \\
& \textit{subject:} \\
& g_j(\boldsymbol{x}) = -2\sum_{i=1}^{5} c_{ij}x_{(10+i)} - 3d_j x_{(10+j)}^2 - e_j + \sum_{i=1}^{10} a_{ij}x_i \le 0 \quad j=1,\ldots,5
\end{aligned}
\tag{3.56}
$$

where:

$$\boldsymbol{b} = [-40,-2,-0.25,-4,-4,-1,-40,-60,5,1]$$

$$n=15 \quad 0 \le x_i \le 10 (i=1,\ldots,15)$$

TABLE 3.1 Parameters for g19

j	1	2	3	4	5
e_j	–15	–27	–36	–18	–12
c_{1j}	30	–20	–10	32	–10
c_{2j}	–20	39	–6	–31	32
c_{3j}	–10	–6	10	–6	–10
c_{4j}	32	–31	–6	39	–20
c_{5j}	–10	32	–10	–20	30
d_j	4	8	10	6	2
a_{1j}	–16	2	0	1	0
a_{2j}	0	–2	0	0.4	2
a_{3j}	–3.5	0	2	0	0
a_{4j}	0	–2	0	–4	–1
a_{5j}	0	–9	–2	1	–2.8
a_{6j}	2	0	–4	0	0
a_{7j}	–1	–1	–1	–1	–1
a_{8j}	–1	–2	–3	–2	–1
a_{9j}	1	2	3	4	5
a_{10j}	1	2	1	1	1

Design objective: Single objective

Function characteristics: Continuous

Dimensions: 15-Dimensional

Optimal value: 32.6555

3.3.2.6 Hmittelman

The mathematical expression of Hmittelman is given in Eq. (3.57).

$$
\begin{aligned}
&\min f(\boldsymbol{x}) = 10y_1 + 7y_2 + y_3 + 12y_4 + 8y_5 + 3y_6 + y_7 + 5y_8 + 3y_9 \\
&s.t. \quad 3y_1 - 12y_2 - 8y_3 + y_4 - 7y_9 + 2y_{10} \le -2, \\
&\qquad y_2 - 10y_3 - 5y_5 + y_6 + 7y_7 + y_8 \le -1, \\
&\qquad 5y_1 - 3y_2 - y_3 - 2y_8 + y_{10} \le -1, \\
&\qquad -4y_3 - 2y_4 - 5y_6 + y_7 - 9y_8 - 2y_9 \le -3, \\
&\qquad 9y_2 - 12y_4 - 7y_5 + 6y_6 + 2y_8 - 15y_9 + 3y_{10} \le -7, \\
&\qquad 5y_2 - 8y_1 + 2y_3 - 7y_4 - y_5 - 5y_7 - 10y_9 \le -1, \\
&\qquad y_1 = x_5x_7x_9x_{10}x_{14}x_{15}x_{16}, \qquad y_6 = x_6x_7x_9x_{14}x_{16}, \\
&\qquad y_2 = x_1x_2x_3x_4x_8x_{11}, \qquad y_7 = x_9x_{10}x_{14}x_{16}, \\
&\qquad y_3 = x_3x_4x_6x_7x_8, \qquad y_8 = x_5x_{10}x_{14}x_{15}x_{16}, \\
&\qquad y_4 = x_3x_4x_8x_{11}, \qquad y_9 = x_1x_2x_{11}x_{12}, \\
&\qquad y_5 = x_6x_7x_8x_{12}, \qquad y_{10} = x_{13}x_{14}x_{15}x_{16}. \\
&\qquad x_i \in \{0,1\}, \quad i = 1,\ldots,16
\end{aligned}
\tag{3.57}
$$

Design objective: Single objective

Function characteristics: Discrete

Dimensions: 16-Dimensional

3.3.2.7 g02

The mathematical expression of g02 is given in Eq. (3.58).

$$
f(x) = -\left| \frac{\sum_{i=1}^{n} \cos^4(x_i) - 2\prod_{i=1}^{n} \cos^2(x_i)}{\sqrt{\sum_{i=1}^{n} i x_i^2}} \right|
$$

$$
\begin{aligned}
&\textit{subject:}\\
&g_1(\boldsymbol{x}) = 0.75 - \prod_{i=1}^{n} x_i \le 0,\\
&g_2(\boldsymbol{x}) = \sum_{i=1}^{n} x_i - 0.75 \quad n \le 0\\
&n = 20 \quad 0 < x_i \le 10 (i = 1, \ldots, n)
\end{aligned} \tag{3.58}
$$

Design objective: Single objective

Function characteristics: Continuous

Dimensions: 20-Dimensional

Optimal value: –0.8036

3.4 ENGINEERING APPLICATION CASES

3.4.1 Tension/Compression Spring Design (TSD)

The design of tension/compression springs (TSD), as shown in Figure 3.29, aims to minimize the spring's weight while being constrained by minimum deflection, shear force, frequency, outer diameter, and lateral constraints.

$$
\begin{aligned}
&\min f(\boldsymbol{x}) = x_1^2 x_2 (x_3 + 2)\\
&s.t. \quad g_1(\boldsymbol{x}) = 1 - \frac{x_2^3 x_3}{71{,}785 x_1^4} \le 0,\\
&\qquad g_2(\boldsymbol{x}) = \frac{4x_2^2 - x_1 x_2}{12{,}566 x_1^3 (x_2 - x_1)} + \frac{1}{5{,}108 x_1^2} - 1 \le 0,\\
&\qquad g_3(\boldsymbol{x}) = 1 - \frac{140.45 x_1}{x_3 x_2^2} \le 0,\\
&\qquad g_4(\boldsymbol{x}) = \frac{x_1 + x_2}{1.5} - 1 \le 0.\\
&\qquad 0.05 \le x_1 \le 2; \quad 0.25 \le x_2 \le 1.3; \quad 2 \le x_3 \le 15
\end{aligned} \tag{3.59}
$$

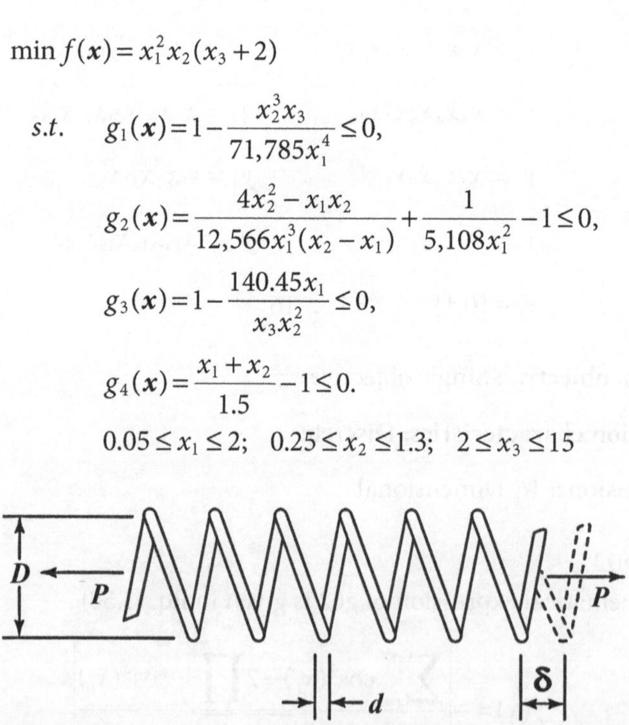

FIGURE 3.29 TSD.

Design objective: Single objective

Function characteristics: Continuous

Dimensions: 3-Dimensional

Optimal value: 0.01267

3.4.2 Welded Beam Design (WBD)

The welded beam design (WBD/WB4), as shown in Figure 3.30, aims to minimize design cost while being constrained by shear force, bending stress within the beam, buckling load on the bar, lateral constraints, and deflection at the end of the beam.

$$\min f(\boldsymbol{x}) = 1.10471 x_1^2 x_2 + 0.04811 x_3 x_4 (14 + x_2)$$

$$\begin{aligned}
s.t. \quad & g_1(\boldsymbol{x}) = \tau(\boldsymbol{x}) - \tau_{\max} \leq 0, \\
& g_2(\boldsymbol{x}) = \sigma(\boldsymbol{x}) - \sigma_{\max} \leq 0, \\
& g_3(\boldsymbol{x}) = x_1 - x_4 \leq 0, \\
& g_4(\boldsymbol{x}) = 0.10471 x_1^2 + 0.04811 x_3 x_4 (14 + x_2) - 5 \leq 0, \\
& g_5(\boldsymbol{x}) = 0.125 - x_1 \leq 0, \\
& g_6(\boldsymbol{x}) = \delta(\boldsymbol{x}) - \delta_{\max} \leq 0, \\
& g_7(\boldsymbol{x}) = P - P_c(\boldsymbol{x}) \leq 0,
\end{aligned}$$

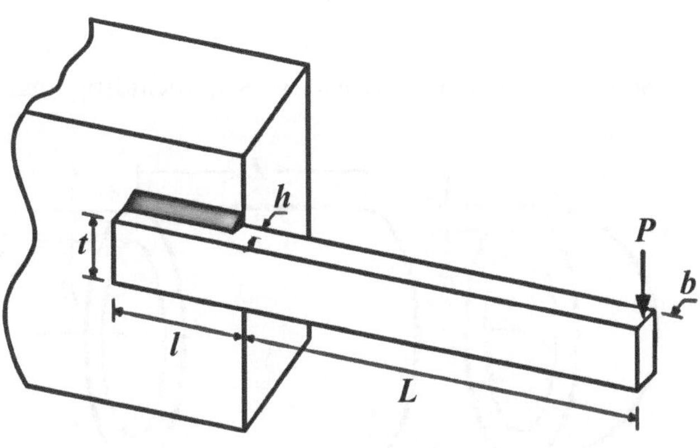

FIGURE 3.30 WBD.

where

$$\tau(\mathbf{x})=\sqrt{(\tau')^2+2\tau'\tau''\frac{x_2}{2R}+(\tau'')^2},\quad \tau'=\frac{P}{\sqrt{2}x_1x_2},\quad \tau''=\frac{MR}{J},$$

$$M=P\left(L+\frac{x_2}{2}\right),\quad R=\sqrt{\frac{x_2^2}{4}+\left(\frac{x_1+x_3}{2}\right)^2},\quad \sigma(x)=\frac{6PL}{x_4x_3^2},$$

$$\delta(\mathbf{x})=\frac{4PL^3}{Ex_3^3x_4},\quad J=2\left\{\sqrt{2}x_1x_2\left[\frac{x_2^2}{12}+\left(\frac{x_1+x_3}{2}\right)^2\right]\right\},\tag{3.60}$$

$$P_c(\mathbf{x})=\frac{4.013E\sqrt{x_3^2x_4^6/36}}{L^2}\left(1-\frac{x_3}{2L}\sqrt{\frac{E}{4G}}\right),G=12\times10^6\text{ psi},$$

$$P=600\text{ lb},\quad L=14\text{ in},\quad \delta_{\max}=0.25\text{ in},\quad E=30\times10^6\text{ psi},$$

$$\tau_{\max}=13{,}600\text{ psi},\quad \sigma_{\max}=30{,}000\text{ psi}.$$

$$0.1\le x_1,x_4\le 2;\quad 0.1\le x_2,x_3\le 10$$

Design objective: Single objective

Function characteristics: Continuous

Dimensions: 4-Dimensional

Optimal value: 1.7249

3.4.3 Pressure Vessel Design (PVD)

The pressure vessel design (PVD), as shown in Figure 3.31, aims to minimize the design cost of a cylindrical vessel, including material cost,

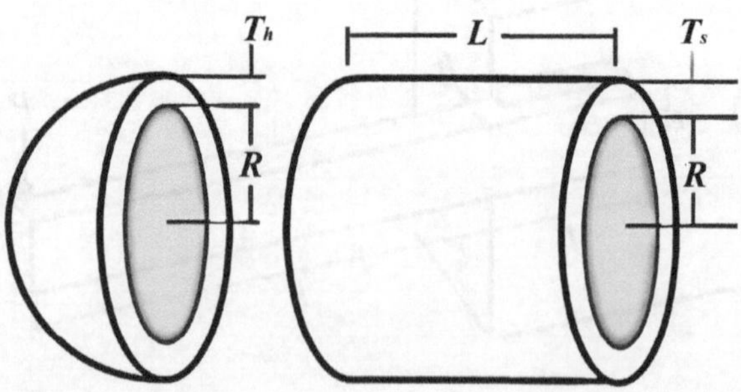

FIGURE 3.31 PVD.

forming cost, and welding cost. The four design variables are the thickness of the pressure vessel, the thickness of the head, the internal radius of the vessel, and the length of the vessel.

$$\min f(\boldsymbol{x}) = 0.6224x_1x_3x_4 + 1.7781x_2x_3^2 + 3.1661x_1^2x_4 + 19.84x_1^2x_3$$

$$\begin{aligned} s.t.\quad & g_1(\boldsymbol{x}) = -x_1 + 0.0193x_3 \le 0, \\ & g_2(\boldsymbol{x}) = -x_2 + 0.00954x_3 \le 0, \\ & g_3(\boldsymbol{x}) = -\pi x_3^2 x_4 - \frac{4}{3}\pi x_3^3 + 1{,}296{,}000 \le 0, \\ & g_4(\boldsymbol{x}) = x_4 - 240 \le 0. \end{aligned} \tag{3.61}$$

$$1 \times 0.0625 \le x_1, x_2 \le 99 \times 0.0625;\quad 10 \le x_3, x_4 \le 200$$

Design objective: Single objective

Function characteristics: Continuous

Dimensions: 4-Dimensional

Optimal value: 5,885.33

3.4.4 Speed Reducer Design (SRD/SR7)

The speed reducer design (SRD/SR7) aims to minimize the total weight of the reducer while being subject to 11 constraints, including limits on the bending stress of gear teeth, surface stress, and the lateral deflection of the shaft.

$$\begin{aligned} \min f(\boldsymbol{x}) = {} & 0.7854x_1x_2^2(3.3333x_3^2 + 14.9334x_3 - 43.0934) - 1.508x_1(x_6^2 + x_7^2) \\ & + 7.4777(x_6^3 + x_7^3) + 0.7854(x_4x_6^2 + x_5x_7^2) \end{aligned}$$

$$\begin{aligned} s.t.\ & g_1(\boldsymbol{x}) = \frac{27}{x_1x_2^2x_3^2} - 1 \le 0,\ g_6(\boldsymbol{x}) = \frac{\left[(745x_5/(x_2x_3))^2 + 157.5\times10^6\right]^{1/2}}{85x_7^3} - 1 \le 0, \\ & g_2(\boldsymbol{x}) = \frac{397.5}{x_1x_2^2x_3^2} - 1 \le 0,\ g_7(\boldsymbol{x}) = \frac{x_2x_3}{40} - 1 \le 0,\ g_8(\boldsymbol{x}) = \frac{5x_2}{x_1} - 1 \le 0, \\ & g_3(\boldsymbol{x}) = \frac{1.93x_4^3}{x_2x_3x_6^4} - 1 \le 0,\ g_9(\boldsymbol{x}) = \frac{x_1}{12x_2} - 1 \le 0,\ g_{10}(\boldsymbol{x}) = \frac{1.5x_6 + 1.9}{x_4} - 1 \le 0, \\ & g_4(\boldsymbol{x}) = \frac{1.93x_5^3}{x_2x_3x_7^4} - 1 \le 0,\ g_{11}(\boldsymbol{x}) = \frac{1.1x_7 + 1.9}{x_5} - 1 \le 0, \end{aligned}$$

$$g_5(x) = \frac{\left[(745x_4/(x_2x_3))^2 + 16.9\times10^6\right]^{1/2}}{110x_6^3} - 1 \le 0,$$

$$2.6 \le x_1 \le 3.6;\quad 0.7 \le x_2 \le 0.8;\quad 17 \le x_3 \le 28;\quad 7.3 \le x_4,\quad x_5 \le 8.3; \quad (3.62)$$

$$2.9 \le x_6 \le 3.9;\quad 5.0 \le x_7 \le 5.5.$$

Design objective: Single objective

Function characteristics: Continuous

Dimensions: 7-Dimensional

Optimal value: 2,994.4711

3.4.5 Stepped Cantilever Beam Design (SCBD)

The stepped cantilever beam design (SCBD), as shown in Figure 3.32, aims to minimize the volume of a five-step cantilever beam with a total length of $L = 500$ cm. The material has an elastic modulus E of 200 GPa, and a concentrated load of 50,000 N is applied at the free end of the beam. There are 11 constraints in total, including 5 bending stress constraints, 1 displacement constraint, and 5 length-to-width ratio constraints.

$$\min V = D(b_1h_1l_1 + b_2h_2l_2 + b_3h_3l_3 + b_4h_4l_4 + b_5h_5l_5)$$

$$s.t.\quad g_1(x) = \frac{6Pl_5}{b_5h_5^2} - 14{,}000 \le 0,\quad E = 2e11\quad D = 1$$

$$g_2(x) = \frac{6P(l_5 + l_4)}{b_4h_4^2} - 14{,}000 \le 0,$$

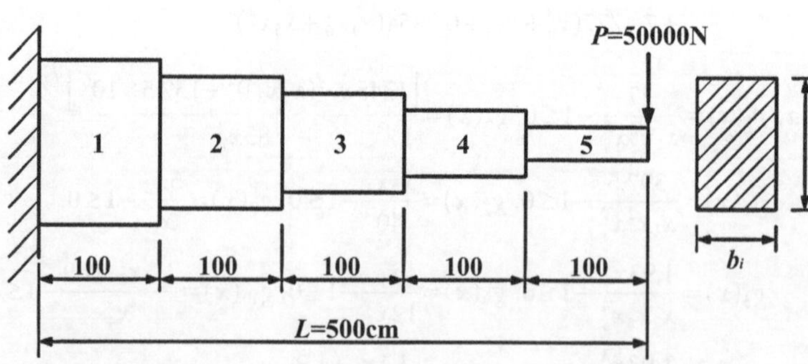

FIGURE 3.32 SCBD.

$$
\begin{aligned}
g_2(\boldsymbol{x}) &= \frac{6P(l_5+l_4)}{b_4h_4^2} - 14{,}000 \le 0,\\
g_3(\boldsymbol{x}) &= \frac{6P(l_5+l_4+l_3)}{b_3h_3^2} - 14{,}000 \le 0,\\
g_4(\boldsymbol{x}) &= \frac{6P(l_5+l_4+l_3+l_2)}{b_2h_2^2} - 14{,}000 \le 0,\\
g_5(\boldsymbol{x}) &= \frac{6P(l_5+l_4+l_3+l_2+l_1)}{b_1h_1^2} - 14{,}000 \le 0,\\
g_6(\boldsymbol{x}) &= \frac{Pl^3}{3E}\left(\frac{1}{I_5}+\frac{7}{I_4}+\frac{19}{I_3}+\frac{37}{I_2}+\frac{61}{I_1}\right) - 2.7 \le 0,\\
g_7(\boldsymbol{x}) &= \frac{h_5}{b_5} - 20 \le 0, \quad I_1 = \frac{b_5h_5^3}{12}\\
g_8(\boldsymbol{x}) &= \frac{h_4}{b_4} - 20 \le 0, \quad I_2 = \frac{b_4h_4^3}{12}\\
g_9(\boldsymbol{x}) &= \frac{h_3}{b_3} - 20 \le 0, \quad I_3 = \frac{b_3h_3^3}{12}\\
g_{10}(\boldsymbol{x}) &= \frac{h_2}{b_2} - 20 \le 0, \quad I_4 = \frac{b_2h_2^3}{12}\\
g_{11}(\boldsymbol{x}) &= \frac{h_1}{b_1} - 20 \le 0, \quad I_5 = \frac{b_1h_1^3}{12}
\end{aligned}
\tag{3.63}
$$

$$(x_1 \sim x_{10}) = (b_1, h_1, b_2, h_2, b_3, h_3, b_4, h_4, b_5, h_5)$$

Design objective: Single objective

Function characteristics: Continuous

Dimensions: 10-Dimensional

Optimal value: 62,791

3.5 CHAPTER SUMMARY

This chapter provides an overview of benchmark test functions for data-driven optimization methods, covering unconstrained low-dimensional cases, unconstrained high-dimensional cases, constrained low-dimensional cases, constrained high-dimensional cases, and engineering application cases. These benchmark test functions are suitable for testing algorithms that solve constrained and unconstrained problems, as well as discrete and high-dimensional problems. They can effectively help researchers verify the efficiency and robustness of their algorithms.

REFERENCES

Adorio, E. P., & Diliman, U. (2005). MVF-multivariate test functions library in c for unconstrained global optimization. *Quezon City, Metro Manila, Philippines, 44*, accessed *January 14*, 2005.

Akbari, H., & Kazerooni, A. (2020). KASRA: A Kriging-based Adaptive Space Reduction Algorithm for Global Optimization of Computationally Expensive Black-Box Constrained Problems. *Applied Soft Computing, 90*, 106154.

Dong, H., Wang, P., Song, B., Zhang, Y., & An, X. (2020). Kriging-Assisted Discrete Global Optimization (KDGO) for Black-Box Problems with Costly Objective and Constraints. *Applied Soft Computing, 94*, 106429.

Jamil, M., & Yang, X.-S. (2013). A Literature Survey of Benchmark Functions for Global Optimisation Problems. *International Journal of Mathematical Modelling and Numerical Optimisation, 4*(2), 150–194.

Li, F.-F., Shoemaker, C. A., Wei, J.-H., & Fu, X.-D. (2013). Estimating Maximal Annual Energy Given Heterogeneous Hydropower Generating Units with Application to the Three Gorges System. *Journal of Water Resources Planning and Management, 139*(3), 265–276.

Liang, J. J., Runarsson, T. P., Mezura-Montes, E., Clerc, M., Suganthan, P. N., Coello, C. C., & Deb, K. (2006). Problem Definitions and Evaluation Criteria for the CEC 2006 Special Session on Constrained Real-Parameter Optimization. *Journal of Applied Mechanics, 41*(8), 8–31.

Liu, C., Wan, Z., Liu, Y., Li, X., & Liu, D. (2021). Trust-Region Based Adaptive Radial Basis Function Algorithm for Global Optimization of Expensive Constrained Black-Box Problems. *Applied Soft Computing, 105*(1), 107233.

Liu, H., Xu, S., Chen, X., Wang, X., & Ma, Q. (2017). Constrained Global Optimization via a DIRECT-Type Constraint-Handling Technique and an Adaptive Metamodeling Strategy. *Structural and Multidisciplinary Optimization, 55*, 155–177.

Mezura-Montes, E., & Cetina-Domínguez, O. (2012). Empirical Analysis of a Modified Artificial Bee Colony for Constrained Numerical Optimization. *Applied Mathematics and Computation, 218*(22), 10943–10973.

Müller, J., Shoemaker, C. A., & Piché, R. (2013). SO-MI: A Surrogate Model Algorithm for Computationally Expensive Nonlinear Mixed-Integer Black-Box Global Optimization Problems. *Computers & operations research, 40*(5), 1383–1400.

Müller, J., Shoemaker, C. A., & Piché, R. (2014). SO-I: A Surrogate Model Algorithm for Expensive Nonlinear Integer Programming Problems Including Global Optimization Applications. *Journal of Global Optimization, 59*(4), 865–889.

Pichitlamken, J., Nelson, B. L., & Hong, L. J. (2006). A Sequential Procedure for Neighborhood Selection-of-the-Best in Optimization via Simulation. *European Journal of Operational Research, 173*(1), 283–298.

Surjanovic, S., & Bingham, D. (2013). Virtual library of simulation experiments: test functions and datasets. *Simon Fraser University, Burnaby, BC, Canada, accessed May, 13*, 2015.

CHAPTER 4

MSSR

Multi-Start Space Reduction Surrogate-Based Global Optimization Method[1]

4.1 INTRODUCTION

Surrogate-based optimization (SBO) is a technique that leverages surrogate models to predict objective and constraint functions, significantly reducing the need for direct evaluations (Edke & Chang, 2011; Queipo et al., 2005). This chapter focuses on applying SBO methods to address black-box optimization problems effectively.

Since surrogate models are typically smooth and continuous functions, directly optimizing them can yield locally optimal solutions. However, these solutions are based on predictions and may significantly deviate from the true solutions. A critical research focus in recent years has been on selecting informative samples to enhance surrogate models and accurately identifying the global optimal region. Numerous scholars have advanced this field, contributing to the ongoing development of surrogate-based global optimization algorithms. Jones et al. (1998) presented a widely cited global optimization algorithm for expensive black-box problems, which is known as EGO. EGO constructs the surrogate model by Kriging and updates the surrogate model by maximizing an expected improvement function. Gary Wang et al. (2001) provided an adaptive response surface method (ARSM),

DOI: 10.1201/9781003636267-4

which creates a quadratic approximation model for the expensive objective function in a reduced space. Gutmann (2001) introduced a global optimization method based on RBF to solve problems with expensive function evaluations. Jin et al. (2001) explored the accuracy of surrogate models and how they affect the sampling strategies. Wang and Simpson (2004) utilized a fuzzy clustering method to get a reduced search space, which can efficiently find the global optimum on nonlinear constrained optimization problems. A stochastic RBF method for the global optimization of expensive functions was proposed by Regis and Shoemaker (2007), who also improved the Gutmann-RBF method by varying the size of the subdomain in different iterations. Younis and Dong (2010) developed a kind of space reduction method called space exploration and unimodal region elimination (SEUMRE), which establishes a unimodal region to speed up the search. SEUMRE has successfully been used for black-box engineering applications. Gu et al. (2012) invented the hybrid and adaptive meta-model-based (HAM) method to divide the design space into several subdomains with different weights. In every iteration, sample points are obtained from these regions based on the size of the weights. At last, HAM performed well on a crash simulation of vehicles. Long et al. (2015) combined a kind of intelligent space exploration strategy with ARSM to provide reduced regions for global optimization. As we can see, the space reduction method is a high-efficiency way to realize global optimization of computationally expensive problems.

In this chapter, a new multi-start space reduction (MSSR) surrogate-based search algorithm is introduced for global optimization problems with computationally expensive black-box objective functions and constraints. The algorithm divides the design space into three regions: global space (GS), medium space (MS) and local space (LS). GS represents the original design region, MS narrows the focus to a promising subset and LS concentrates on the vicinity of the current best solution. The search process employs a Kriging-based multi-start optimization method for local optimization, sample selection and exploration. Latin hypercube sampling is used to generate starting points, while sequential quadratic programming (SQP) refines local solutions. A newly introduced selection strategy identifies high-quality sample points to enhance the Kriging model, and the estimated mean square error guides the exploration of unexplored regions in the design space. The search alternates among GS, MS and LS until the global optimum is located.

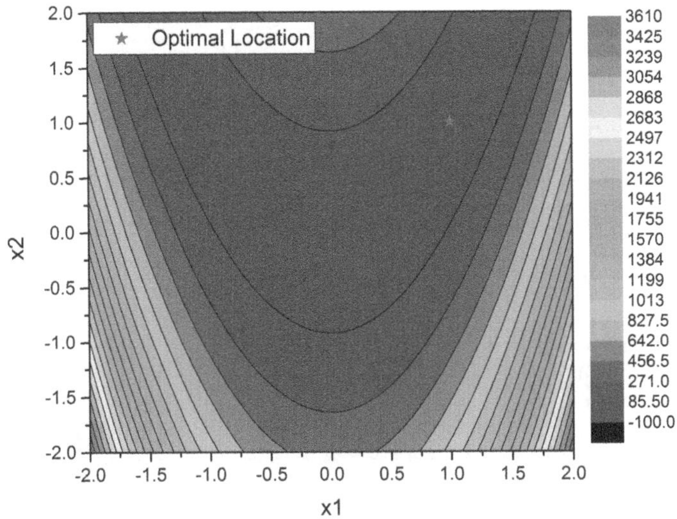

FIGURE 4.1 Original Banana function.

4.2 KRIGING-BASED MODEL

To validate the accuracy of the surrogate model, this chapter uses the Banana function as an example. Fifteen experimental design points are generated using optimal Latin hypercube sampling (OLHS), and a Kriging surrogate model is constructed. The detailed formulas of Kriging are provided in Section 2.2. As shown in Figures 4.1 and 4.2, the 15 triangular markers represent the experimental design points. Overall, the Kriging model closely aligns with the original function, though minor deviations are observed in some regions.

4.3 THE PROPOSED MULTI-START OPTIMIZATION PROCESS

The proposed multi-start optimization process for the Kriging-based model comprises three key components: local optimization using the surrogate model, selection of high-potential sample points and exploration of uncharted areas within the design space.

To ensure randomly selected starting points that adequately cover the search space, Latin hypercube sampling (LHS) is employed. These selected starting points are used iteratively during the search process. Sequential quadratic programming (SQP) is applied to the Kriging surrogate model to identify local optimal solutions, which are stored in a database of

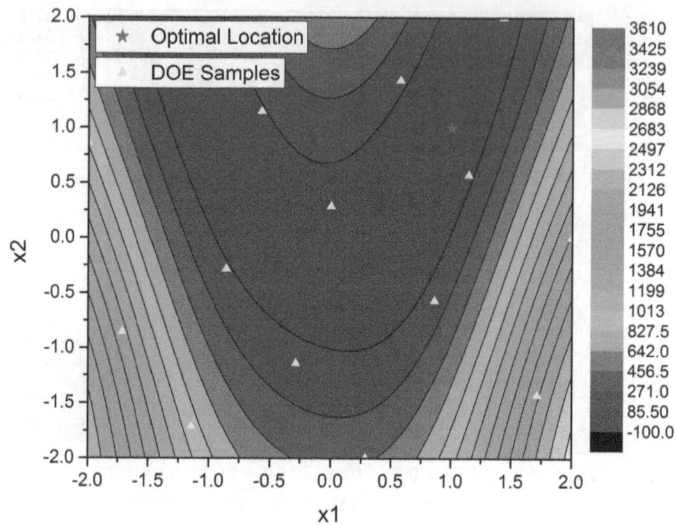

FIGURE 4.2 Kriging prediction with 15 samples.

"Potential Sample Points." However, SQP may converge to the same local optimum from different starting points, resulting in duplicates in the database. Additionally, local optima may coincide with existing sample points. To mitigate these issues, new sample points are required to maintain a defined distance from previously obtained points. Furthermore, in cases where no suitable local optima exist within the defined space, the multi-start optimization process maximizes the Kriging model's estimated mean squared error (MSE) to explore uncharted areas. A special selection strategy is employed to extract the most promising results from the "Potential Sample Points." The pseudo-codes summarizing the processes of optimization, selection and exploration of unknown areas are presented as follows.

4.1 *Optimization*:

(01) Begin

(02) **Initialize Dimension *n***, Database "Potential Samples," Design Space, Kriging Predictor, MSE;

(03) Acquire ***m*** starting points by LHS; (Here, it is suggested that ***m*** can be defined in the range [20, 40] on two-dimensional problems, [6***n***, 8***n***] when the dimension of the problem is 2<***n***<10 and [50, 70] on high-dimensional problems.)

(04) **for** ***i*=1: m**
(05) Employ SQP algorithm;
(06) Optimize the Kriging Predictor from the ***i***th starting point;
(07) Store the local optimal solutions and their predicted values in the database "Potential Samples;"
(08) **end**
(09) "Potential Samples" is a matrix with $m \times (n+1)$ elements;
(10) **end**

/* The design space is selected among GS, MS and LS, which will change with the iteration going on. The Kriging predictor and its estimated MSE can be obtained by the DACE toolbox (Lophaven et al. 2002). The "fmincon" function of MATLAB® can be employed to realize the SQP algorithm (The Mathworks 2015). */

<u>4.2 Selection:</u>

(01) **Begin**
(02) **Sort** the predicted values in "Potential Samples" and get the maximum ***(Xpsmax Ypsmax)*** and minimum ***(Xpsmin, Ypsmin)***; (The sample and the predicted value in "Potential Samples" are expressed as ***(Xps, Yps)***)
(03) Initialize parameters ***k***=1, ***flag_repeat***=0, ***flag_stop***=0, ***e_error***=0.00001 (If ***n***>= 10, ***e_error***=0.0001), ***MAXK***; /****MAXK*** is a parameter that decides how many points can be sampled at most in one iteration. Here, ***MAXK*** equals to 3 on two-dimensional problems and equals to 4 on higher-dimensional problems. For nonlinear constrained optimization problems, ***MAXK*** equals to 3. */
(04) **Acquire** the size of the expensive samples set **S** as ***m_size***;
(05) **While** ***k*** < ***MAXK*** *and* ***flag_stop***==0
(06) **for** i=1: ***m_size***
(07) **if** square of the distance between ***Xpsmin*** and the sample **S**(*i*) <= ***e_error***
(08) ***flag_ repeat*** =1;
(09) **end**
(10) **end**

/* Here, the new promising samples that go much close to the existing points will be flagged. */

```
        if flag_repeat==0
            record the current sample Xpsmin; k=k+1;
        end
        for i=1: m
            if |Yps (i) -Ypsmin | <=0.0001
                Yps (i)= Ypsmax+10;
            end
        end
```

/* At each iteration, just one local optimal solution from the Kriging model is selected and the same results are covered by a big value "*Ypsmax*+10." When the next iteration comes, the bigger values are ignored. */

```
        Sort the predicted values Yps in "Potential Samples" again and
        update (Xpsmin, Ypsmin);
        If Ypsmin == Ypsmax+10
            flag_stop= 1
        end
        flag_repeat =0;
    end
    if k>1
        Store the selected samples and evaluate the true function values.
    end
end
```

4.3 *Explore Unknown Area:*

```
Begin
    if k==1
        Implement the above-mentioned Optimization method to get
        the local maximums of the MSE function.
        Get two new samples and evaluate the true function values.
    end
end
```

/* If the algorithm cannot find a satisfactory solution by the above-mentioned selection process, the estimated MSE can be maximized to acquire new samples which must be located in an unexplored area. */

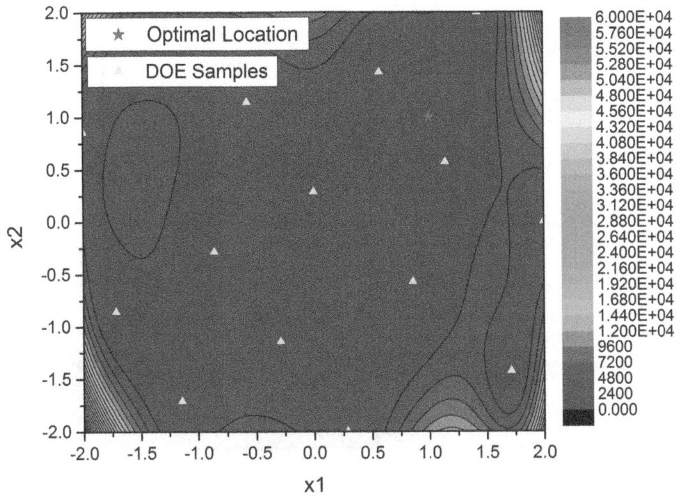

FIGURE 4.3 Estimated MSE of Kriging.

The estimated MSE function of the Kriging model is illustrated in Figure 4.3, where local maxima of the MSE consistently appear in unexplored areas. The MSE value increases with distance from known sample points and approaches zero at the locations of these points. Selecting one of the locally maximal MSE solutions for sample updates typically enhances space-filling. An optimization process that effectively leverages these properties of the Kriging model can fully exploit its potential. Figure 4.4 demonstrates the multi-start optimization process on a Kriging model, starting with 30 initial points. Eventually, two local optimal solutions are selected, both situated in the valley of the Banana function—a region associated with better solutions.

4.4 SPACE REDUCTION APPROACH

A sample set obtained using the design of experiments (DOE) method is used to store the data from expensive evaluations. Based on the values of these samples, three spaces—GS, MS and LS—are defined for the multi-start optimization process. GS represents the entire region of the original design space. MS is based on the portion of design space of the current better samples. LS is the neighborhood area of the best current sample point. During the iterative search process, the sample set

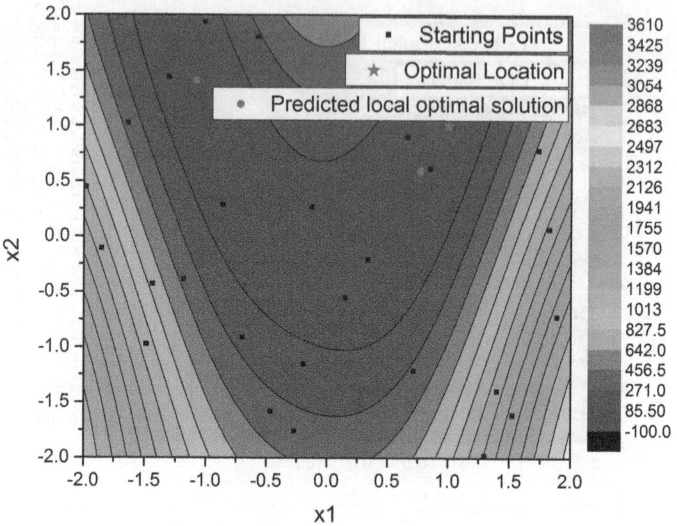

FIGURE 4.4 Multi-start process on Kriging.

is continuously updated with new samples, and the better-performing samples are refined. MS and LS dynamically adjust as iterations progress until the optimization process concludes. The detailed definitions of MS and LS are provided below.

$$dis_i = \left| \max\left(\mathbf{S}(1:k)_i\right) - \min\left(\mathbf{S}(1:k)_i\right) \right|, \quad i = 1,2,\ldots,n$$

$$Lob_i = \begin{cases} S_i^{best} - dis_i, & S_i^{best} - dis_i \geq \min(range_i) \\ \min(range_i), & S_i^{best} - dis_i \leq \min(range_i) \end{cases}$$

$$Ub_i = \begin{cases} S_i^{best} + dis_i, & S_i^{best} + dis_i \leq \max(range_i) \\ \max(range_i), & S_i^{best} + dis_i \geq \max(range_i) \end{cases} \tag{4.1}$$

$$range\_local_i = [Lob_i, Ub_i]$$

$$Lob_i = \min\left(\mathbf{S}_i(1:p)\right)$$

$$Ub_i = \max\left(\mathbf{S}_i(1:p)\right) \tag{4.2}$$

$$range\_medium_i = [Lob_i, Ub_i], \quad i = 1,2,\ldots,n$$

where n is the dimension of a problem. $\boldsymbol{S}(1{:}k)_i$ is the i-th dimension of the top k samples selected from the ranked sample set. $range_i$ is the i-th dimension of the original design range. S_i^{best} is the i-th dimension of the current best sample. Equations (4.3) and (4.4) define the LS and MS, respectively. If the distance of Lob_i and Ub_i in Eq. (4.3) or (4.4) is smaller than 1e-5, it is suggested that setting a smaller space to search:

$$\begin{aligned} Lob_i &= Lob_i - 0.025 \times \left(\max\left(range_i\right) - \min\left(range_i\right)\right) \\ Ub_i &= Ub_i + 0.025 \times \left(\max\left(range_i\right) - \min\left(range_i\right)\right) \end{aligned} \tag{4.3}$$

Meanwhile, the new range should also be the subset of the original design range. Both of the two spaces change their scopes based on the better samples acquired from the design space. Here, k and p are two user-defined parameters, which represent the number of the better samples. In MSSR, we define k and p as follows:

$$k = \begin{cases} 3, n \leq 2 \quad and \quad CS \leq 150 \\ round(CS/30), \quad n \leq 2 \quad and \quad CS > 150 \\ 5, n > 2 \quad and \quad CS \leq 150 \\ round(CS/30), \quad n > 2 \quad and \quad CS > 150 \end{cases} \tag{4.4}$$

$$p = \begin{cases} round(CS/3), \quad n \leq 2 \\ 3n, \quad n > 2 \quad and \quad CS \leq 60 \\ round(CS/3), \quad n > 2 \quad and \quad CS > 60 \end{cases} \tag{4.5}$$

where CS is the number of current sample points. k will be smaller than p with continuous iterations. According to Eq. (4.1) to (4.5), MS can give a reduced region that may include several promising solutions and LS can make the search focus on one of them quickly. In some cases, when LS turns into a tiny space or the search in LS, MS or GS repeats around a local optimal solution, there are no appropriate locations that can be selected as new samples. Or if new samples cannot be found after optimization and selection, the estimated MSE of Kriging can be used to explore the unknown area. The ranges for getting the local maximums of MSE in local, medium and global searches are defined as follows:

$$dis_i = \left|\max(range_i) - \min(range_i)\right|, \quad i = 1,2,\ldots,n$$

$$Lob_i = \begin{cases} S_i^{best} - 0.5 \times dis_i, & S_i^{best} - 0.5 \times dis_i \geq \min(range_i) \\ \min(range_i), & S_i^{best} - 0.5 \times dis_i \leq \min(range_i) \end{cases}$$

$$Ub_i = \begin{cases} S_i^{best} + 0.5 \times dis_i, & S_i^{best} + 0.5 \times dis_i \leq \max(range_i) \\ \max(range_i), & S_i^{best} + 0.5 \times dis_i \geq \max(range_i) \end{cases} \tag{4.6}$$

$$range\_mse\_local_i = [Lob_i, Ub_i]$$

$$range\_mse\_medium_i = range\_medium_i$$

$$range\_mse\_global_i = range_i$$

The parameters in Eq. (4.6) share the same definitions as those in Eqs. (4.1) and (4.2). Intuitively, the defined ranges enclose the current best solution and dynamically adjust as iterations progress. The algorithm effectively combines GS, MS and LS to fully leverage the Kriging predictor, accelerating convergence toward the global optimum. Simultaneously, it explores unknown areas, enabling the current best solution to escape potential local optima and improve the overall search performance.

4.5 THE ENTIRE OPTIMIZATION PROCESS

The complete MSSR global optimization process is illustrated by the flowchart in Figure 4.5. The key steps in this process are summarized as follows:

/* The initial process */

1. Apply OLHS to generate DOE sample points over the entire design space.
2. Evaluate the expensive function using the DOE sample points and store the results in the sample set. (For nonlinear constrained problems, expensive functions include objective and constraint functions.)
3. Rank all expensive samples based on their function values. (Here, if a sample point does not satisfy the true constraints, the sample values should add a large penalty factor of 1e6.)

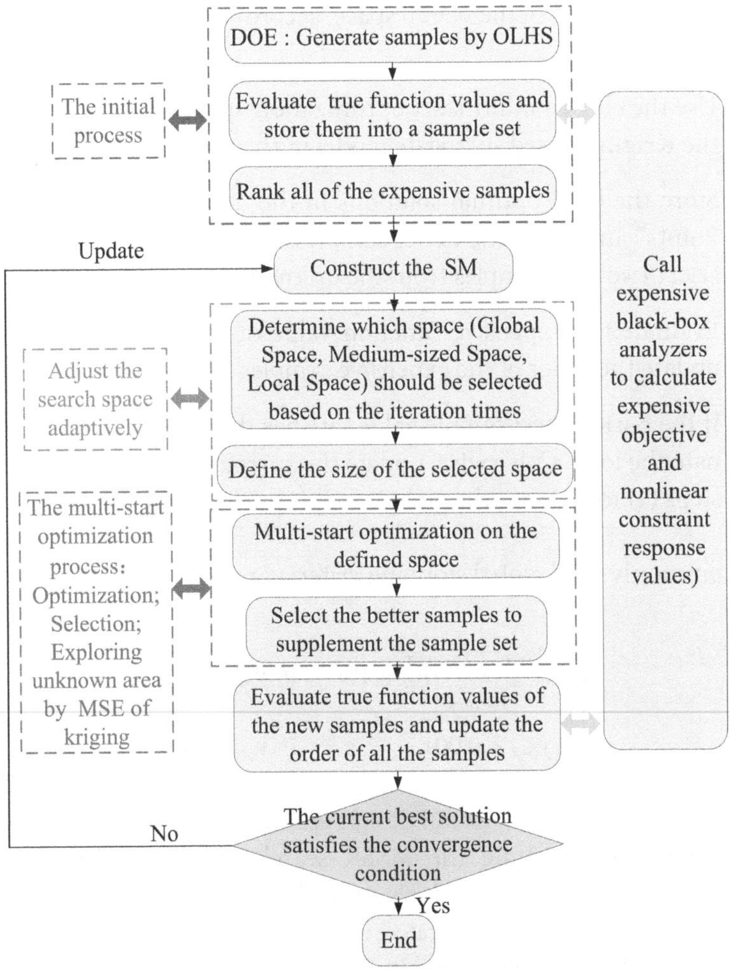

FIGURE 4.5 Flowchart of the MSSR optimization process.

/* The search loop */

4. Construct the Kriging-based surrogate model. (For nonlinear constrained problems, surrogate models of objective and constraint functions are built, respectively. Here, sample values use the true objective values without the additional penalty factor.)

5. Determine which space should be explored based on the present number of iterations. The global search, medium-sized search and local search will be implemented alternatively in the process.

6. Define the size of the search space, according to the expensive sample set.
7. Use the chosen multi-start optimization approach, SQP, to optimize the Kriging-based surrogate model in the defined space.
8. Store the local optimal solutions in the database "Potential Sample Points" and select the better samples. If there is not a better sample, select two new samples from the unknown area.
9. Evaluate the expensive function value of the selected samples and update the order of the expensive samples like step (3).
10. If the current best sample value satisfies the stopping criteria, terminate the loop. Otherwise, update the surrogate model and repeat the steps (4) to (9) until the global stopping criteria are satisfied.

The commonly used global stopping criteria are:

$$\begin{cases} \dfrac{\left|y_{best} - y_{optimal}\right|}{\left|y_{optimal}\right|} < 1\% & \text{if } y_{optimal} \neq 0 \\ y_{best} < 0.001 & \text{if } y_{optimal} = 0 \end{cases} \tag{4.7}$$

Figure 4.5 illustrates the overall design optimization process for MSSR:

To better demonstrate the MSSR search process, generations and updates of the sample points during the global optimization on a Banana function are graphically illustrated using Figure 4.6a–e. Each figure contains three iterations which involve the GS, MS and LS. At the start, the region of LS is larger than that of MS. As the iteration goes on and the expensive sample points increase, LS quickly shrinks to focus on the region around the global optimum. MS always provides a medium-sized region that includes the current best solution. The MS and LS are getting smaller and smaller when more and more points are supplemented. Intuitively, LS can make the search concentrate on the current most promising region and accelerate the convergence. MS can provide a promising region that may include several true local optimal solutions. And GS can guarantee that the multi-start optimization process will explore the entire design space. As Figure 4.6a, c shows, sometimes, LS may not include the true global optimal position, but LS will eventually get close to it with the current best sample point moving. Ultimately, 15 iterations and 37 expensive

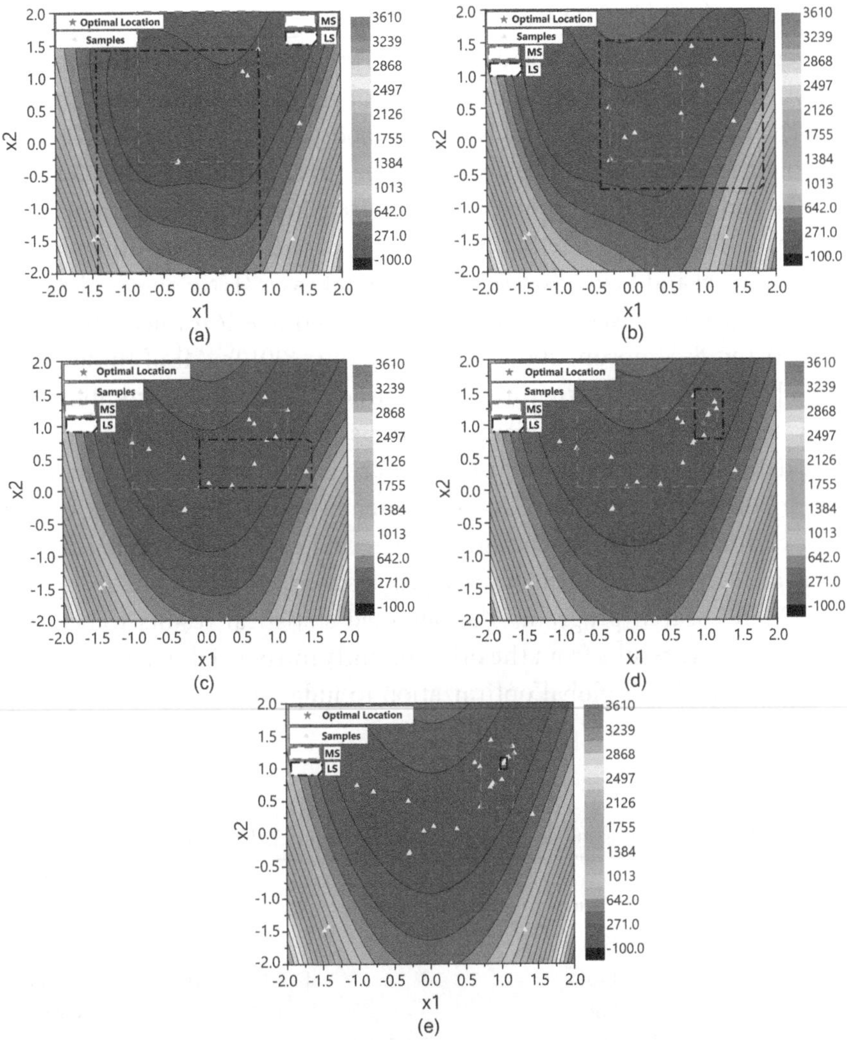

FIGURE 4.6 (a–e) MSSR optimization process on benchmark Banana function.

sample points are used to find a satisfactory global solution. Initially, with only eight DOE samples, the basic shape of the surrogate model was quite different from the real situation, but with the addition of new samples, the surrogate model gradually approached the real function, especially near the global optimum region.

4.6 TEST CASES AND RESULTS

To verify the capabilities and demonstrate the advantages of the new MSSR algorithm, various commonly used global optimization benchmarks which encompass bound and nonlinear constrained problems were used during the tests. The dimensions of these problems range from 2 to 16. For bound-constrained problems, there are eight two-dimensional cases (Banana, Peaks, GP, SC, Shub, GF, HM, Leon), two four-dimensional cases (Shekel and Levy), two six-dimensional cases (HN6 and Trid6), two ten-dimensional cases (Sphere and Trid10) and one 16-dimensional case F16 (Wang & Simpson, 2004; Younis & Dong, 2010). All of these problems have their own structures and characteristics, and in combination they can better represent many situations in engineering optimization. The detailed forms of these functions are given in Table 4.1. For nonlinear constrained problems, two representative mathematical cases and four commonly used benchmark engineering cases were employed. Ten runs on each of these benchmark problems have been made using the new MSSR search program. The obtained statistical results were compared with the results from the other recently introduced space reduction search methods for global optimization to judge their relative efficiency and robustness.

TABLE 4.1 Bound-Constrained Benchmark Problems for Global Optimization

Category	Func.	Number of Dims.	Design Space	Analytic Global Minimum
Low-dimensional problems (n=2–6)	Banana	2	$[-2, 2]^2$	0.0000
	Peaks	2	$[-3\ 3]\times[-4\ 4]$	−6.5511
	GP	2	$[-2, 2]^2$	3.0000
	SC	2	$[-2, 2]^2$	−1.0320
	Shub	2	$[-10, 10]^2$	−186.7309
	GF	2	$[-2, 2]^2$	0.5233
	HM	2	$[-6, 6]^2$	0.0000
	Leon	2	$[-10, 10]^2$	0.0000
	Shekel	4	$[0, 10]^4$	−10.1532
	Levy	4	$[-10, 10]^4$	0.0000
	HN6	6	$[0, 1]^6$	−3.3220
	Trid6	6	$[-36, 36]^6$	−50.0000
High-dimensional problems (n >= 10)	Sphere	10	$[-5.12, 5.12]^{10}$	0.0000
	Trid10	10	$[-100, 100]^{10}$	−210.0000
	F16	16	$[-1, 1]^{16}$	25.8750

4.6.1 The Algorithmic Test

At first, Harmony Search (Yang, 2010) and Differential Evolution (Storn & Price, 1997) algorithms were selected as reference cases to demonstrate that nature-inspired global optimization methods commonly have larger computation costs on expensive black-box problems. An effective space reduction algorithm DIRECT (Björkman & Holmström, 1999), and a widely cited surrogate-based space exploration method MPS (Wang et al., 2004) were also employed for comparison. Meanwhile, EGO that uses the Kriging model for expensive functions was compared to prove the advantage of the proposed algorithm. Here, Mueller's surrogate model toolbox (Mueller, 2012) was used to realize the "Expected Improvement" strategy in the EGO algorithm. Finally, a comparison with a multi-start optimization algorithm that does not use a spatial reduction strategy is made to demonstrate the importance of spatial reduction.

For MSSR, $3n+2$ DOE sample points have been generated to construct the initial surrogate model. Seven representative functions from Table 4.1 were used as test cases, and the seven algorithms have been used to run the tests for ten times. Table 4.2 shows the collected median values of the number of function evaluations (NFE) and obtained minimum values (Min). The seven algorithms tried to get the values that satisfy the condition of Eq. (4.8). It is worth mentioning that EGO has much higher CPU time than other algorithms when the samples and dimensions increase. Hence,

TABLE 4.2 Preliminary Comparison Results on Seven Representative Benchmark Functions

Algorithms		**Banana**	**GP**	**SC**	**Shub**	**Shekel**	**HN6**	**F16**
HS	**NFE**	9,122	512	310	450	10,000	698	915
	Min	8.84e-4	3.0164	−1.0276	−185.6736	−2.6829	−3.3033	26.1207
DE	**NFE**	1,390	830	450	3,070	3,730	3,660	3,690
	Min	4.05e-4	3.0075	−1.0299	−185.3988	−10.0930	−3.3085	26.1022
DIRECT	**NFE**	603	101	117	2,883	103	213	6,439
	Min	3.01e-4	3.0073	−1.0248	−185.5823	−10.0934	−3.2975	26.0884
MPS	**NFE**	145	134	35	545	680	783	3,319
	Min	0.0358	3.0014	−1.0311	−186.7119	−5.0473	−3.3205	29.7177
EGO	**NFE**	216	167	35	227	250	54	200
	Min	9.67e-4	3.0323	−1.0297	−181.0324	−7.5345	−3.3152	27.4815
MS	**NFE**	61	124	25	117	289	121	161
	Min	2.51e-4	3.0065	−1.0299	−186.4286	−10.0863	−3.2973	26.1116
MSSR	**NFE**	41	82	22	115	197	83	138
	Min	3.45e-4	3.0049	−1.0303	−186.4203	−10.0829	−3.2967	26.1257

a maximum allowable NFE (250 for low-dimensional problems. 200 for high-dimensional problems) was given when EGO was tested. As shown by the results listed in Table 4.2, HS and DE consistently had larger NFE than the other algorithms. DIRECT performed well on most cases except for the Banana, Shub and F16 functions. Basically, EGO and MPS could easily get the approximate global optimal values on simpler cases like GP and SC, but most of the time they needed larger NFE on complex cases like Shub, Shekel and F16. From Table 4.2, it can be found that the proposed MS and MSSR algorithms had better performance in all these cases. Moreover, MSSR used fewer NFE than MS and has shown its advantage. Obviously, the "Space Reduction" strategy improves the presented multi-start optimization algorithm. In summary, nonsurrogate-based methods generally have larger NFE, since they directly call the exact function when searching the optimal solutions. Surrogate-based methods are guided by predictive models to explore the design space, which effectively decrease NFE.

In summary, nonsurrogate-based methods generally have larger NFE, since they directly call the exact function when searching the optimal solutions. Surrogate-based methods are guided by predictive models to explore the design space, which effectively decrease NFE. Upon comparison with nature-inspired global optimization methods as well as existing optimization methods for classical agent models, it can be initially seen that the MSSR algorithm proposed in this chapter has some superiority.

However, once these surrogate models focus on the same region, the algorithm will converge to a local optimal location and can hardly explore other promising areas. In this chapter, SEUMRE and HAM used Eq. (4.8) as the termination criteria, and all the user-defined parameters were assigned based on the two original papers (Gu et al., 2012; Younis & Dong, 2010). Since grid sampling can find the global optimal positions of GP and Banana by luck before the iteration process of SEUMRE begins, the DOE ranges of SEUMRE were changed as 95% of the original ranges on the two problems. To deal with the randomness of these methods, each of the experiment tests was done ten times.

Table 4.3 provides the mean values of NFE and the range of the obtained best values. Table 4.4 shows the statistical results of NFE, which involve the minimum NFE, maximum NFE and the median. The NFE values with the ">" sign indicate that at least one of the tests could not satisfy the stopping criteria within 500 function evaluations, and the numbers in the brackets represent how many failures it had. As indicated by Tables 4.3 and 4.4, the MSSR method has successfully found the global optimum in all cases

TABLE 4.3 Mean Values of NFE and Ranges of Optimal Values Obtained by the Three Algorithms

	MSSR		SEUMRE		HAM	
Func.	**NFE**	**Obtained Value**	**NFE**	**Obtained Value**	**NFE**	**Obtained Value**
Banana	42.8	[1.91e-5, 7.32e-4]	90.9	[4.75e-5, 6.37e-4]	68.3	[1.27e-5, 6.34e-4]
Peaks	28.1	[−6.5477, −6.5007]	42.7	[−6.5509, −6.4868]	>228.5	[−6.5510, −3.0498]
GP	87.1	[3.0001, 3.0273]	133.6	[3.0002, 3.0191]	122	[3.0001, 3.0227]
SC	22.5	[−1.0316, −1.0274]	48.8	[−1.0307, −1.0241]	33.9	[−1.0316, −1.0259]
Shub	122.9	[−186.7259, −184.9656]	>329.5	[−186.4404, −117.0721]	168.4	[−186.7209, −185.9839]
GF	34.2	[0.5233, 0.5277]	>208.4	[0.5259, 0.5350]	94.1	[0.5238, 0.5283]
HM	40.3	[7.79e-5, 7.56e-4]	>266.8	[1.04e-5, 0.0028]	120	[1.01e-4, 9.08e-4]
Leon	181.7	[8.88e-5, 9.68e-4]	>253.9	[1.12e-4, 0.3207]	239.4	[1.21e-4, 9.58e-4]
Shekel	207.1	[−10.1486, −10.0716]	>471.7	[−10.0546, −2.6303]	>458.1	[−10.1472, −2.6166]
Levy	218.5	[3.96e-4, 8.04e-4]	>358.1	[6.63e-4, 0.1103]	>341.7	[2.96e-5, 2.26e-2]
HN6	84.8	[−3.3119, −3.2890]	>282.5	[−3.3009, −3.1046]	93.5	[−3.3194, −3.2967]
Trid6	92.1	[−49.9021, −49.5544]	>500	[−47.5255, −7.9626]	127.5	[−49.9614, −49.6379]
Sphere	115.4	[4.57e-4, 9.98e-4]	>500	[1.8147, 17.2568]	>288.3	[4.20e-4, 0.1847]
Trid10	142.4	[−208.9614, −208.0416]	>500	[−83.0087, 990.0295]	>500	[−166.6914, −48.9592]
F16	137.7	[26.1053, 26.1307]	>500	[27.5243, 29.5178]	>249.8	[26.0410, 26.6333]

within 500 function evaluations and used the least NFE. SEUMRE could perform well on Banana, Peaks, GP and SC, but it had difficulties in solving the multimodal and high-dimensional problems. As Table 4.4 shows, SEUMRE just succeeded one time on Shekel, four times on Levy and six times on HN6, but it failed all the ten runs on Trid6, Sphere, Trid10 and F16. The best value SEUMRE obtained on F16 is 27.5243 with 500 function evaluations, which is much larger than the results from MSSR and HAM. HAM is an effective method that could perform better on Banana, GP, SC,

TABLE 4.4 Specific Statistical Results of NFE Obtained by MSSR, SEUMRE and HAM

	MSSR			SEUMRE			HAM		
Func.	**Min**	**Max**	**Median**	**Min**	**Max**	**Median**	**Min**	**Max**	**Median**
Banana	24	66	41	72	114	86	45	104	62
Peaks	18	50	24	37	44	44	34	>500(4)	73
GP	51	141	82	79	359	93	82	195	117
SC	18	27	22	44	58	49	26	52	29
Shub	24	215	115	68	>500(3)	377	86	315	160
GF	15	64	29	65	>500(2)	100	46	281	76
HM	22	95	32	65	>500(4)	157	46	288	66
Leon	67	408	146	142	>500(2)	194	102	433	233
Shekel	68	415	197	217	>500(9)	>500	269	>500(8)	>500
Levy	89	376	181	119	>500(6)	>500	104	>500(5)	>370
HN6	52	117	83	125	>500(4)	149	87	108	91
Trid6	63	146	85	>500	>500(10)	>500	106	144	130
Sphere	94	145	117	>500	>500(10)	>500	180	>500(3)	198
Trid10	125	162	139	>500	>500(10)	>500	>500	>500(10)	>500
F16	103	168	138	>500	>500(10)	>500	136	>500(2)	201

Shub, GF, HM, Leon, HN6 and Trid6, but it had a poor performance on Shekel and Trid10. For high-dimensional problems Sphere and F16, HAM could get satisfactory solutions most of the time.

Figure 4.7a–f provide the main iterative results of the three methods on the high-dimensional problems with the obtained best objective function value and increasing NFE. To improve the readability, two adjacent iterative results have a small interval that is basically more than two units of NFE. Figure 4.7a, c, e shows the entire search process within the 200 function evaluations, and Figure 4.7b, d, f gives a clearer comparison on the results of HAM and MSSR from the NFE values of 100–200. It can be found that MSSR got closer to the true global optimal solutions quicker than HAM and SEUMRE. In addition, only MSSR could satisfy the stopping criteria of Eq. (4.8) within 200 function evaluations. All these methods were run on a computer with a Core i7–4720HQ CPU (2.60 GHZ) and 16 GB memory. The execution time the three algorithms averagely spent on these test functions has also been recorded. Figure 4.8 shows that MSSR and HAM spent more time than SEUMRE on two-dimensional problems. This is due to the fact that MSSR needs to call the SQP solver many times in one loop and HAM needs to construct three surrogate models in each iteration. Furthermore, the three methods have the common feature that

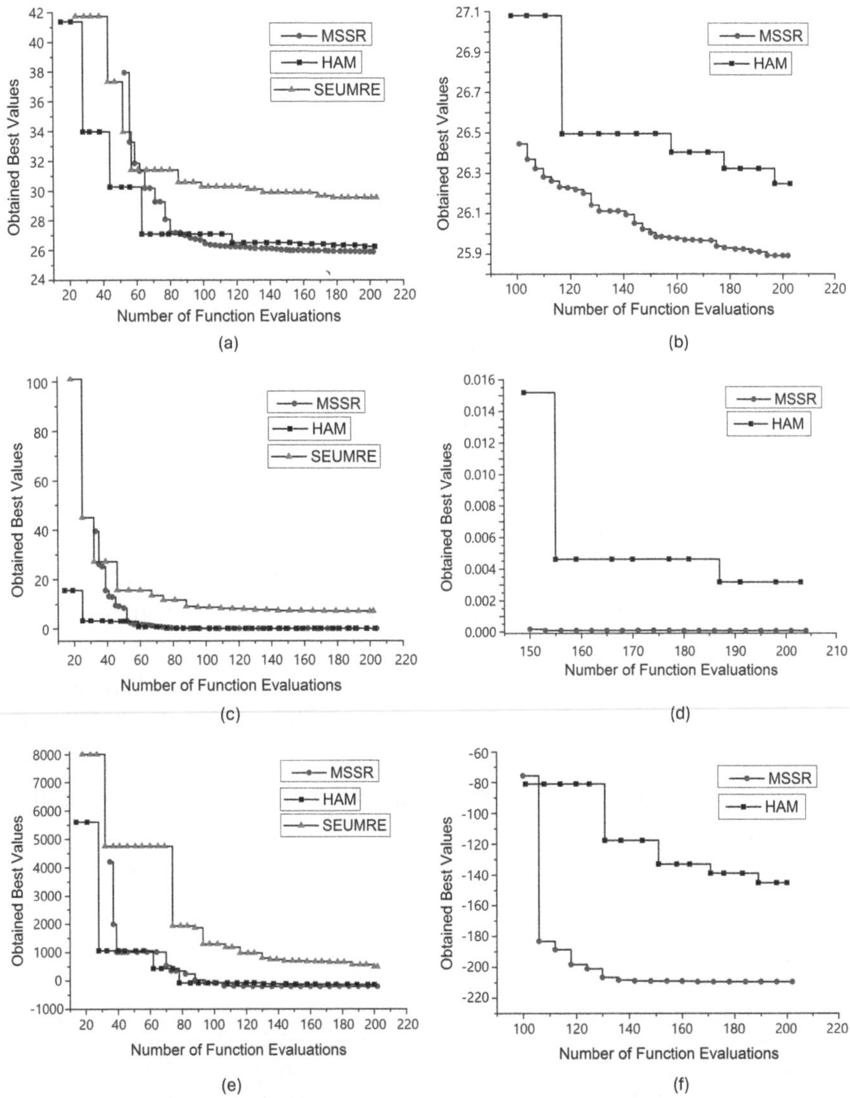

FIGURE 4.7 (a–f) Iterative results on high-dimensional problems.

they will be more time-consuming on higher-dimensional and multimodal problems.

In summary, NFE is always the most important evaluation indicator for the algorithm's performance on expensive black-box problems. HAM presents a good performance most of the time, but it may be trapped around some local optima sometimes. SEUMRE can perform better on low-dimensional problems, but it cannot work well on multimodal and high-dimensional problems. The new MSSR method satisfies the given stopping criteria with the least NFE and has the highest robustness, presenting to be the most promising black-box global optimization technique.

4.6.2 Engineering Case Testing

In this chapter, six classical nonlinear constrained problems were used to test the MSSR method. One of the test problems (G6) comes from the well-known constrained optimization problems that were used by Coello Coello (2002), Abdel-Rahman (2004) and Egea (2008). Another one comes from the widely used Himmelblau's nonlinear problems (Gen & Cheng, 1999; Himmelblau, 1972). Four structural engineering applications are Tension/Compression Spring Design (TSD), Welded Beam Design (WBD), Pressure Vessel Design (PVD) and Speed Reducer Design (SRD), respectively (Coello Coello, 2002; Gen & Cheng, 1999). All of these six test problems' objectives and constraints were regarded as expensive black-box functions. The dimensions of these test cases (G6, TSD, WBD, PVD, Him, SRD) range from 2 to 7, and their numbers of constraints are 2, 4, 7, 4, 6 and 11.

Figure 4.9a, b, d, f shows that MSSR usually could not find the feasible solutions at the beginning, but it would eventually acquire the global optimum. According to the references in these test cases, the obtained values in Table 4.5 and Figure 4.9 are sufficiently accurate and the corresponding NFEs are much smaller.

To verify the robustness of MSSR in dealing with nonlinearly constrained optimization problems, the results of ten independent runs of the

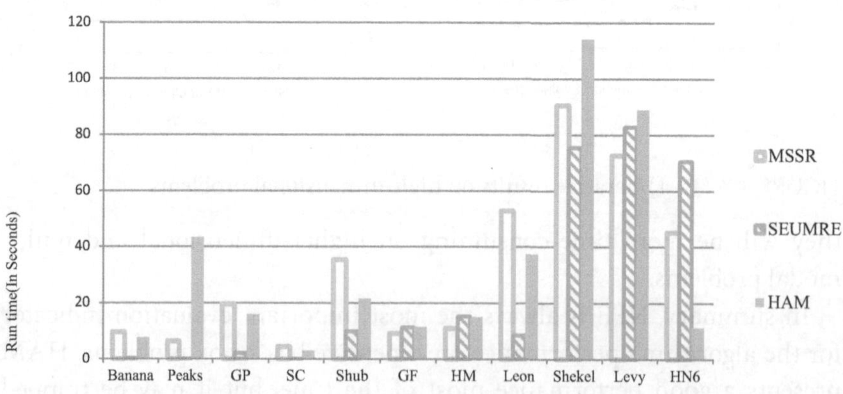

FIGURE 4.8 Execution time of MSSR, SEUMRE and HAM on benchmark functions.

TABLE 4.5 Global Optimal Results Obtained by MSSR on Nonlinear Constrained Problems

	Design Variables							
Problems	***x*1**	***x*2**	***x*3**	***x*4**	***x*5**	***x*6**	***x*7**	***f*(*x*)**
G6	14.097149	0.847352						−6,956.8719
TSD	0.0516827	0.3565636	11.2980133					0.0126652
WBD	0.2056902	3.4683028	9.0445203	0.2056904				1.7256
PVD	0.778187	0.384658	40.320586	199.986548				5,885.3653
Him	78.000000	33.000000	27.072136	45.000000	44.967954			−31,025.3139
SRD	3.500177	0.700000	17.000000	7.332558	7.715387	3.350284	5.286657	2,994.8487

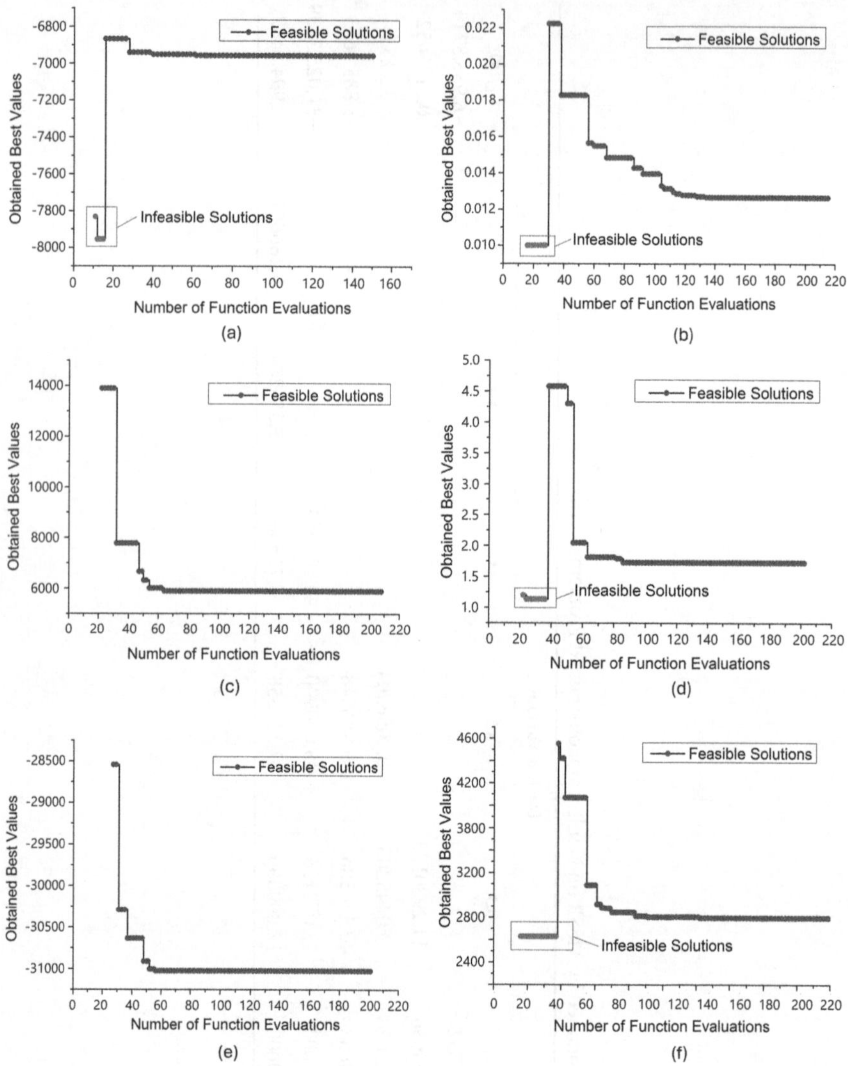

FIGURE 4.9 (a–f) Iterative results obtained by MSSR on constrained optimization problems.

computation are given in this chapter, and it is clear from Tables 4.6 and 4.7 that each time the results are very close to the true optimal solution and the number of NFEs is sufficiently small.

Overall, MSSR can not only perform well on bound-constrained expensive black-box optimization problems but also efficiently and robustly obtain global optimal solutions on nonlinearly constrained problems.

TABLE 4.6 Summary of Results Obtained by MSSR on G6, TSD and Him

	G6		TSD		Him	
Exp.	NFE	Opt. Value	NFE	Opt. Value	NFE	Opt. Value
No.1	62	−6,957.3896	81	0.0126664	60	−31,025.5575
No.2	79	−6,958.4628	213	0.0126817	61	−31,025.2482
No.3	19	−6,955.8152	139	0.0126817	48	−31,025.5270
No.4	44	−6,958.2769	97	0.0126654	93	−31,025.0141
No.5	20	−6,958.2769	97	0.0126653	100	−31,021.3633
No.6	53	−6,957.8394	140	0.0126655	69	−31,023.6350
No.7	39	−6,958.0899	114	0.0126670	57	−31,023.9933
No.8	40	−6,961.2597	66	0.0126698	63	−31,025.5595
No.9	29	−6,955.2008	109	0.0126654	51	−31,025.5557
No.10	63	−6,955.2106	108	0.0126652	51	−31,023.2053

TABLE 4.7 Summary of Results Obtained by MSSR on WBD, PVD and SRD

	WBD		PVD		SRD	
Exp.	NFE	Opt. Value	NFE	Opt. Value	NFE	Opt. Value
No.1	110	1.7253	88	5,885.4051	131	2,994.8493
No.2	133	1.7253	87	5,885.3782	164	2,996.4051
No.3	99	1.7249	75	5,885.3427	189	2,995.5840
No.4	162	1.7253	125	5,885.3979	134	2,994.4745
No.5	167	1.7560	107	5,885.3576	102	2,997.4988
No.6	201	1.7535	97	5,885.3658	102	2,997.4988
No.7	113	1.7256	91	5,885.3778	111	2,994.6535
No.8	100	1.7256	112	5,885.4247	96	2,997.0597
No.9	153	1.7256	98	5,885.3408	69	2,995.4729
No.10	105	1.7254	73	5,885.3993	71	2,997.3088

4.7 CHAPTER SUMMARY

In this work, a new multi-start optimization strategy is introduced to search the three spaces. This strategy applies OLHS to provide multiple starting points and then employs an SQP solver to explore the surrogate model using these starting points in the defined space. The other two varying spaces, namely, MS and LS, are two reduced regions that include the promising solutions and adjust their positions and boundaries automatically during the search. Each of the three spaces has its own functions. GS ensures that the true global solution will not be missed. MS plays an important role in providing a promising region that involves several current best solutions. And LS is an accelerator to finish the search around a true local optimum quickly. In this work, a new multi-start optimization strategy is introduced to search the three spaces. This strategy applies LHS to provide multiple starting points and then employs an SQP solver to explore the surrogate model using these starting points in the defined space. In each of the iterative search loops, a new selection scheme is used to obtain several promising samples. This selection scheme ensures that the Kriging-based surrogate model is sufficiently exploited, and the unknown area of the surrogate model can be gradually explored. The estimated MSE of the Kriging-based surrogate model is used as an important tool to explore the unknown area of the design space.

The new algorithm has been applied to 15 benchmark bound-constrained optimization examples, two nonlinear constrained optimization problems and four structural engineering applications. All the benchmark test results showed MSSR's superior performance and robustness in dealing with expensive black-box optimization problems.

NOTE

1 Based on "Multi-start Space Reduction (MSSR) Surrogate-based Global Optimization Method," published in [Structural and Multidisciplinary Optimization], [2016]. Permission obtained from [Springer].

REFERENCES

Abdel-Rahman, A. (2004). *Studies on metaheuristics continuous global optimization problems.* Kyoto University, Japan.

Björkman, M., & Holmström, K. (1999). Global Optimization Using DIRECT Algorithm in Matlab. *Advanced Model Optimization, 1*(2), 17–37.

Coello Coello, C. A. (2002). Theoretical and Numerical Constraint-Handling Techniques Used with Evolutionary Algorithms: A Survey of the State of the Art. *Computer Methods in Applied Mechanics and Engineering, 191*(11), 1245–1287. https://doi.org/10.1016/S0045-7825(01)00323-1

Edke, M. S., & Chang, K.-H. (2011). Shape Optimization for 2-D Mixed-Mode Fracture Using Extended FEM (XFEM) and Level Set Method (LSM). *Structural and Multidisciplinary Optimization, 44*(2), 165–181. https://doi.org/10.1007/s00158-010-0616-5

Egea, J. (2008). *New heuristics for global optimization of complex bioprocesses.* Universidade de Vigo, Galiza.

Gary Wang, G., Dong, Z., & Aitchison, P. (2001). Adaptive Response Surface Method - A Global Optimization Scheme for Approximation-Based Design Problems. *Engineering Optimization, 33*(6), 707–733. https://doi.org/10.1080/03052150108940940

Gen, M., & Cheng, R. (1999). *Genetic algorithms and engineering optimization* (Vol. 7). John Wiley & Sons.

Gu, J., Li, G. Y., & Dong, Z. (2012). Hybrid and Adaptive Meta-Model-Based Global Optimization. *Engineering Optimization, 44*(1), 87–104. https://doi.org/10.1080/0305215x.2011.564768

Gutmann, H. M. (2001). A Radial Basis Function Method for Global Optimization. *Journal of Global Optimization, 19*(3), 201–227.

Himmelblau, D. (1972). *Applied nonlinear programming*. McGraw-Hill.

Jin, R., Chen, W., & Simpson, T. W. (2001). Comparative studies of metamodelling techniques under multiple modelling criteria. *Structural and Multidisciplinary Optimization, 23*, 1–13.

Jones, D. R., Schonlau, M., & Welch, W. J. (1998). Efficient Global Optimization of Expensive Black-Box Functions. *Journal of Global Optimization, 13*(4), 455–492. https://doi.org/10.1023/A:1008306431147

Long, T., Wu, D., Guo, X., Wang, G. G., & Liu, L. (2015). Efficient Adaptive Response Surface Method Using Intelligent Space Exploration Strategy. *Structural and Multidisciplinary Optimization, 51*(6), 1335–1362. https://doi.org/10.1007/s00158-014-1219-3

Lophaven, S. N., & Nielsen, H. B., & Søndergaard J. (2002). DACE-A Matlab Kriging toolbox. version 2.0.

Mueller, J. (2012). *User guide for modularized surrogate model toolbox*. T. U. o. T. Department of Mathematics.

Queipo, N. V., Haftka, R. T., Shyy, W., Goel, T., Vaidyanathan, R., & Kevin Tucker, P. (2005). Surrogate-Based Analysis and Optimization. *Progress in Aerospace Sciences, 41*(1), 1–28. https://doi.org/10.1016/j.paerosci.2005.02.001

Regis, R. G., & Shoemaker, C. A. (2007). A Stochastic Radial Basis Function Method for the Global Optimization of Expensive Functions. *Informs Journal on Computing, 19*(4), 497–509.

Storn, R., & Price, K. (1997). Differential Evolution – A Simple and Efficient Heuristic for global Optimization over Continuous Spaces. *Journal of Global Optimization, 11*(4), 341–359. https://doi.org/10.1023/A:1008202821328

Wang, G. G., & Simpson, T. (2004). Fuzzy Clustering Based Hierarchical Metamodeling for Design Space Reduction and Optimization. *Engineering Optimization, 36*(3), 313–335.

Wang, L., Shan, S., & Wang, G. G. (2004). Mode-Pursuing Sampling Method for Global Optimization on Expensive Black-Box Functions. *Engineering Optimization, 36*(4), 419–438.

Yang, X. (2010). *Engineering optimization: An introduction with metaheuristic applications*. John Wiley & Sons.

Younis, A., & Dong, Z. (2010). Metamodelling and Search Using Space Exploration and Unimodal Region Elimination for Design Optimization. *Engineering Optimization, 42*(6), 517–533. https://doi.org/10.1080/03052150903325540

CHAPTER 5

SOCE

Surrogate-Based Optimization with Clustering-Based Space Exploration for Expensive Multimodal Problems[1]

5.1 INTRODUCTION

In complex multidisciplinary designs, there exist a large number of computationally intensive black-box problems involving expensive hardware or software resources (Zeng et al., 2016). Commonly, response outputs from an expensive analyzer form the objective and/or constraint functions of an EBOP. Intuitively, the total number of objective or constraint function evaluations (NFE) reflects the computation load in an EBOP. Especially, when the EBOP is nonconvex (Deshmukh & Allison, 2016; Yin et al., 2016), that is, the expensive black-box problem has multiple locally optimal solutions, the NFE will become larger (Alexandrov et al., 1998; Leifsson & Koziel, 2016; Toropov et al., 1993). Traditional global optimization algorithms, such as nature-inspired methods (Sadollah et al., 2015; Wang, 2010; Yang, 2009), need to create a diverse population and meanwhile update generations to explore the design space. In genetic algorithms (GA), the "promising parents" have a bigger opportunity to pass their genetic information to the children, which is inspired by evolutionary concepts. GA can find the optimal fitness function value generation by generation with four main steps, which are reproduction, crossover, mutation and selection (Al-Sultan & Nizami, 1996). The particle swarm optimization (PSO) algorithm uses

DOI: 10.1201/9781003636267-5

simple formulas to imitate the social behavior patterns of organisms like swarms, bats, bees and ants that can work in a team (Shi & Eberhart, 1998). Recently, a remarkable algorithm called GWO was presented by Mirjalili et al. (2014), which was inspired by gray wolves' leadership hierarchy and hunting strategies. Due to its efficiency and robustness, GWO has been widely used in engineering applications. These nature-inspired algorithms can effectively solve highly nonlinear, discrete, nonconvex optimization problems; therefore, considerable contributions have been made in this field. However, all the above-mentioned algorithms have difficulties in dealing with EBOPs, because stochastic search produces substantial function evaluation (Weise et al., 2016).

To control the NFE in an expensive black-box optimization process, surrogate-assisted global optimization algorithms have been developed (Haftka et al., 2016; Zadeh et al., 2009). Jones et al. (1998) introduced an efficient global optimization algorithm called EGO, which has shown its excellent performance in comparison with other classical algorithms. EGO combines the prediction uncertainty of Kriging and the current best value to create an "expected improvement (EI) function," and updates the sample set by maximizing the EI function. Gutmann (2001) utilized the radial basis function (RBF) to construct a surrogate model and measured the bumpiness of the surrogate model. This algorithm updates the sample set by selecting a new position with a hypothetical value that yields the "least bumpiness" of these surrogate models. Regis and Shoemaker (2013) provided a quasi-multi-start response surface framework (AQUARS) for global optimization of EBOPs. This proposed framework not only focuses on the current best local optimal region of the surrogate model but also explores the neighborhoods of the least explored local optimum. Finally, AQUARS was employed to solve a watershed calibration problem and had a remarkable performance. Jie et al. (2015) provided an adaptive meta-model-based global optimization algorithm (AMGO) for unconstrained EBOPs. AMGO employs Kriging and augmented RBF for modeling, and their weight factors are dynamically selected with iterations increasing. With tests on different benchmark examples, AMGO showed satisfactory precision and low computation cost.

When it comes to constrained EBOPs, the state of the art is relatively weak (Zhou et al., 2016). A lot of work has been done for constrained evolutionary optimization algorithms (Coello Coello, 2002), but the huge NFE makes them hard to deal with constrained EBOPs.

The previously introduced surrogate-based algorithms had better performances on unconstrained EBOPs, but they were not tested on benchmark constrained problems. Regis (2014) developed two algorithms (COBRA and Extended ConstrLMSRBF) for constrained EBOPs. The two algorithms follow a two-phase approach, in which the first one guarantees the algorithm to find feasible solutions quickly and the second one makes the feasible solution go close to the true global optimal location. Cutbill and Wang (2016) introduced a probabilistic method to reduce the redundant constraints for black-box optimization problems. They defined a series of rules to express the relationships among constraints, but the accuracy of these rules depended on the number of samples in a particular region.

In this chapter, a new surrogate-based global optimization algorithm with clustering-based space exploration (SOCE) for multimodal and/or constrained EBOPs is presented. This proposed algorithm uses QRS and Kriging to construct two surrogate models. Based on the characteristics of QRS and Kriging, two different optimizers (a multi-start local optimizer and the GWO global optimizer) are connected to the two models, respectively. In the employed multi-start local search, collected samples need to keep a defined distance from each other to satisfy the diversity of predicted local optima. Besides, SOCE suggests a local convergence criterion to judge when to carry out space exploration. The presented space exploration approach employs the *k*-means clustering algorithm to create multiple subspaces and defines an iterative process to select the promising samples that are far away from the clustering centers. In addition, two penalty function methods are proposed to make the algorithm applicable to constrained optimization.

5.2 SOCE ALGORITHM

5.2.1 Surrogate Modeling and Optimization

In SOCE, Kriging and QRS models are constructed separately to approximate the true model. Each has distinct predictive characteristics. Kriging is an interpolation method that commonly generates an approximation model with multiple local optima. Owing to its remarkable capacity in predicting nonconvex problems, Kriging has been widely used for complex engineering applications. QRS belongs to one of the regression methods that generally can reflect the overall trend of a true model. Especially, if it is a convex problem, QRS can accurately predict the global optimum. However, it has difficulties in dealing with multimodal problems.

Sequential quadratic programming (SQP) is a well-known local optimization algorithm that can search for the optimal solution from one given starting point. The success rate that SQP finds the global optimum depends on positions of starting points and the complexity of this problem. For a multimodal function, it is hard for SQP to directly find the global optimal solution. Hence, we utilized a multi-start SQP (MSSQP) algorithm to realize the global optimization process. The MSSQP algorithm includes two parts: the pre- and post-treatment parts. In the pre-treatment process, Symmetric Latin Hypercube Sampling (SLHS) is employed to capture the initial starting points. SLHS can make starting points have a random and centro-symmetric distribution in a design space. On one hand, the random nature increases the success rate of MSSQP to find the global optimum when the main loop keeps running. On the other hand, a better coverage rate can improve the probability of obtaining the global optimum in one iteration. In the post-treatment process, the key point is how to get new samples with diversity, which can avoid supplementing samples around the same local optima. Here, we define an allowable minimum distance between these promising local optimal locations as follows.

$$Dis = w\left\|\max\left(\boldsymbol{Range}\right) - \min\left(\boldsymbol{Range}\right)\right\| \tag{5.1}$$

In Eq. (5.1), ***Range*** is a vector representing the range of a design space. The default value of the coefficient w is 0.005 in SOCE and it affects the length of Dis.

GWO is a recently presented nature-inspired global optimizer, which has been widely used. GWO divides the gray wolves into four types (alpha, beta, delta and omega) based on their leadership hierarchy. The four types correspond to different fitness values. Additionally, GWO simulates the gray wolves' hunting mechanism that includes encircling prey, hunting and attacking, to get new samples. In summary, GWO is appropriate for multimodal problems and can explore the QRS model efficiently. It is worth noting that global optimization on QRS sometimes may produce repeated samples in multiple iterations, thus the algorithm needs to delete redundant samples in time. Besides, QRS needs at least $0.5n^2 + 1.5n + 1$ samples to guarantee the predictive accuracy. Here, n represents the design dimension. Equation (5.2) describes the optimization on surrogate models.

$$\begin{aligned} &MSSQP \rightarrow \min \hat{f}_{Krg}(\boldsymbol{x}) \quad \boldsymbol{lb} \le \boldsymbol{x} \le \boldsymbol{ub} \\ &GWO \quad\ \rightarrow \min \hat{f}_{QRS}(\boldsymbol{x}) \quad \boldsymbol{lb} \le \boldsymbol{x} \le \boldsymbol{ub} \end{aligned} \tag{5.2}$$

where $\hat{f}_{Krg}(\boldsymbol{x})$ and $\hat{f}_{QRS}(\boldsymbol{x})$ represent the Kriging and QRS models of an objective function, respectively.

5.2.2 Initialization and Loop of SOCE

The optimization flow of SOCE includes initialization and loop. In SOCE, the initialization process mainly defines some basic parameters and carries out the design of experiments (DOEs). After the expensive sample values are evaluated at these DOE sample points, two initial surrogate models (Kriging and QRS) are constructed, respectively. Furthermore, the initial expensive samples are sorted to get the current best value for the following loop. Algorithms 5.1(a) and (b) show the main details of this process.

In the loop process, the Kriging and QRS models are optimized by the MSSQP and GWO algorithms, respectively. After a series of detections and selections, the new samples are added into the sample set. At this moment, if the algorithm satisfies a local convergence criterion, it will go on exploring the unknown space. Finally, sample ranks, two surrogate models and the design range will be updated for the next iteration.

Algorithm 5.1(a) is shown below.

Algorithm 5.1(a) The Proposed Optimization Flow—Initialization

(01) **Begin**
(02) Initialize Kriging and quadratic polynomial parameters, and set the population size and max iterations of the gray wolf optimizer.
(03) Carry on the initial DOE process, evaluate the expensive function values and construct the initial surrogate models.
(04) Set the structure variables of the Kriging and QRS predictors as global variables for the subsequent optimization.
(05) ***n*** ← Get the dimension of design variables
(06) ***Iteration*** ← Count the iteration number
(07) ***Current_NFE*** ← Count the number of function evaluations
(08) ***Y_best*** ← Sort the initial sample values
(09) ***Range_new*** ← Set the new range as the initial space.
(10) **End**

Algorithm 5.1(b) is shown below.

Algorithm 5.1(b) The Proposed Optimization Flow—Loop

(01) **Begin**
(02) **while** ***Y_best*** does not reach the target value **and** ***Current_NFE***<300
(03) ***M*** ← Carry on SLHS to obtain multiple starting points.
(04) ***A*** ← Call SQP optimizer at ***M*** to obtain multiple locally optimal solutions from the Kriging model.
(05) ***S_Kriging*** ← Find two promising locations from ***A*** that keep a defined distance with each other and meanwhile cannot go close to existing samples. The distance is *w*|| Max (***Range_new***) – Min (***Range_new***) ||.
(06) **if** ***Current_NFE***$>0.5n^2+1.5n+1$
(07) ***S_QRS*** ← Call GWO optimizer to obtain the global optimal location from the QRS model.
(08) **end if**
(09) ***S*** ← Promise that ***S_Kriging*** and ***S_QRS*** are not repeated samples and store them into the sample set.
(10) ***Y*** ← Evaluate the expensive objective function values.
(11) ***Local_error*** ← Sort the current sample values ***Y*** and obtain the local convergence error.
(12) **if** ***Local_error*** satisfies the local convergence criteria
(13) ***S_explore*** ← Call **Algorithm 5.2** to get several samples from the unknown design space.
(14) ***Y_explore*** ← Evaluate the expensive function values.
(15) **end if**
(16) **if** ***Iteration***>3
(17) **if** REM (***Iteration***, 2) == 0
(18) ***Range_new*** ← Keep the new range for Kriging as the original design range.
(19) **else**
(20) ***Range_new*** ← The new range for Kriging is reduced to a region that encloses the top 50% of samples. The minimum and maximum **X** coordinates in each dimension are selected to create this region. If this region focuses on a point or a line, the new range is defined as the original design range.
(21) **end if**
(22) **end if**
(23) Update and obtain the algorithm parameters, Kriging and QRS models.

```
(24)        Y_best ← Get the current best function value
(25)    end while
(26) End
```

5.2.3 Clustering-Based Space Exploration

As previously discussed, if the search falls into a local optimal region for multiple iterations, the algorithm needs to jump out and explore the sparsely sampled region. In the current work, the sparsely sampled regions are defined based on the *k*-means clustering algorithm.

First, a local convergence criterion is proposed. In each iteration, the mean value of the current top *m* samples is stored. When the iteration number is more than 3, a local error is obtained by the difference of the present and last mean values. If the error equals to zero, it means that no better samples are added into the top *m* samples in this iteration. The algorithm allows this case but it cannot continue for too many consecutive times. If this situation continues for the maximal times (here, the threshold value is defined as 10), the algorithm will use Latin hypercube sampling (LHS) to get new samples in a promising region. When the local error drops below a user-defined small value, the clustering-based space exploration is activated. In this work, a small value of 0.001 is defined as the maximum error. The main steps are summarized below.

- Utilize the *k*-means clustering algorithm (Hartigan & Wong, 1979) to produce multiple clustering centers.
- Evaluate the total length of the design range in each dimension and set a small percentage. Create multiple small regions around these clustering centers. The specific expressions can be found in Lines (14) and (15) of Algorithm 5.2(a).
- Count the number of the samples being located in these created small regions. If the proportion of the counted samples in the total samples exceeds a user-defined value ***Ratio***, the loop ends. Otherwise, the percentage ***w*** gets increased and the loop continues.
- Generate new samples by LHS in the whole design range and delete those samples located in the clustering-based regions.
- Finally, evaluate the expensive sample values and update the sample set.

Algorithms 5.2(a) and (b) show the details of the clustering-based space exploration.

Algorithm 5.2(a) Clustering-based Space Exploration—Search Strategy

(01) **Begin**
(02) **if** 0 < ***Local_error*** < 0.001
(03) ***S_explore_number*** ← Define the number of initial samples for exploration.
(04) ***S_explore*** ← Call LHS to obtain the initial samples for exploration.
(05) ***S_ number*** ← Count the number of the current samples.
(06) ***w*** ← Set the initial parameters for the size of clustering regions.
(07) ***Center_number*** ← Define the number of clustering centers.
(08) ***Center*** ← Employ the K-means algorithm to obtain clustering centers.
(09) ***Ratio*** ← The defined percentage of ***S_ number.***
(10) ***Sum_ratio*** ← 0
(11) ***dis_range*** ← Get the distance between low and up bounds in the design range.
(12) **while** ***Sum_ratio*** < ***Ratio***
(13) **for** each clustering center *i*
(14) ***Range_clusters*** (*i*) ← [***Center*** (*i*)-***w**dis_range***; ***Center*** (*i*) + ***w**dis_range***]
(15) Keep the ***Range_clusters*** enclosed by the original design space.
(16) **end for**
(17) Call **Algorithm 5.2(b)** to sign the Samples of ***S*** located in ***Range_clusters.***
(18) ***Sum_in*** ← Count the number of the samples of ***S*** in these ***Range_clusters.***
(19) ***Sum_ratio*** ← ***Sum_in***/ ***S_ number.***
(20) ***w*** ← ***w*** + 0.025.
(21) **end while**
(22) Call **Algorithm 5.2(b)** to sign the Samples of ***S _explore*** located in ***Range_clusters***
(23) ***S_explore_save*** ← Save the ***S_explore*** samples outside the ***Range_clusters.***
(24) Make sure that ***S_explore_save*** keeps a small distance with existing samples ***S.***

(25) ***Y_explore_save*** ← Evaluate the function values at samples ***S_explore_save***.
(26) [**S**, **Y**] ← Update the expensive samples set.
(27) **else if** ***Local_error*** == −1
(28) ***Range_promising*** ← Utilize the top 50% of samples to create a promising range.
(29) ***S_explore*** ← Call LHS to obtain 3***n***+2 samples that cannot go close to existing samples.
(30) ***Y_explore*** ← Evaluate the function values at samples ***S_explore***.
(31) [**S**, **Y**] ← Update the expensive samples set.
(32) **end if**
(33) **end if**
(34) **End**

Algorithm 5.2(b) is a function of Algorithm 5.2(a). It describes how to sign the samples in a particular region. Algorithm 5.2(b) will return a vector with logical values to identify whether the samples are in a particular region. The details are as follows:

Algorithm 5.2(b) Check Samples in the Clustering Ranges or Not

(01) **Begin**
(02) ***S_Number*** ← Input the number of samples.
(03) ***Range_clusters*** ← Input the defined range.
(04) ***S_test*** ← Create a zero vector with the length of ***S_Number***.
(05) **for** each clustering range *k*
(06) **for** each existing expensive samples *i*
(07) **if** ***S_test*** (*i*)== 0
(08) ***IN*** ← Check the **S** (*i*) in the ***Range_clusters*** (*k*) or not.
(09) **if** ***IN*** == 1
(10) ***S_test*** (*i*) ← 1
(11) **end if**
(12) **end if**
(13) **end for**
(14) **end for**
(15) **return** ***S_test***
(16) **End**

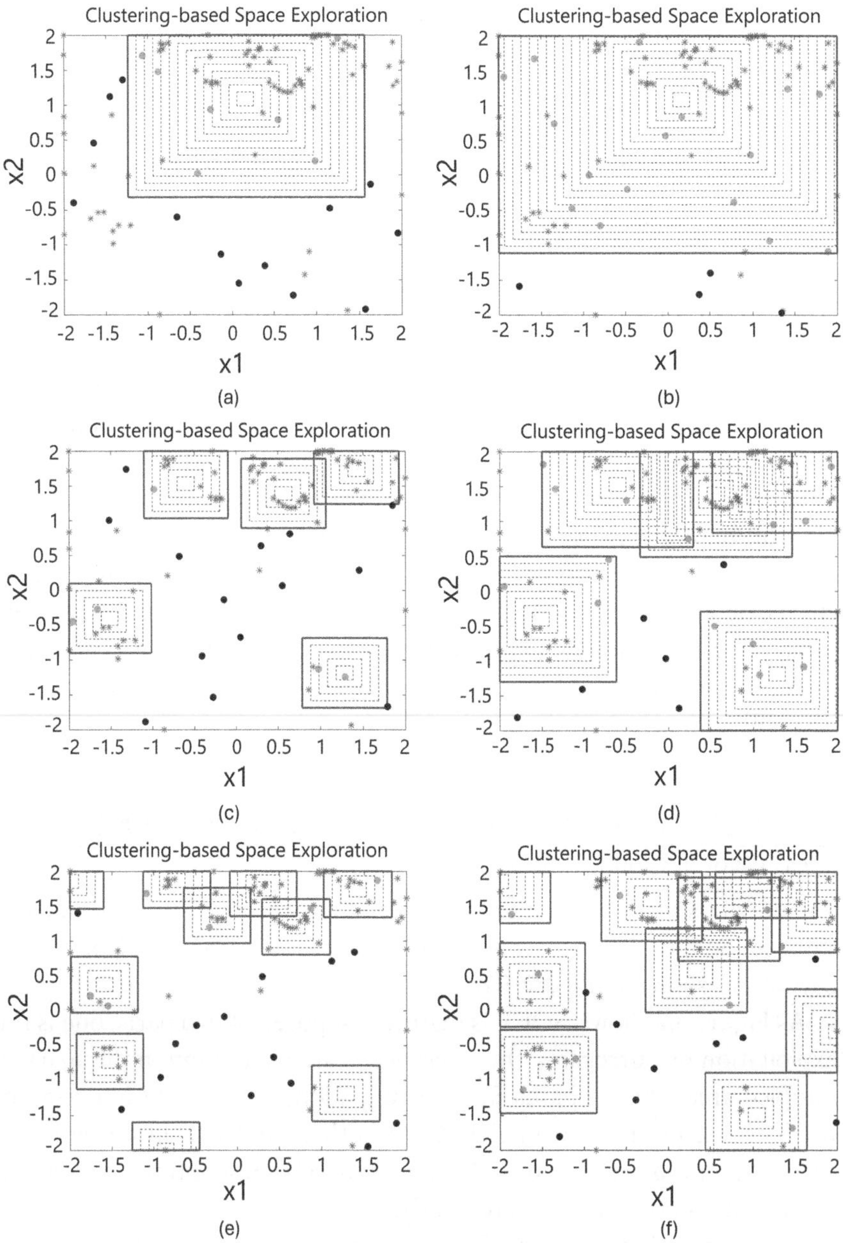

FIGURE 5.1 (a–f) Clustering-based exploration on sparsely sampled regions.

To demonstrate it clearly, several graphic examples are provided. Figure 5.1 shows six groups of results with different parameters. The dots in Figure 5.1 are generated by LHS, among which light ones are located in these clustering regions and black ones are outside of these regions. There are 20 dots in each following figure. Besides, the dashed rectangles describe the dynamic updates of these regions, and the solid-line rectangles are the final boundaries of the clustering regions. According to the previous discussion, the number of dashed rectangles reflects the number of iterations in Algorithm 5.2a. Figures 5.1a and b just have one clustering center and their parameters Ratio are 0.7 and 0.9, respectively. It is clear that the final region will get bigger and the number of iterations will increase if Ratio is larger. What is more, one clustering center cannot describe the distribution of samples well. Figures 5.1c and d show the results when there are five centers. Compared with Figures 5.1a and b, the clustering regions in Figures 5.1c and d can better cover the clustering samples and the new samples can also fill the sparsely sampled space well. Figures 5.1e and f show a similar phenomenon. As observed in Figure 5.1, the number of clustering centers and ***Ratio*** affect the exploration process. Intuitively, more clustering centers can make this strategy more accurate. In summary, if the number of clustering centers is too small, this strategy cannot explore the sparsely sampled space accurately. On the contrary, too many clustering centers will increase the number of loops in Algorithm 5.2(a) and the computational cost will get larger.

5.3 OVERALL OPTIMIZATION FRAMEWORK OF SOCE

5.3.1 Overall Optimization Process

The previous section described the specific process of the algorithm, and this section describes the optimization process of the SOCE algorithm as a whole, as shown in Figure 5.2.

As Figure 5.2 shows, SOCE is mainly composed of two parts: one is the exploitation on surrogates; the other one is an exploration in the sparsely sampled area. On one hand, SOCE can quickly identify a local optimum with the help of surrogate models. On the other hand, the clustering-based space exploration can make SOCE jump out of a local optimal region and begin a new optimization search in the unexplored area.

The termination criterion of SOCE is suggested as Eq. (5.3).

$$\begin{cases} \dfrac{|y_{optimal} - y_{best}|}{|y_{optimal}|} \leq 1\% \quad or \quad NFE > 300, & if\ y_{optimal} \neq 0 \\ y_{best} < 0.001 \quad or \quad NFE > 300, & if\ y_{optimal} = 0 \end{cases} \tag{5.3}$$

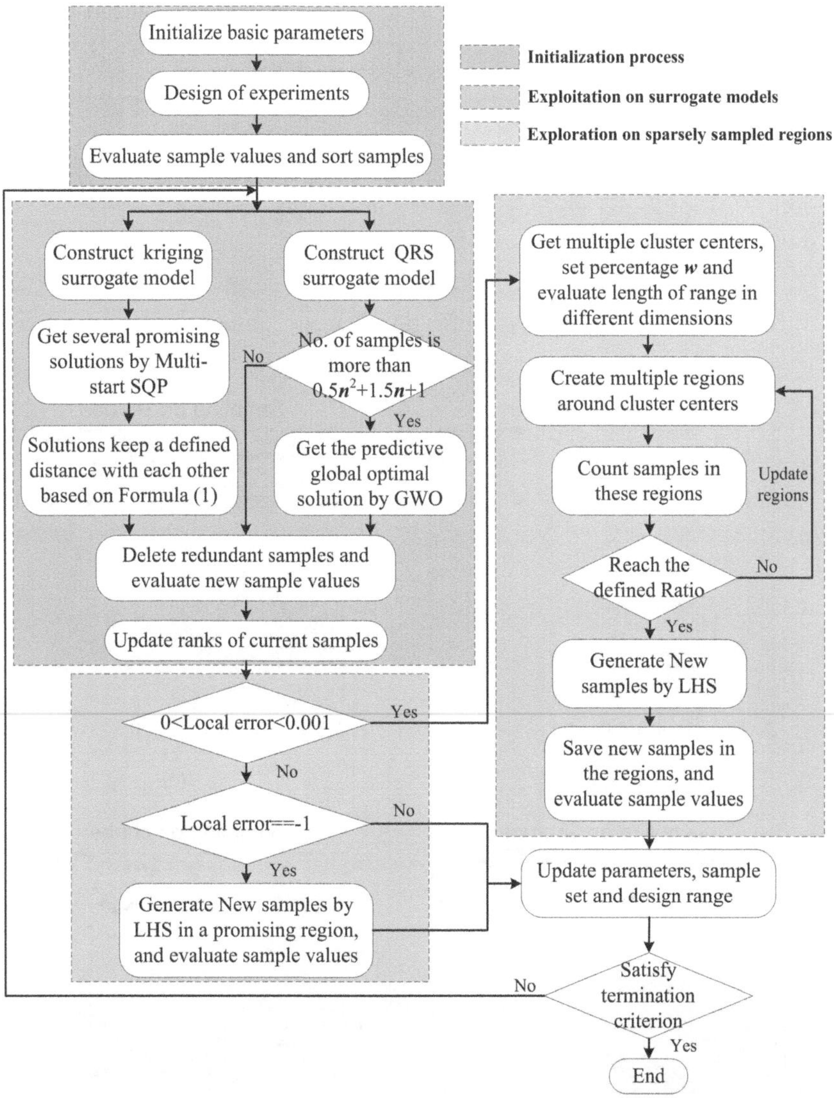

FIGURE 5.2 Flowchart of SOCE.

Here, $y_{optimal}$ is the analytical global optimum, y_{best} is the present best value and *NFE* is the number of objective function evaluations.

To make SOCE easy to understand, a graphic example is shown to demonstrate the capacity of the proposed algorithm. Figures 5.3 and 5.4 illustrate the search process of SOCE on a variety of nonlinear multi-peak problems.

To increase difficulty, a group of DOE samples [−2, −0.857], [2, −0.286], [0.286, 0.286], [0.857, −1.429], [1.429, 1.429], [−0.857, −2], [−1.429, 0.857],

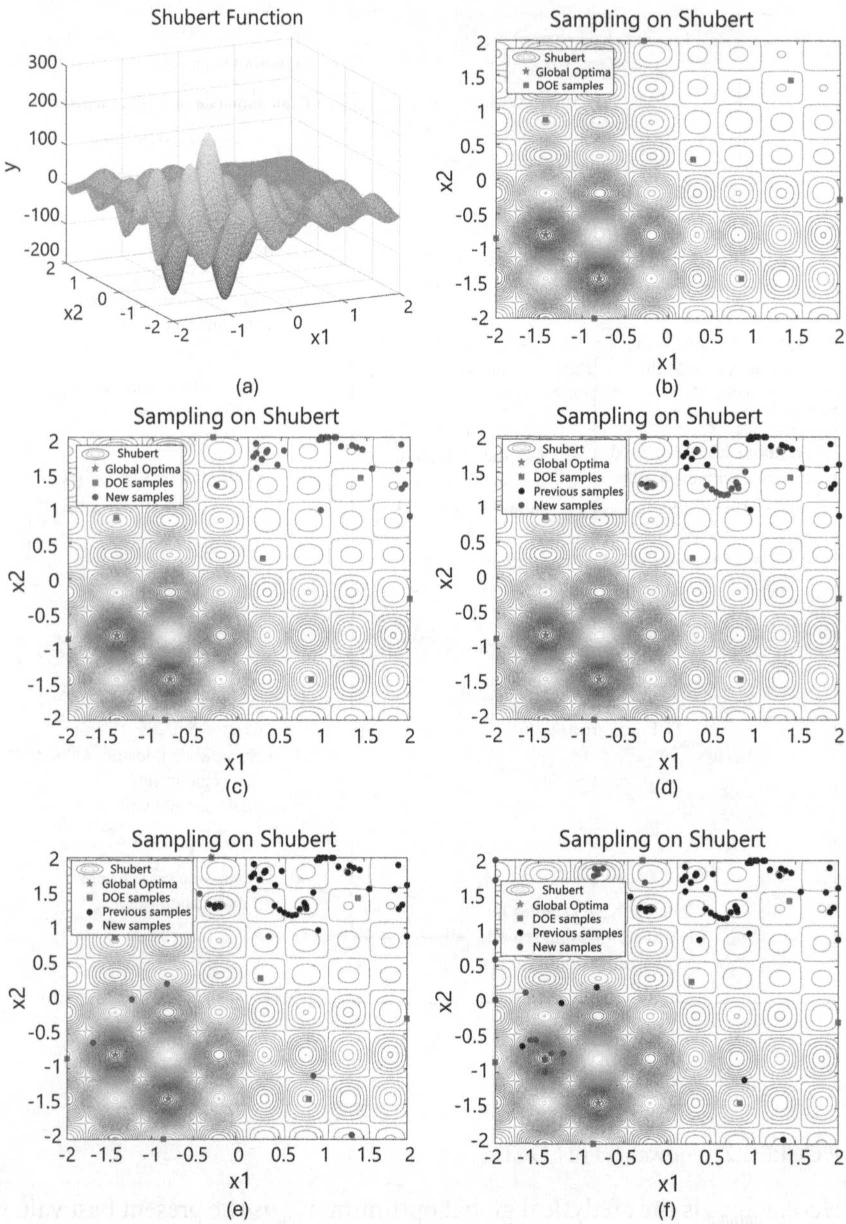

FIGURE 5.3 Optimization process of SOCE on Shubert.

[−0.286, 2] that do not locate in the neighborhoods of the global optimum are selected. Figure 5.3a shows the true surface of Shubert and Figure 5.3b presents the initial DOE samples on its contour plot. Figure 5.3c–f describes the dynamic process of sample updating. Since the initial samples cannot

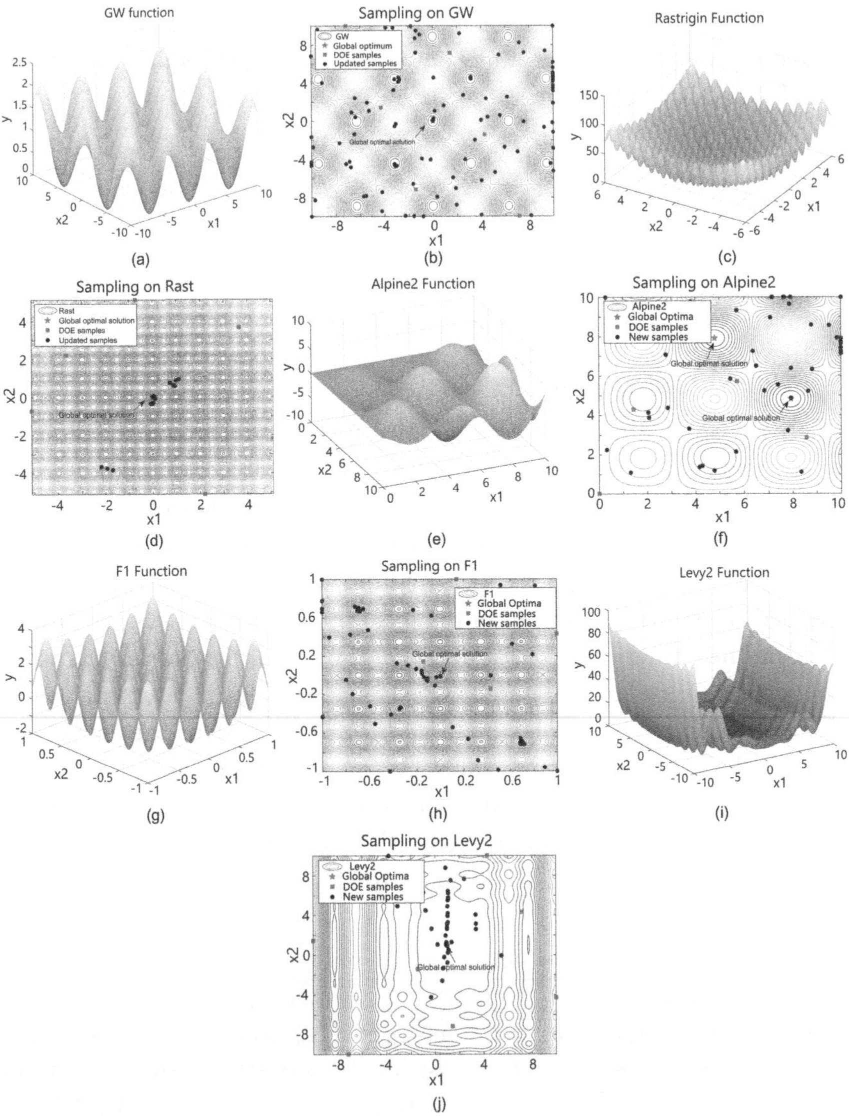

FIGURE 5.4 Optimization process in other multimodal arithmetic cases.

provide an accurate surrogate model, the first search focuses on the regions far away from the true global optimum. From Figure 5.3e, it can be found that new samples are filled in unexplored regions by the clustering-based strategy when the algorithm gets trapped in local minima. Owing to the newly supplemented samples in Figure 5.3e, the accuracy of local regions around the global optimum gets improved. Finally, 89 samples are used in total to find the global optimum.

In addition, the optimization search process of SOCE on some classical nonlinear multi-peak problems is given in Figure 5.4, which shows that SOCE has a strong global search capability.

5.3.2 Parameters Analysis of SOCE

Section 5.1 has mentioned some algorithm parameters, among which the number of clustering centers (NCC), the percentage ***Ratio***, the number of starting points (NSP) and the max allowable number of supplementary samples obtained from multi-start optimization (MANS) may have significant effects on the whole algorithm. Therefore, the sensitivity of these parameters will be analyzed in this section.

This chapter utilizes a representative multimodal function, Shubert, which has a global optimum of −186.7309. To avoid the randomness associated with DOEs, a DOE sample such as the one in Figure 5.3 is used as the initial sample point. Since the initial surrogate models constructed by the eight samples have poor approximation accuracy, SOCE is easy to get trapped in the neighborhood of the local optimum −10.9786, which can activate the clustering-based strategy and make all the parameters work adequately. Considering the stochastic behavior of SOCE, each case needs to be tested for ten times. For NCC and ***Ratio***, 12 cases are given and the statistical results are shown in Table 5.1, where the results with the symbol ">" indicate that at least one test cannot find target values within 300 NFE. In addition, the numbers in brackets reflect the failure times. In Table 5.1, Cases 4, 8 and 12 show that the bigger ***Ratio*** always brings the worse result.

TABLE 5.1 Parametric Analysis of NCC and Ratio on Shubert

	Parameters		Test Results			
Cases	**NCC**	***Ratio***	**Min NFE**	**Mean NFE**	**Max NFE**	**Obtained Values**
Case 1	1	0.6	74	>159.9	>300 (1)	[−186.710, −123.580]
Case 2	1	0.7	50	>117.9	>300(1)	[−186.730, −123.580]
Case 3	1	0.8	50	124.9	274	[−186.730, −185.610]
Case 4	1	0.9	82	>176.1	>300(2)	[−186.720, −79.330]
Case 5	5	0.6	100	143.2	217	[−186.640, −185.100]
Case 6	5	0.7	87	146.5	214	[−186.720, −185.070]
Case 7	5	0.8	94	146.4	263	[−186.730, −185.030]
Case 8	5	0.9	109	>184.4	>300(1)	[−186.730, −79.330]
Case 9	10	0.6	54	121.7	197	[−186.720, −185.080]
Case 10	10	0.7	75	140.8	219	[−186.720, −185.480]
Case 11	10	0.8	77	127.6	171	[−186.720, −185.450]
Case 12	10	0.9	80	148	275	[−186.660, −185.120]

It is clear that Case 4 fails to find satisfactory results within 300 NFE for two times. Cases 9–12 where NCC equals to 10 have relatively smaller NFE values in all the 12 cases. Furthermore, with NCC increasing, the failure times decrease significantly. In conclusion, a ***Ratio*** with a value between 0.6 and 0.8 can make SOCE more efficient. Additionally, a bigger NCC can make the exploration strategy more accurate, but the number of loops in Algorithm 5.2(a) will increase. Therefore, the recommended parameter ranges for NCC and ***Ratio*** are [5, 10] and [0.6, 0.8], respectively.

For the parameter NSP, seven cases are provided and the statistical results are listed in Table 5.2. In this test, NCC is defined as 10 and ***Ratio*** is 0.6. It is easy to find that all the cases can find satisfactory solutions within 300 NFE. Mean NFE does not change too much but it gets smaller gradually when NSP increases. It is worth noting that the CPU time is significantly affected by NSP. This is because a bigger NSP can increase the number of running the SQP optimizer. Eventually, we suggest the parameter range [30, 50] for NSP.

MANS is a main factor to affect the parallelism of SOCE. If MANS gets bigger, the number of supplementary samples in each iteration may increase. Analysis results of MANS are shown in Table 5.3. In this test,

TABLE 5.2 Parametric Analysis of NSP on Shubert

		Test Results				
Cases	**NSP**	**Min NFE**	**Mean NFE**	**Max NFE**	**Obtained Values**	**CPUt**
Case 1	5	67	149.2	227	[−186.73, −185.03]	23.11s
Case 2	10	78	138.4	192	[−186.72, −185.00]	20.33s
Case 3	20	81	140.2	199	[−186.69, −185.34]	25.77s
Case 4	30	63	125.2	174	[−186.67, −185.53]	27.83s
Case 5	40	58	135.7	213	[−186.69, −184.96]	36.07s
Case 6	50	67	130.4	244	[−186.73, −184.87]	41.57s
Case 7	60	48	128.3	243	[−186.71, −185.12]	49.94s

TABLE 5.3 Parametric Analysis of MANS on Shubert

		Test Results				
Cases	**MANS**	**Min NFE**	**Mean NFE**	**Max NFE**	**Obtained Values**	**Iteration**
Case 1	2	64	130.2	183	[−186.72, −184.88]	36.9
Case 2	3	85	144.8	248	[−186.73, −185.31]	31.9
Case 3	4	88	149.8	309	[−186.67, −185.47]	28.5
Case 4	5	121	185.7	323	[−186.72, −185.39]	31.4
Case 5	6	106	184	319	[−186.62, −185.08]	27.6
Case 6	7	121	183.8	248	[−186.67, −185.13]	27.3

NCC, ***Ratio*** and NSP are defined as 10, 0.6 and 30, respectively. As Table 5.3 shows, Mean NFE increases remarkably when MANS changes from 2 to 7. Meanwhile, it can be found that the mean number of iterations has a negative correlation with MANS. In a parallel computing environment, users can increase MANS properly to improve the parallelism of SOCE. To sum up, in this work, the recommended parameter range for MANS is [2, 4].

To verify the recommended parameters, tests are carried out on a more challenging problem (two-dimensional Griewank function with the range $X_1 \in [-5, 15]$ and $X_2 \in [-15, 5]$). Similarly, eight samples [15, −6.429], [3.571, 5], [9.286, −12.143], [0.714, −15], [−5, −9.286], [12.143, 2.143], [−2.143, −0.714], [6.429, −3.571] are given to construct the initial surrogate models that make SOCE easily get trapped around a local optimal value 7.40e-3 at the beginning. The same cases in Tables 5.1–5.3 are tested on Griewank and the specific results are shown in Tables 5.4–5.6. From Table 5.4, it can be seen that Cases 1–4 have the worst performance and Cases 10, 9 and 5 can go close to the global optimum with fewer function evaluations. Additionally, all the better parameter groups in Table 5.4 are located in the recommended parameter ranges.

In Table 5.5, Case 3 has the smallest NFE value and the CPU time gradually gets longer from Case 1 to Case 7. Besides, the results in Table 5.6 show that Cases 1 and 2 have the best performance. Like Tables 5.2 and 5.3, Tables 5.5 and 5.6 also give the laws that a larger NSP value can cause longer CPU time and a larger MANS value may bring more function evaluations. In summary, the parametric analyses on the two representative multimodal problems Shubert and Griewank get similar laws, and meanwhile

TABLE 5.4 Parametric Analysis of NCC and Ratio on GW

Cases	Parameters		Test Results			
	NCC	*Ratio*	Min NFE	Mean NFE	Max NFE	Obtained Values
Case 1	1	0.6	206	>285.5	>300(8)	[1.32e-5, 7.40e-3]
Case 2	1	0.7	166	>286.6	>300(9)	[1.31e-4, 7.40e-3]
Case 3	1	0.8	>300	>300	>300(10)	[7.40e-3, 7.40e-3]
Case 4	1	0.9	>300	>300	>300(10)	[7.40e-3, 7.40e-3]
Case 5	5	0.6	112	>216.2	>300(3)	[1.90e-6, 7.40e-3]
Case 6	5	0.7	184	>275.4	>300(6)	[7.89e-6, 7.40e-3]
Case 7	5	0.8	107	>260.5	>300(6)	[3.53e-5, 7.40e-3]
Case 8	5	0.9	269	>296.9	>300(9)	[4.73e-5, 7.40e-3]
Case 9	10	0.6	80	>203.9	>300(2)	[7.95e-6, 7.40e-3]
Case 10	10	0.7	82	>152.7	>300(1)	[3.38e-5, 7.40e-3]
Case 11	10	0.8	133	>235.4	>300(5)	[2.77e-6, 7.40e-3]
Case 12	10	0.9	151	>257.8	>300(6)	[1.36e-5, 7.40e-3]

TABLE 5.5 Parametric Analysis of NSP on GW

Cases	NSP	Test Results				
		Min NFE	Mean NFE	Max NFE	Obtained Values	CPUt
Case 1	5	147	>225.8	>300(2)	[1.25e-5, 7.40e-3]	25.75s
Case 2	10	118	>192.2	>300(1)	[4.79e-5, 7.40e-3]	23.75s
Case 3	20	88	>183	>300(2)	[1.50e-5, 7.40e-3]	28.64s
Case 4	30	86	>179.8	>300(1)	[5.36e-6, 7.40e-3]	35.82s
Case 5	40	72	>224.5	>300(3)	[1.42e-6, 7.40e-3]	51.80s
Case 6	50	126	>221.2	>300(1)	[2.60e-6, 7.40e-3]	58.55s
Case 7	60	101	>245.9	>300(2)	[1.78e-6, 7.40e-3]	78.46s

TABLE 5.6 Parametric Analysis of MANS on GW

Cases	MANS	Test Results				
		Min NFE	Mean NFE	Max NFE	Obtained Values	Iteration
Case 1	2	91	212.3	370	[1.17e-6, 5.28e-4]	46.0
Case 2	3	113	225	478	[6.54e-6, 9.43e-4]	41.3
Case 3	4	124	258.3	435	[4.51e-6, 9.72e-4]	38.1
Case 4	5	149	281.2	468	[2.26e-6, 6.96e-4]	36.9
Case 5	6	147	311.1	579	[2.34e-6, 8.56e-4]	36.0
Case 6	7	166	293.7	620	[1.51e-5, 9.83e-4]	30.2

the recommended parameter ranges can make SOCE work well. In the subsequent comparison experiments, the four parameters NCC, ***Ratio***, NSP and MANS of SOCE are defined as 10, 0.6, 30 and 2, respectively.

5.4 EXPERIMENTS ON BENCHMARK EXAMPLES

5.4.1 Comparison Test on Bound-Constrained Examples

Considering that SOCE is a multi-point global optimization algorithm, MSEGO supplements multiple samples in each cycle based on different surrogate models and is tested as the preliminary comparison. The test cases and the results of MSEGO come from Long et al. (2015). When ten independent tests are finished, the statistical results are given in Tables 5.7 and 5.8. It is easy to find that SOCE has a better performance than MSEGO on most cases (SE, PK, SC, BR, RS, GN and HN). SOCE can quickly go close to the true global optimum within 40 function evaluations on SE, PK, SC and BR, but MSEGO needs more than 100 function evaluations. Furthermore, SOCE can get better values than MSEGO on RS, GN and HN with fewer function evaluations. Although both SOCE and MSEGO

TABLE 5.7 Obtained Values of SOCE and MSEGO

	SOCE		MSEGO	
Func.	Var. Range	Median	Var. Range	Median
SE	[−1.456, −1.448]	−1.456	[−1.456, −1.454]	−1.456
PK(Peaks)	[−6.551, −6.494]	−6.544	[−6.498, −5.979]	−6.498
SC	[−1.032, −1.030]	−1.032	[−1.024, −0.987]	−1.024
BR	[0.398, 0.399]	0.399	[0.398, 0.431]	0.398
RS(F1)	[−2.000, −1.980]	−1.994	[−1.874, −1.636]	−1.874
GF	[0.003, 0.009]	0.007	[0.001,0.035]	0.001
GP	[3.000, 3.029]	3.008	[3.002, 3.014]	3.002
GN	[3.33e-15, 4.81e-3]	7.33e-4	[0.176, 0.627]	0.177
HN(HN6)	[−3.317, −3.290]	−3.306	[−3.208, −3.052]	−3.145

TABLE 5.8 NFE of SOCE and MSEGO

	SOCE		MSEGO	
Func.	Var. Range	Mean	Var. Range	Mean
SE	[29, 55]	33.4	[70, 123]	109.6
PK(Peaks)	[29, 46]	37.3	[129, 132]	130.4
SC	[26, 47]	34.9	[130, 132]	131.2
BR	[22, 29]	25.9	[36, 132]	112.6
RS(F1)	[29, 242]	108.5	[131, 132]	131.4
GF	[47, 162]	113.5	[132, 132]	132.0
GP	[68, 239]	145.9	[101, 132]	120.4
GN	[11, 130]	95.7	[132, 132]	132.0
HN(HN6)	[55, 149]	89.1	[176, 176]	176.0

can find a value close to 3 on GP, MSEGO has a smaller computation cost. In summary, SOCE is more efficient and robust.

To further verify the algorithm's efficiency and robustness, 15 representative benchmark problems are provided for comparison testing. Among them, there are 12 low-dimensional problems and three high-dimensional problems. More details about them are listed in Table 5.9. Since the stochastic nature of SOCE, ten tests are carried out on these examples. In this work, three surrogate-based algorithms EGO, HAM and KMS are tested as contrast. EGO and HAM are two well-known global optimization algorithms that have been widely cited. KMS is a Kriging-based multi-start global optimization method, which employs the MSSQP algorithm proposed in SOCE. In addition, KMS uses the same multi-start optimization strategy as SOCE.

TABLE 5.9 Bound-Constrained Benchmark Problems for Global Optimization

Category	Func.	Number of Dims.	Design Space	Analytic Global Minimum
Low-dimensional problems (most of them are multimodal problems with lots of local minima)	Shub	2	$[-2, 2]^2$	−186.731
	GW2	2	$[-10, 10]^2$	0.000
	SE	2	$[0, 5]^2$	−1.457
	Peaks	2	$[-3\ 3]\times[-4\ 4]$	−6.551
	Beale	2	$[-4.5, 4.5]^2$	0.000
	Alp	2	$[0, 10]^2$	−6.130
	F1	2	$[-1, 1]^2$	−2.000
	Rast	2	$[-5.12, 5.12]^2$	0.000
	Levy	2	$[-10, 10]^2$	0.000
	Zakh	2	$[-5, 10]^2$	0.000
	Shek10	4	$[0, 10]^4$	−10.536
	HN6	6	$[0, 1]^6$	−3.322
High-dimensional problems (n=10–16)	GW10	10	$[-600, 600]^{10}$	0.000
	Sphere	15	$[-5.12, 5.12]^{15}$	0.000
	F16	16	$[-1, 1]^{16}$	25.875

Table 5.10 shows the best values obtained by SOCE, KMS, EGO and HAM within 300 function evaluations. Table 5.11 presents the NFE used by the four algorithms in the experiments.

It is clear that the best values from SOCE mostly go much closer to the true global optima with the fewest function evaluations and meanwhile it has the fewest failure times. Since GW10 is a high-dimensional and multimodal problem, SOCE cannot find a value that is smaller than 0.001 within 300 NFE. However, SOCE can get satisfactory accuracy on GW10. KMS can perform well on some problems that have fewer local optima, such as SE, Peaks, Beales, levy and HN6, but most of the time it misses the global optimum on more complex problems. EGO can solve low-dimensional multimodal problems well, except Rast, GW2 and Beale. Moreover, EGO has the worst performance on high-dimensional problems. Although the hybrid meta-model technology improves the robustness of HAM, HAM may still miss the global optimum. This is because HAM does not provide a search strategy to explore the sparsely sampled area. As observed in Tables 5.10 and 5.11, HAM sometimes can just find a local optimum on the multimodal problem, but HAM has a relatively robust performance on most of the examples.

To improve readability, the mean values of NFE for the four algorithms are given as a histogram in Figure 5.5. Meanwhile, the total ranks of the

TABLE 5.10 Best Values Obtained by SOCE, KMS, EGO and HAM

	SOCE			KMS			EGO			HAM		
Func	**Best**	**Med.**	**Worst**	**Best**	**Med.**	**Worst**	**Best**	**Med.**	**Worst**	**Best**	**Med.**	**Worst**
Shub	−186.701	−186.053	−185.342	−186.203	−101.456	−39.589	−186.664	−186.109	−184.941	−186.720	−119.826	−39.589
GW2	2.86e-6	2.14e-4	8.28e-4	7.40e-3	8.63e-3	1.97e-2	1.03e-6	5.73e-4	3.53e-3	2.30e-6	7.40e-3	9.86e-3
SE	−1.456	−1.456	−1.448	−1.457	−1.454	2.866	−1.457	−1.455	−1.451	−1.457	−1.453	−1.447
Peaks	−6.551	−6.544	−6.494	−6.550	−6.524	−6.492	−6.551	−6.549	−6.511	−6.551	−6.542	−3.050
Beale	4.19e-5	3.15e-4	8.37e-4	1.18e-4	7.09e-4	9.51e-4	2.35e-3	2.12e-2	8.27e-2	3.37e-7	2.54e-4	2.87e-3
Alp	−6.127	−6.115	−6.084	−6.123	−6.080	−2.854	−6.129	−6.116	−6.089	−6.126	−6.121	−2.854
F1	−2.000	−1.994	−1.980	−1.997	−1.879	−0.660	−2.000	−1.997	−1.985	−2.000	−1.991	−1.879
Rast	4.26e-14	1.78e-4	8.79e-4	1.40e-12	0.995	3.980	2.02e-3	1.00e-2	7.92e-2	1.40e-5	1.05e-4	8.84e-4
Levy	9.83e-6	3.10e-4	6.85e-4	1.07e-5	3.90e-4	9.20e-4	5.13e-5	4.48e-4	1.23e-3	6.03e-7	5.33e-5	9.29e-4
Zakh	7.58e-6	2.72e-4	8.47e-4	2.32e-5	5.04e-4	2.80e-3	6.31e-6	7.68e-4	7.71e-3	3.31e-6	7.51e-5	2.35e-4
Shekel	−10.523	−10.486	−2.871	−10.507	−9.998	−5.126	−10.523	−10.169	−5.029	−10.517	−9.753	−2.427
HN6	−3.317	−3.306	−3.290	−3.312	−3.308	−3.291	−3.318	−3.298	−3.201	−3.316	−3.2945	−3.159
GW10	1.18e-2	3.28e-2	9.75e-2	0.912	1.093	1.367	13.968	28.083	56.223	9.88e-3	2.39e-2	0.585
Sphere	2.09e-10	2.33e-8	5.74e-5	1.29e-3	3.06e-3	7.00e-3	0.180	0.562	0.863	5.31e-4	3.24e-3	0.154
F16	26.073	26.109	26.130	26.096	26.122	26.876	26.356	26.668	27.270	26.061	26.129	26.323

TABLE 5.11 Specific Statistical Results of NFE Obtained by SOCE, KMS, EGO and HAM

	SOCE			KMS			EGO			HAM		
Func	Min	Max	Mean	Min	Max	Mean	Min	Max	Mean	Min	Max	Mean
Shub	14	138	68	14	>300(8)	>243.7	29	111	71	34	>300(5)	>208.7
GW2	32	273	140.3	>300	>300(10)	>300	24	>300(3)	>157.3	58	>300(7)	>244
SE	29	55	33.4	20	>300(1)	>57.4	30	123	54.2	21	73	41.5
Peaks	29	46	37.3	17	92	39.8	21	45	34	21	>300(1)	>67
Beale	58	249	151.9	80	237	156.1	>300	>300(10)	>300	114	>300(1)	>185.2
Alp	23	68	38.4	23	>300(4)	>176.5	15	43	23.8	21	>300(1)	>75.4
F1	29	242	108.5	56	>300(7)	>235.5	39	105.1	155	30	>300(1)	>93.7
Rast	10	64	25.6	11	>300(7)	>214.9	>300	>300(10)	>300	46	171	102.4
Levy	16	74	38.6	19	71	41.2	18	>300(1)	>103.1	37	76	50.9
Zakh	63	231	134.8	54	>300(1)	>198	28	>300(4)	>157.5	42	63	48.2
Shekel	107	>300(2)	>166.1	245	>300(9)	>294.5	240	>300(5)	>282.5	108	>300(8)	>263.3
HN6	55	149	89.1	70	112	87.3	37	>300(3)	>123.4	68	>300(3)	>151.1
GW10	>300	>300(10)	>300	>300	>300(10)	>300	>300	>300(10)	>300	>300	>300(10)	>300
Sphere	138	141	138.6	>300	>300(10)	>300	>300	>300(10)	>300	261	>300(5)	>286.5
F16	111	265	176.8	114	>300(1)	>182.4	>300	>300(10)	>300	187	>300(4)	>252.8

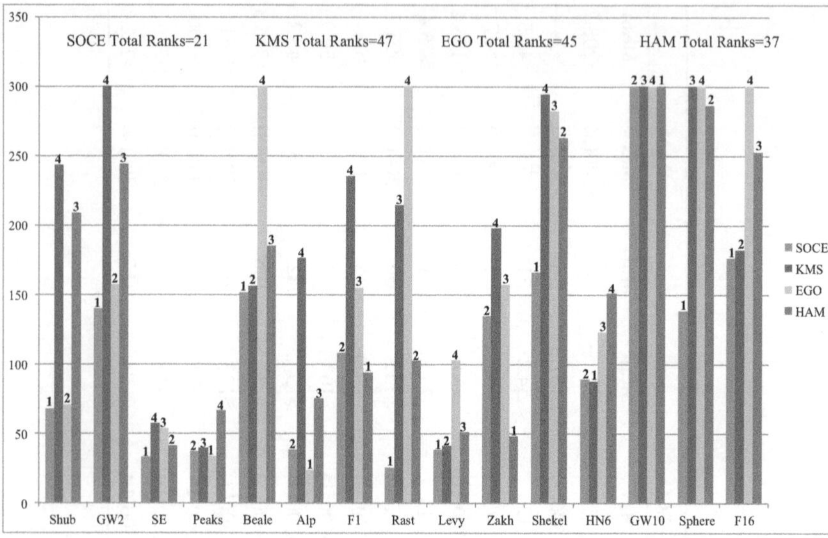

FIGURE 5.5 Histogram of mean values of NFE.

four algorithms are evaluated. Among them, SOCE performs the best. It is worth noting that all the four algorithms have the same NFE on GW10. In Table 5.10, the median values of SOCE, KMS, EGO and HAM on GW10 are 3.28e-2, 1.093, 28.083 and 2.39e-2, respectively. Hence, the accuracy of their results is used for ranking. In summary, SOCE is a promising global optimization algorithm for expensive black-box multimodal problems.

Although Kriging-based optimization technologies commonly cannot work well on high-dimensional problems, the proposed SOCE is still tested on the 50-dimensional Rosenbrock function in this work. Here, the maximal NFE is defined as 1,500. Figure 5.6 shows the true function values obtained by SOCE within 1,500 function evaluations. As discussed previously, QRS begins to work after 1,326 ($0.5n^2+1.5n+1$) samples are added.

In Figure 5.6, the best value changes slightly between 2.5e6 and 2e7 within 1,326 function evaluations. It demonstrates that Kriging cannot efficiently guide SOCE to search the design space anymore. The combination of QRS and Kriging makes the overall trend begin to decrease after 1,400 samples. However, the obtained best value 1.81e6 after 1,500 samples is still far away from the true global optimum 0. The essential reason for the poor performance is that the approximation accuracy of the two surrogates is not good enough on 50-dimensional problems. Hence, SOCE

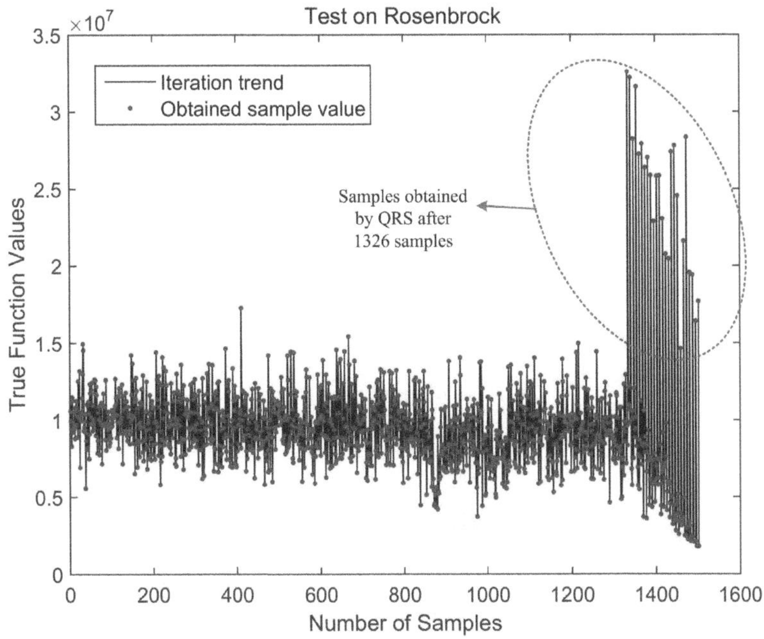

FIGURE 5.6 Test results of SOCE on 50-dimensional Rosenbrock.

is not appropriate for high-dimensional problems (with 50+ design variables) and its better scope of application is lower than 20 dimensions.

5.4.2 Comparison Test on Nonlinear-Constrained Examples

Based on the overall optimization flow in Figure 5.2, a penalty strategy is suggested to make SOCE applicable for nonlinear-constrained optimization. This strategy involves two penalty functions as follows.

$$F = Y_{obj} + 10^{10} \cdot \sum_{i=1}^{m} \max(Z_i, 0) \quad i = 1,2,\ldots,m \tag{5.4}$$

$$F = \begin{cases} Y_{obj} & \text{if } \forall Z_i \le 0 \quad i = 1,2,\ldots m \\ Y_{obj} + 10^{10} & \text{if } \exists Z_i > 0 \quad i = 1,2,\ldots,m \end{cases} \tag{5.5}$$

Y_{obj} refers to the true objective value and Z_i is one of the constraint values. As previously discussed, SOCE needs to sort samples to get the local convergence criterion and update promising regions in Algorithm 5.1. Hence, Eq. (5.4) is employed to get the actual function value with the penalty term for ranking. Besides, Eq. (5.4) is also used to select promising solutions in the multi-start optimization process. On the other side, Eq. (5.5) is used to get the present best value for the termination judgment. For nonlinear-constrained optimization problems, SOCE constructs surrogate models for objective and constraint functions, respectively. Equations (5.6) and (5.7) describe the optimization process on surrogate models.

$$GWO \rightarrow Minimize \quad \hat{f}_{QRS}(\boldsymbol{x}) + 10^{10} \sum_{i=1}^{m} \max\left(\hat{g}_{Krg}^{i}(\boldsymbol{x}), 0\right) \tag{5.6}$$

$$\begin{aligned} & MSSQP \rightarrow Minimize \quad \hat{f}_{Krg}(x) \\ & \qquad \qquad \hat{g}_{Krg}^{1}(\boldsymbol{x}) \le 0 \\ & s.t. \quad \hat{g}_{Krg}^{2}(\boldsymbol{x}) \le 0 \\ & \qquad \qquad \vdots \\ & \qquad \qquad \hat{g}_{Krg}^{m}(\boldsymbol{x}) \le 0 \quad i = 1, 2, \ldots, m \end{aligned} \tag{5.7}$$

where $\hat{g}_{Krg}^{i}(\boldsymbol{x})$ is the Kriging model of the ith constraint. $\hat{f}_{QRS}(\boldsymbol{x})$ and $\hat{f}_{Krg}(x)$ are the QRS and Kriging models of objective functions, respectively. In this section, SOCE is tested on nonlinearly constrained problems (Zhang et al., 2008; Regis, 2014), which include five complex mathematical examples (G6, G7, G8, G9, G10) and two engineering problems (welded beam design (WB4) and speed reducer design (SR7)). The target values are given in Table 5.12. In the same way, all the tests are repeated ten times. Additionally, as contrast, KMS, EGO and HAM utilize Eq. (5.4) to deal with these constrained benchmark examples. Figure 5.7 presents the representative results of SOCE on these problems. Since G7 is a high-dimensional problem, the clear results close to the present best value are

TABLE 5.12 Nonlinear-Constrained Benchmark Problems for Global Optimization

Problems		Number of Design Variables	Number of Constraints	Best Known Value	Target Value
Benchmark mathematical examples	G6	2	2	−6,961.8139	−6,800
	G7	10	8	24.3062	25
	G8	2	2	−0.0958	−0.09
	G9	7	4	680.6301	1,000
	G10	8	6	7,049.3307	8,000
Engineering examples	WB4	4	7	1.7250	2.5
	SR7	7	11	2,994.42	2,995

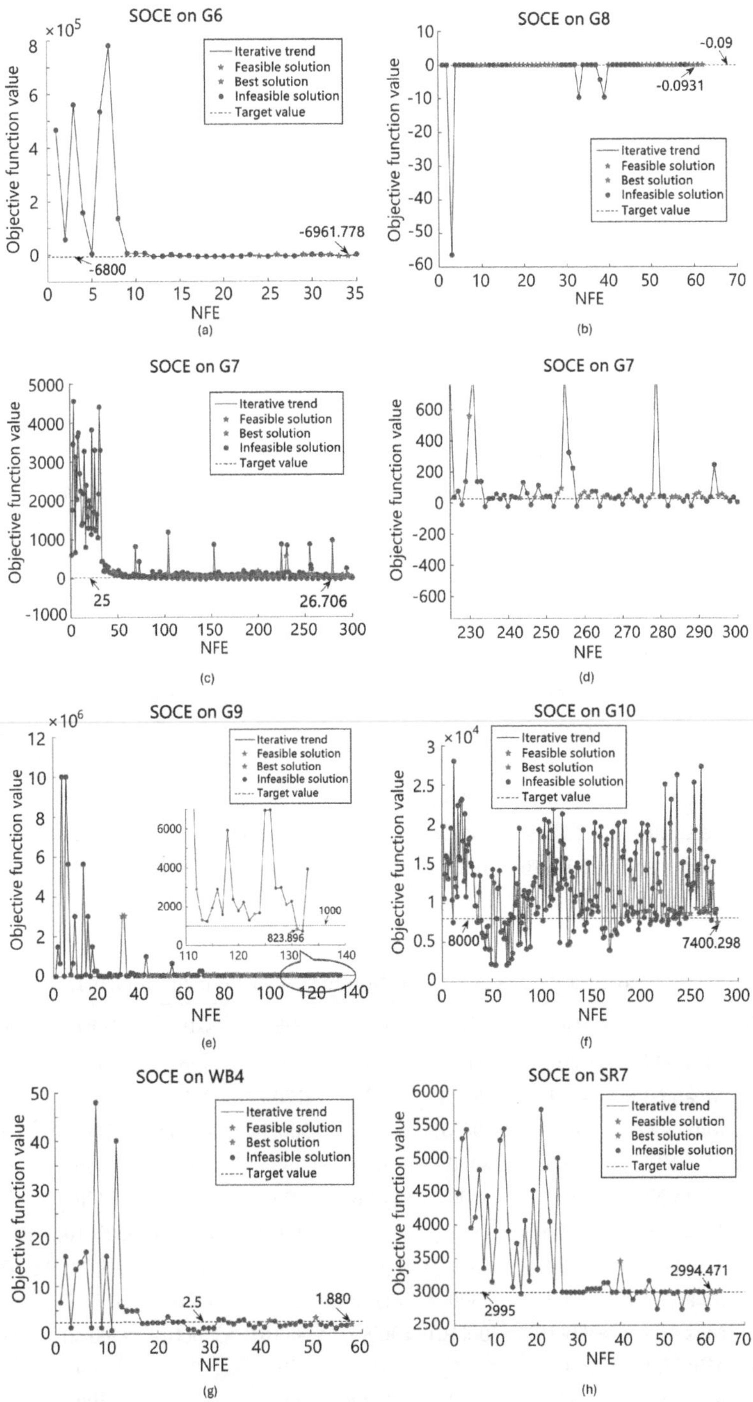

FIGURE 5.7 SOCE on nonlinearly constrained problems.

also provided. It is clear that SOCE can quickly focus on the boundaries of constraints and find feasible solutions. Especially, SOCE can get accurate results with fewer function evaluations on G6, G8, WB4 and SR7. For high-dimensional examples G7, G9 and G10, SOCE can mostly find their target values but higher computation cost is required.

Tables 5.13 and 5.14 give the statistical results including the obtained function values and NFE. SOCE has excellent performances on G6, G8, WB4 and SR7. For G9 and G10, SOCE just fails two times within 300 function evaluations, but it still gets the acceptable results 1,012.288 and 8,260.758 that are quite close to their target values. Additionally, it can

TABLE 5.13 Statistical Results of Function Values on Nonlinear-Constrained Problems

Problems		G6	G7	G8	G9	G10	WB4	SR7
SOCE	Best Value	−6,961.813	24.644	−0.0958	772.220	7,109.074	1.726	2,994.471
	Med. value	−6,953.338	26.208	−0.0937	927.255	7,767.577	2.205	2,994.471
	Worst value	−6,872.775	28.571	−0.0902	1,012.288	8,260.758	2.349	2,994.657
HAM	Best Value	−6,339.926	602.933	−0.0950	966.166	1.28e11	2.934	3,124.147
	Med. value	−3,338.948	1.54e10	−0.0940	1,294.343	2.21e11	3.191	3,222.033
	Worst value	−1,356.719	4.13e10	−0.0912	1,740.238	3.71e11	8.272	3,401.861
KMS	Best value	−6,073.916	169.209	−0.0943	825.588	1.65e11	2.314	3,041.883
	Med. value	−1,500.888	398.865	−0.0738	1,403.688	2.66e11	5.097	3,112.072
	Worst value	1.15e10	2,088.117	−0.0579	3,083.203	3.87e11	1.45e9	3,191.210
EGO	Best value	1.36e9	385.207	−0.091	763.358	6.30e10	2.654	3,051.468
	Med. value	1.16e10	622.641	−0.057	1,033.078	1.88e11	5.142	3,070.588
	Worst value	3.52e10	1,178.897	−0.015	1,295.343	2.58e11	7.511	3,151.450

TABLE 5.14 Statistical Results of NFE on Nonlinear-Constrained Problems

Problems		G6	G7	G8	G9	G10	WB4	SR7
SOCE	Min NFE	20	117	35	79	156	34	39
	Mean NFE	43.9	>266.4	57	>170.7	>241.2	57.3	62.9
	Max NFE	65	>300(8)	114	>300(1)	>300(1)	100	85
HAM	Min NFE	>300	>300	69	295	>300	>300	>300
	Mean NFE	>300	>300	115.8	>299.5	>300	>300	>300
	Max NFE	>300(10)	>300(10)	192	>300(9)	>300(10)	>300(10)	>300(10)
KMS	Min NFE	>300	>300	16	71	>300	17	>300
	Mean NFE	>300	>300	>271.6	>218.3	>300	>271.7	>300
	Max NFE	>300(10)	>300(10)	>300(9)	>300(6)	>300(10)	>300(9)	>300(10)
EGO	Min NFE	>300	>300	10	29	>300	>300	>300
	Mean NFE	>300	>300	>243.6	>207.9	>300	>300	>300
	Max NFE	>300(10)	>300(10)	>300(8)	>300(6)	>300(10)	>300(10)	>300(10)

be seen that SOCE cannot find its target value within 300 function evaluations on G7 in most cases. However, the range of the optimal values [24.644, 28.571] obtained by SOCE on G7 is satisfactory. HAM, EGO and KMS sometimes can reach the target value within 300 NFE on G8 and G9, but all of them cannot go close to the target value on G7 and G10. Especially on G10, HAM, EGO and KMS can hardly find feasible solutions within 300 NFE. Although KMS and HAM also failed on G6, they have almost reached the target. HAM, EGO and KMS have acceptable performance on WB4 and SR7, but few of them can complete the mission within the maximal NFE. After the comparison test, it can be found that SOCE is also an efficient and robust algorithm for nonlinear-constrained EBOPs.

5.5 CHAPTER SUMMARY

In SOCE, a surrogate-based global optimization algorithm SOCE is presented, which can solve multimodal EBOPs and constrained EBOPs. SOCE employs Kriging and QRS to construct two surrogate models. Since Kriging models can always generate multiple predictive optimal locations, an MSSQP is suggested to find them as supplementary samples. To guarantee the diversity of the new samples, MSSQP defines an allowable distance to eliminate redundant samples. QRS models can predict the overall trend of a true model, thus the nature-inspired global optimization algorithm GWO is utilized to capture the global optimum of QRS models. When the optimization process gets trapped in a local optimum, a clustering-based space exploration strategy is activated to make the search focus on unexplored regions. This proposed strategy includes four steps: (1) Generate multiple clustering centers; (2) Create small regions around these centers; (3) Count the current samples located in these regions and update regions until a defined ratio is reached; (4) Generate new samples and delete the samples outside of these regions. In this work, the specific pseudo is provided and a graphic example is shown to demonstrate the remarkable capacity of SOCE on multimodal EBOPs. To verify the robustness of SOCE, tests were repeated ten times on 15 benchmark examples. Besides, three surrogate-based global optimization algorithms EGO, HAM and KMS were compared with SOCE. The results showed the powerful capacity of SOCE in dealing with multimodal EBOPs. Finally, two penalty functions were proposed to make SOCE applicable for constrained optimization. In the tests of seven nonlinear-constrained examples, SOCE successfully found satisfactory solutions with fewer function evaluations. In summary, SOCE is a promising global optimization algorithm for multimodal EBOPs and constrained EBOPs.

NOTE

1 Based on "Surrogate-based Optimization with Clustering-based Space Exploration for Expensive Multimodal Problems," published in [Structural and Multidisciplinary Optimization], [2018]. Permission obtained from [Springer].

REFERENCES

Al-Sultan, K. S., & Nizami, M. F. H. S. (1996). A Genetic Algorithm for the Set Covering Problem. *Journal of the Operational Research Society, 47*(5), 702–709.

Alexandrov, N. M., Dennis, J. E., Lewis, R. M., & Torczon, V. (1998). A Trust-Region Framework for Managing the Use of Approximation Models in Optimization. *Structural Optimization, 15*(1), 16–23. https://doi.org/10.1007/BF01197433

Coello Coello, C. A. (2002). Theoretical and Numerical Constraint-Handling Techniques Used with Evolutionary Algorithms: A Survey of the State of the Art. *Computer Methods in Applied Mechanics and Engineering, 191*(11), 1245–1287. https://doi.org/10.1016/S0045-7825(01)00323-1

Cutbill, A., & Wang, G. G. (2016). Mining Constraint Relationships and Redundancies with Association Analysis for Optimization Problem Formulation. *Engineering Optimization, 48*(1), 115–134. https://doi.org/10.1080/0305215X.2014.995177

Deshmukh, A. P., & Allison, J. T. (2016). Multidisciplinary Dynamic Optimization of Horizontal Axis Wind Turbine Design. *Structural and Multidisciplinary Optimization, 53*(1), 15–27. https://doi.org/10.1007/s00158-015-1308-y

Gutmann, H. M. (2001). A Radial Basis Function Method for Global Optimization. *Journal of Global Optimization, 19*(3), 201–227. https://doi.org/10.1023/A:1011255519438

Haftka, R. T., Villanueva, D., & Chaudhuri, A. (2016). Parallel Surrogate-Assisted Global Optimization with Expensive Functions – A Survey. *Structural and Multidisciplinary Optimization, 54*(1), 3–13. https://doi.org/10.1007/s00158-016-1432-3

Hartigan, J. A., & Wong, M. A. (1979). Algorithm AS 136: A k-Means Clustering Algorithm. *Journal of the Royal Statistical Society, Series C (Applied Statistics), 28*(1), 100–108.

Jie, H., Wu, Y., & Ding, J. (2015). An Adaptive Metamodel-Based Global Optimization Algorithm for Black-Box Type Problems. *Engineering Optimization, 47*(11), 1459–1480.

Jones, D. R., Schonlau, M., & Welch, W. J. (1998). Efficient Global Optimization of Expensive Black-Box Functions. *Journal of Global Optimization, 13*(4), 455–492. https://doi.org/10.1023/A:1008306431147

Leifsson, L., & Koziel, S. (2016). Surrogate Modelling and Optimization Using Shape-Preserving Response Prediction: A Review. *Engineering Optimization, 48*(3), 476–496. https://doi.org/10.1080/0305215X.2015.1016509

Long, T., Wu, D., Guo, X., Wang, G. G., & Liu, L. (2015). Efficient Adaptive Response Surface Method Using Intelligent Space Exploration Strategy. *Structural and Multidisciplinary Optimization, 51*(6), 1335–1362. https://doi.org/10.1007/s00158-014-1219-3

Mirjalili, S., Mirjalili, S. M., & Lewis, A. (2014). Grey Wolf Optimizer. *Advances in Engineering Software, 69*, 46–61. https://doi.org/10.1016/j.advengsoft.2013.12.007

Regis, R. G. (2014). Constrained Optimization by Radial Basis Function Interpolation for High-Dimensional Expensive Black-Box Problems with Infeasible Initial Points. *Engineering Optimization, 46*(2), 218–243. https://doi.org/10.1080/0305215X.2013.765000

Regis, R. G., & Shoemaker, C. A. (2013). A Quasi-Multistart Framework for Global Optimization of Expensive Functions Using Response Surface Models. *Journal of Global Optimization, 56*(4), 1719–1753. https://doi.org/10.1007/s10898-012-9940-1

Sadollah, A., Eskandar, H., & Kim, J. H. (2015). Water Cycle Algorithm for Solving Constrained Multi-objective Optimization Problems. *Applied Soft Computing, 27*, 279–298. https://doi.org/10.1016/j.asoc.2014.10.042

Shi, Y., & Eberhart, R. C. (1998). Parameter selection in particle swarm optimization. *Evolutionary Programming VII: 7th International Conference, EP98*, San Diego, California, USA, March 25–27, 1998 Proceedings 7.

Toropov, V. V., Filatov, A. A., & Polynkin, A. A. (1993). Multiparameter Structural Optimization Using FEM and Multipoint Explicit Approximations. *Structural Optimization, 6*(1), 7–14. https://doi.org/10.1007/BF01743169

Wang, X. (2010). A New Metaheuristic Bat-Inspired Algorithm. In J. R. González, D. A. Pelta, C. Cruz, G. Terrazas, & N. Krasnogor, (Eds.), *Nature inspired cooperative strategies for optimization. Studies in computational intelligence* (Vol. 10, pp. 65–74). Springer-Verlag Berlin Heidelberg.

Weise, T., Wu, Y., Chiong, R., Tang, K., & Lässig, J. (2016). Global versus Local Search: The Impact of Population Sizes on Evolutionary Algorithm Performance. *Journal of Global Optimization, 66*(3), 511–534. https://doi.org/10.1007/s10898-016-0417-5

Yang, X. S. (2009). *Harmony search as a metaheuristic algorithm*. Springer Berlin Heidelberg.

Yin, H., Fang, H., Wen, G., Wang, Q., & Xiao, Y. (2016). An Adaptive RBF-Based Multi-objective Optimization Method for Crashworthiness Design of Functionally Graded Multi-cell Tube. *Structural and Multidisciplinary Optimization, 53*(1), 129–144. https://doi.org/10.1007/s00158-015-1313-1

Zhang, M., Luo, W., & Wang, X. (2008). Differential Evolution with Dynamic Stochastic Selection for CONSTRAINED OPTIMIZATION. *Information Sciences, 178*(15), 3043–3074.

Zadeh, P. M., Toropov, V. V., & Wood, A. S. (2009). Metamodel-Based Collaborative Optimization Framework. *Structural and Multidisciplinary Optimization, 38*(2), 103–115. https://doi.org/10.1007/s00158-008-0286-8

Zeng, F., Xie, H., Liu, Q., Li, F., & Tan, W. (2016). Design and Optimization of a New Composite Bumper Beam in High-Speed Frontal Crashes. *Structural and Multidisciplinary Optimization, 53*(1), 115–122. https://doi.org/10.1007/s00158-015-1312-2

Zhou, Y., Haftka, R. T., & Cheng, G. (2016). Balancing Diversity and Performance in Global Optimization. *Structural and Multidisciplinary Optimization, 54*(4), 1093–1105. https://doi.org/10.1007/s00158-016-1434-1

CHAPTER 6

HSOSR

Hybrid Surrogate-Based Optimization Using Space Reduction for Expensive Black-Box Functions[1]

6.1 INTRODUCTION

Expensive black-box problems (EBPs) are prevalent in modern engineering design (Craven et al., 2016). Traditional optimization techniques, like swarm intelligence and evolutionary computation (Mirjalili et al., 2014; Park & Kim, 2017), are difficult to get EBPs' global optima, primarily because of the substantial number of expensive function evaluations. Surrogate-based global optimization (SBGO) plays an important role in today's simulation-based industrial design (Queipo et al., 2005; Wang & Shan, 2007).

In the past two decades, plenty of researchers focused on the development of SBGO algorithms and their applications (Forrester & Keane, 2009; Gutmann, 2001; Kleijnen, 2009; Myers et al., 2004; Tang et al., 2013; Younis & Dong, 2010). Ong et al. (2003) introduced a hybrid approach that combines surrogate modes with evolutionary algorithms to solve the global optimization of EBPs. Wild et al. (2008) presented a derivative-free optimization algorithm called ORBIT, which employs RBF and a trust-region framework to solve unconstrained expensive optimization problems.

DOI: 10.1201/9781003636267-6

ORBIT was tested on two engineering applications, calibration of a watershed and optimization of a bioremediation plan, and the final results suggested that ORBIT could use fewer function evaluations to complete optimization. Park and Kim (2017) combined a generalized regression neural network with the particle swarm optimizer (PSO) to develop a new surrogate-assisted global optimization algorithm (MUGPSO). Compared to the original PSO, MUGPSO showed improvement in solution quality and computational efficiency. Li et al. (2017) presented a Kriging-based constrained global optimization (KCGO) algorithm to solve expensive nonlinear constrained problems. KCGO includes two phases: (1) "How to find the feasible solutions" and (2) "How to find the better solutions," which can help KCGO find the global optimum even if the initial samples are infeasible.

In recent years, many researchers have been paying attention to hybrid surrogate-based optimization approaches. Zhou et al. (2011) combined different independent surrogates into an ensemble model to improve the prediction accuracy. A recursive process was proposed to obtain the updated weights for each stand-alone surrogate. Through tests on five numerical cases, the ensemble technique showed its advantages in saving sampling costs. Gu et al. (2012) developed a hybrid and adaptive meta-model-based (HAM) global optimization algorithm that employed PR, Kriging, and RBF to estimate the exact objective function, respectively. According to predictive results from the three surrogate models, HAM created seven candidate sets to generate supplementary sample points. HAM was verified by various numerical examples and used for the crashworthiness simulation of a vehicle, and the results showed remarkable computation efficiency and robustness performance. Viana et al. (2013) proposed the multiple surrogates EGO (MSEGO) algorithm that improves the parallelism of EGO. MSEGO can add several new sample points in each optimization cycle based on the predictions of these surrogates, which can considerably reduce the number of iterations required for convergence.

In order to solve unconstrained EBPs, a new algorithm—"Hybrid Surrogate-based Optimization using Space Reduction" (HSOSR) is proposed in this chapter. Since Kriging and RBF have advantages in predicting nonlinear problems, they are employed to construct surrogate models of objective functions, respectively. Generally, different approximation techniques may get different promising regions in a design space. We present a space reduction approach that fuses "Potentially Better Regions" from Kriging and RBF to create two subspaces for exploration. Additionally, a multi-start optimization strategy is employed to search these subspaces on Kriging and RBF. The supplementary samples in each cycle will be selected to avoid repetition. Once the algorithm gets trapped in a local optimal location, a proposed strategy will be triggered to make HSOSR explore the sparsely sampled area of the design space.

6.2 HSOSR ALGORITHM

6.2.1 Surrogate Models – Radial Basis Function

The RBF approximation technique was originally developed by Hardy (1971) and then modified by Dyn et al. (1986). As its name shows, RBF is composed of multiple radial basis functions and it can also be understood as a single-layer neural network. The general expression of RBF is summarized as follows:

$$\hat{f}(\boldsymbol{x}) = \boldsymbol{w}^T \psi = \sum_{i=1}^{m} w_i \psi\left(\left\|\boldsymbol{x} - \boldsymbol{c}^{(i)}\right\|\right) \tag{6.1}$$

In Eq. (6.1), $\boldsymbol{w}$ is a weight vector, $\boldsymbol{x}$ is the to-be-tested location, and $\boldsymbol{c}$ is the center vector. Besides, m denotes the number of input samples and $\psi(\bullet)$ represents the basis functions that have multiple forms and they are shown in Eq. (6.2).

$$\begin{aligned} &\textit{linear}\ \psi(r) = r \\ &\textit{cubic}\ \psi(r) = r^3 \\ &\textit{thin plate spline}\ \psi(r) = r^2 \ln r \end{aligned} \tag{6.2}$$

where r refers to the Euclidean distance between input vector $\boldsymbol{x}$ and center vector $\boldsymbol{c}$. In HSOSR, the cubic basis function is employed to construct RBF.

6.2.2 HSOSR Construction Process

In this section, the proposed HSOSR will be explained in detail. HSOSR is different from the traditional hybrid surrogate-based (or meta-model-based) methods that commonly fuse the results from all the surrogates with weights or construct an ensemble model to combine the advantages of all the surrogates. HSOSR employs Kriging and RBF to construct surrogates for the same expensive black-box objective function, respectively. The better design regions predicted by Kriging and RBF are identified and two promising design subspaces are created for optimization exploration. The specific demonstrations are summarized in the following contents.

6.2.2.1 Space Reduction

For nonlinear multimodal problems, space reduction techniques can make optimization search focus on the potentially optimal regions and avoid unnecessary computation costs in non-significant areas. In this chapter, a large number of samples are first generated to get the predictive values of the Kriging and RBF models. According to the predictive results, HSOSR selects top M sample points from Kriging and RBF, respectively. Subsequently, two promising regions are identified based on these sample points: one is from Kriging and the other comes from RBF.

$$\begin{aligned}
&\boldsymbol{S}_{krg}^{topM} \Leftarrow \begin{bmatrix} S_{krg1}^{Rank1}, & S_{krg2}^{Rank1}, & \cdots, & S_{krgd}^{Rank1}, & Y_{krg}^{Rank1} \\ S_{krg1}^{Rank2}, & S_{krg2}^{Rank2}, & \cdots, & S_{krgd}^{Rank2}, & Y_{krg}^{Rank2} \\ \vdots & \vdots & \ddots & & \vdots \\ S_{krg1}^{RankM}, & S_{krg2}^{RankM}, & \cdots, & S_{krgd}^{RankM}, & Y_{krg}^{RankM} \end{bmatrix} \\
&\qquad Lb_{krgi} = \min\left[S_{krgi}^{Rank1}, S_{krgi}^{Rank2}, \ldots, S_{krgi}^{RankM} \right]^T \\
&\qquad Ub_{krgi} = \max\left[S_{krgi}^{Rank1}, S_{krgi}^{Rank2}, \ldots, S_{krgi}^{RankM} \right]^T \\
&\qquad Range_krging_i = \left[Lb_{krgi}, Ub_{krgi} \right]^T \\
&\qquad i = 1,2,\ldots,d \\
&\boldsymbol{S}_{rbf}^{topM} \Leftarrow \begin{bmatrix} S_{rbf1}^{Rank1}, & S_{rbf2}^{Rank1}, & \cdots, & S_{rbfd}^{Rank1}, & Y_{rbf}^{Rank1} \\ S_{rbf1}^{Rank2}, & S_{rbf2}^{Rank2}, & \cdots, & S_{rbfd}^{Rank2}, & Y_{rbf}^{Rank2} \\ \vdots & \vdots & \ddots & & \vdots \\ S_{rbf1}^{RankM}, & S_{rbf2}^{RankM}, & \cdots, & S_{rbfd}^{RankM}, & Y_{rbf}^{RankM} \end{bmatrix} \\
&\qquad Lb_{rbfi} = \min\left[S_{rbfi}^{Rank1}, S_{rbfi}^{Rank2}, \ldots, S_{rbfi}^{RankM} \right]^T \\
&\qquad Ub_{rbfi} = \max\left[S_{rbfi}^{Rank1}, S_{rbfi}^{Rank2}, \ldots, S_{rbfi}^{RankM} \right]^T \\
&\qquad Range_RBF_i = \left[Lb_{rbfi}, Ub_{rbfi} \right]^T \\
&\qquad i = 1,2,\ldots,d
\end{aligned} \tag{6.3}$$

In Eq. (6.3), *Range_Kriging* and *Range_RBF* denote the two promising regions, respectively. $\boldsymbol{S}_{krg}^{topM}$ and $\boldsymbol{S}_{rbf}^{topM}$ are the ranked top M samples from Kriging and RBF, M is the number of good samples, and d denotes the dimension. In addition, Y_{krg}^{Ranki} and Y_{rbf}^{Ranki} are the predictive values from the two surrogate models, and Lb_i and Ub_i represent the lower and upper bounds, respectively. Based on *Range_Kriging* and *Range_RBF*, two subspaces *Range_union* and *Range_intersection* are defined. The detailed pseudo codes are summarized in Algorithm 6.1.

Algorithm 6.1(a) Create the Promising Ranges—*Range_union*

(01) **Begin**
(02) ***Sp*** ← Carry on Latin Hypercube Sampling (LHS) to obtain multiple starting points.
(03) ***Yrbf*** ← Evaluate the RBF values at these starting points ***Sp***
(04) ***Ykrg*** ← Evaluate the Kriging values at these starting points ***Sp***
(05) ***Goodpoints_r*** ← Sort the RBF values ***Yrbf*** and find the corresponding points
(06) ***Goodpoints_k*** ← Sort the Kriging values ***Ykrg*** and find the corresponding points
(07) ***Num_rank*1** ← Define the number of the selected good points.
(08) ***Range_r*, *Range_k*** ← Define the promising range by the top ***Num_rank*1** good points, respectively.
(09) ***for i*** ← 1 to D (The number of dimensions)
(10) ***Range_union_lb*(*i*)** ← Get the minimum boundary from ***Range_r*** and ***Range_k*** at the ***i***th dimension.
(11) ***Range_union_ub*(*i*)** ← Get the maximum boundary from ***Range_r*** and ***Range_k*** at the ***i***th dimension.
(12) ***end for***
(13) ***Range_union*** ← Get the range [***Range_union_lb*; *Range_union_ub***].
(14) **End**

Intuitively, *Range_union* is the union set of *Range_Kriging* and *Range_RBF*, and *Range_intersection* is the overlap of *Range_Kriging* and *Range_RBF*. Both *Range_union* and *Range_intersection* include better regions from Kriging and RBF, which decreases the risk of missing the global optimum. In Algorithm 6.1(b), after the top ***Num_rank*2** points of Kriging and RBF are obtained, the same points in ***Goodpoints_r_inter*** and ***Goodpoints_k_inter*** will be recorded in ***Points_intersection***. These points in ***Points_intersection*** may gather in a small region or may be distributed in several local optimal regions. Finally, the subspace is created to enclose the points

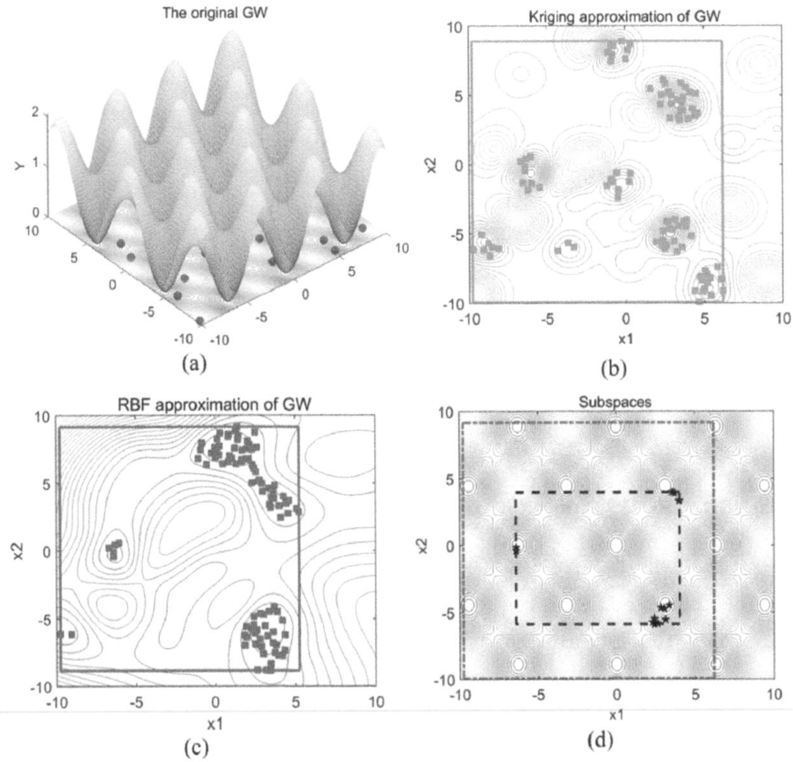

FIGURE 6.1 (a–d) Graphic demonstration of GW function.

set ***Points_intersection***. Figures 6.1 and 6.2 show how to create the proposed subspaces, which involve two cases on Griewank (GW) and Ackley functions, respectively. Intuitively, GW and Ackley have many local valleys, but Ackley has a clearer convergence trend to its global minimum. Since GW has multiple similar valleys in the whole space, the top M samples from Kriging and RBF scatter in different local optimal regions. It can be seen from Figure 6.1b that *Range_Kriging* and *Range_RBF* enclose several promising local regions, respectively. Besides, *Range_intersection* focuses on the common better regions from the two surrogates. From Figure 6.1d, it is clear that *Range_intersection* encloses the exact global minimum and three other local minima and *Range_union* covers all the predicted promising regions from Kriging and RBF. Due to the characteristic of Ackley, the top M samples from the two surrogates are located in a concentrated area around the true global optima. Therefore, *Range_intersection* identifies an accurate reduced space that encloses the global optimal solution. In conclusion, *Range_union* can contain more local optimal regions to

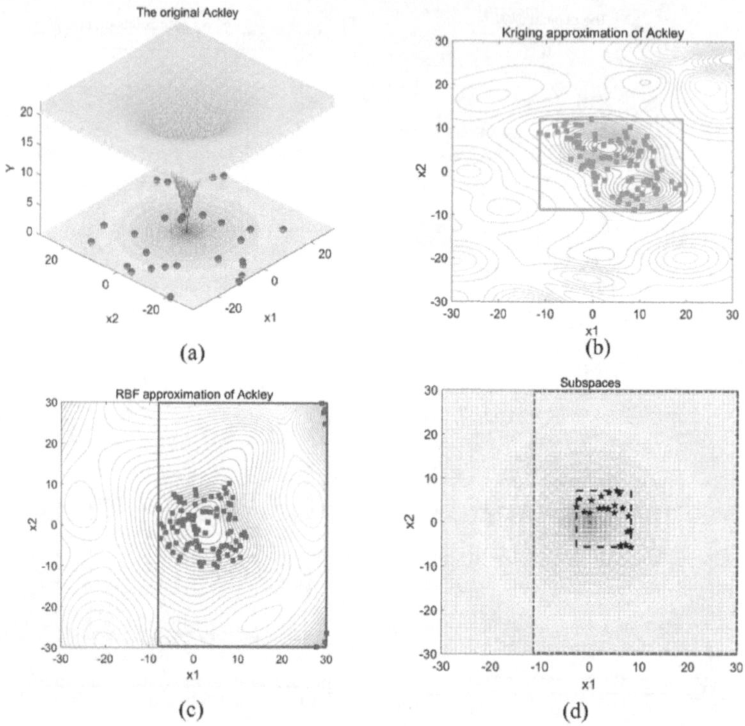

FIGURE 6.2 Graphic demonstration of Ackley function.

avoid missing the global optimum, and Range_intersection can focus on the best joint regions of Kriging and RBF to accelerate convergence. In this work, the total number of cheap points generated by LHS is 1,000, ***Num_rank*1** = 100 and ***Num_rank*2** = 50. It is worth noting that the two subspaces are alternately employed for optimization search with iterations going on. The detailed pseudo codes are summarized in Algorithm 6.1(b).

Algorithm 6.1(b) Create the Promising Ranges—*Range_intersection*

Begin
Goodpoints_r ← Sort the RBF values ***Yrbf*** and find the corresponding points
Goodpoints_k ← Sort the Kriging values ***Ykrg*** and find the corresponding points
***Num_rank*2** ← Define the number of the selected good points.
Goodpoints_r_inter ←***Goodpoints_r*** (1: ***Num_rank*2**, :)

(06) ***Goodpoints_k_inter*** ←***Goodpoints_k*** (1: ***Num_rank2***, :)
(07) ***for i*** ← 1 to ***Num_rank2***
(08)   ***for j*** ← 1 to ***Num_rank2***
(09)   ***Error*** ← Get the error of | ***Goodpoints_r_inter(i***, :)-***Goodpoints_k_inter(j***, :) |
(10)   ***if Error*** is small enough
(11)   Record ***i*** and ***j***
(12)   ***end if***
(13)   ***end for***
(14) ***end for***
(15) ***Points_intersection*** ← Select the points that have good performance on both Kriging and RBF.
(16) ***for i*** ← 1 to D
(17)   ***Range_intersection_lb***(***i***) ← Min(***Points_intersection***(:, ***i***))
(18)   ***Range_intersection_ub***(***i***) ← Max(***Points_intersection***(:, ***i***))
(19) ***end for***
(20) ***Range_intersection*** ← Get the range [***Range_intersection_lb***; ***Range_intersection_ub***]
ith dimension.
(21) **End**

6.2.2.2 *Multi-Start Optimization on Kriging and RBF*

As Figures 6.1 and 6.2 show, surrogate models like Kriging and RBF, always produce multiple approximate local optima, especially for highly nonlinear multimodal problems. Some of these local optima are in the neighborhood of the true local or global optimal solutions but some are not. Compared with traditional global optimization algorithms, like genetic algorithm or particle swarm optimization, multi-start optimization can capture multiple local optima from surrogate models more easily. On one hand, supplementing multiple sample points in each cycle can improve the algorithm's parallelism. On the other hand, multi-start optimization can increase the probability of successfully capturing the global optimum.

In each iteration, HSOSR utilizes Latin hypercube sampling (LHS) to generate several starting points, where the local optimizer–sequential quadratic programming (SQP) is employed to perform optimization search. Considering the demand for sample diversity, two different distance values are defined to select new samples alternately. The multi-start optimization exploration is carried out in the subspaces of the Kriging and RBF models, respectively. Then, all the to-be-supplemented samples are collected and added to a database for further selection. HSOSR promises that the final supplementary samples should keep a distance between each other. Once the algorithm gets trapped in a local optimal region, a proposed strategy begins to explore the sparsely sampled area. The specific pseudo code is listed in Algorithm 6.2.

Algorithm 6.2(a) Exploitation on Surrogates

Begin

if the remainder of (***iteration***/3)== 0

dis ← Δ_1*sqrt(***range_legnth***(1)^2+ ***range_legnth***(2)^2). Here, ***range_length*** denotes the length vector of the design range (1e-3)

else

dis ← Δ_2*sqrt(***range_legnth***(1)^2+ ***range_legnth***(2)^2).(1e-6)

end if

Gn ← If D (Dimension) is smaller than 7, the number is 5D. Otherwise, the number is 2D

M ← Call Latin Hypercube Sampling to get ***Gn*** starting points in the defined subspace.

for i=1: ***Gn***

S_rbf ← Call SQP to perform optimization search at the *i*th starting points ***M***(*i*) on RBF. Save the obtained local optimal solution in the defined subspace.

end for

S_rbf_select ← Select the better samples from ***S_rbf*** and guarantee that the selected sample points keep a distance (bigger than ***dis***) between the existing samples.

for i=1: ***Gn***

S_Kriging ← Call SQP to perform optimization search at the *i*th starting points ***M***(*i*) on Kriging. Save the obtained local optimal solution in the defined subspace.

end for

S_krg_select ← Select the better samples from ***S_Kriging*** and guarantee that the selected sample points keep a distance ***dis*** between the existing samples.

S_new ← [***S_rbf_select***; ***S_krg_select***]

End

In Algorithm 6.2(a), a distance ***dis*** is provided to make samples have the diversity. When ***iteration*** is the multiple of 3, the coefficient of ***dis*** is Δ_1. Otherwise, the coefficient is defined as Δ_2. In the subsequent tests, Δ_1 is 1e-3 and Δ_2 is 1e-6. In the loop, different sizes of ***dis*** affect the selection of samples from the predicted sets ***S_rbf*** and ***S_Kriging***. The larger the parameter ***dis*** is, the more rigorous the selection will be. Eventually, the selected promising samples from Kriging and RBF are gathered into a sample set ***S_new***.

When ***dis*** gets larger (Employ Δ_1), multi-start optimization sometimes may hardly find a satisfactory solution from Kriging and RBF, which makes ***S_new*** empty. Sometimes, the algorithm may get stuck near a local valley, and the present best value cannot be improved for multiple iterations. Once the above-mentioned cases happen, Algorithm 6.2(b) is activated to explore the sparsely sampled area. Since the estimated MSE of Kriging has the maxima at the sparsest area as Figure 6.3 shows, the proposed multi-start optimization is employed to get the updating samples in a randomly generated range. Figure 6.4 shows the captured new samples located in the sparse area. Algorithm 6.2(b) is shown below.

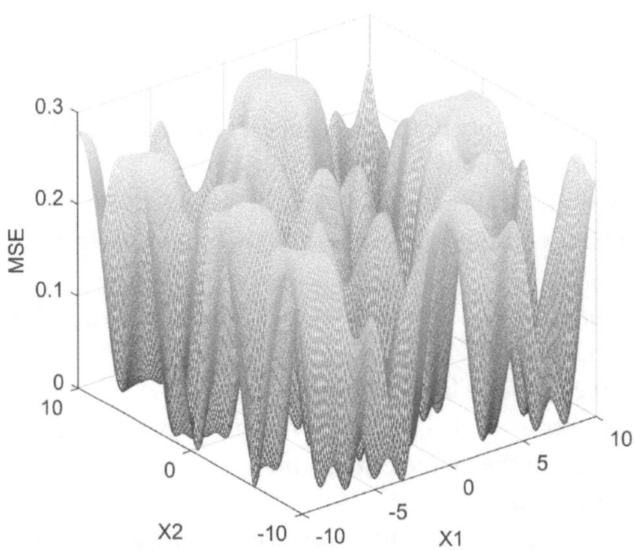

FIGURE 6.3 MSE of Kriging.

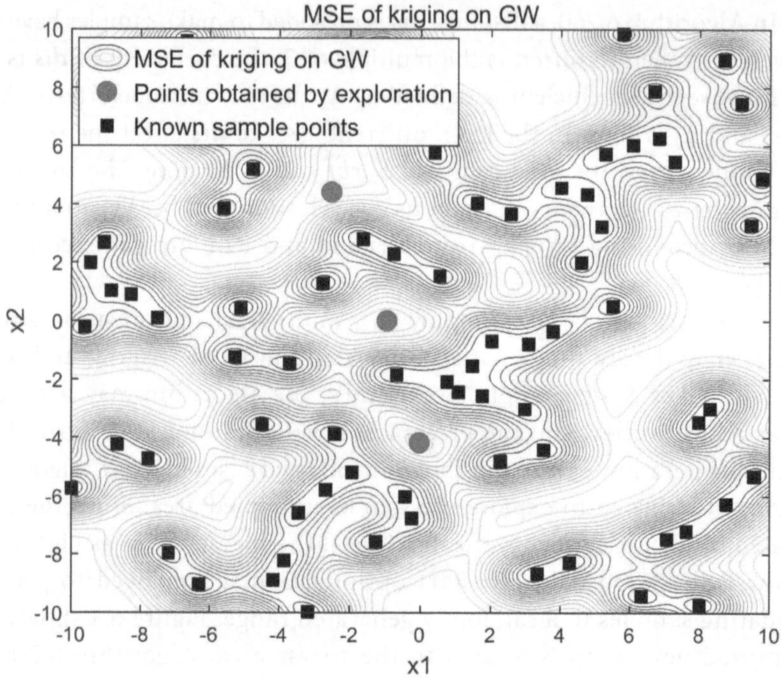

FIGURE 6.4 Samples updating by maximizing MSE.

Algorithm 6.2(b) Exploration of Sparsely Sampled Area

(01) **Begin**

(02) ***if S_new*** is empty or the present best value does not change remarkably for continuous 10 iterations.

(03) ***w*** ← Generate a random value between 0 to 1.

(04) ***lbmse*** ← (range_lb+range_ub)/2-***w***(range_ub-(range_lb+range_ub)/2).

(05) ***ubmse*** ← (range_lb+range_ub)/2+***w***(range_ub-(range_lb+range_ub)/2).

(06) ***range_mse*** ← [***lbmse;ubmse***]

(07) ***M_mse*** ← Call Latin Hypercube Sampling to get ***Gn*** starting points in the defined ***range_mse***.

(08) ***for*** *i*=1: ***Gn***

(09) ***S_mse*** ← Call SQP to perform optimization search at the *i*th starting points ***M_mse***(*i*) on the MSE function of Kriging. Save the samples with local maximal MSE values in ***range_mse***.

(10) ***end for***
(11) ***S_new*** ← [***S_new***; ***S_mse***].
(12) ***end if***
(13) ***S_new_checked*** ← Check the repeated points in ***S_new*** and delete them.
(14) **END**

6.2.2.3 Optimization Flow

The whole optimization process is shown in Figure 6.5, where "Exploitation" and "Exploration" affect each other and jointly search the global optimum. The termination criterion is proposed for the subsequent comparison tests as below,

$$\begin{cases} \dfrac{|y_{optimal} - y_{best}|}{|y_{optimal}|} \leq 1\% \quad or \quad NFE > 300, & if\ y_{optimal} \neq 0 \quad and \quad dim < 8 \\ \dfrac{|y_{optimal} - y_{best}|}{|y_{optimal}|} \leq 1\% \quad or \quad NFE > 500, & if\ y_{optimal} \neq 0 \quad and \quad dim \geq 8 \\ y_{best} < 0.001 \quad or \quad NFE > 300, & if\ y_{optimal} = 0 \quad and \quad dim < 8 \\ y_{best} < 0.001 \quad or \quad NFE > 500, & if\ y_{optimal} = 0 \quad and \quad dim \geq 8 \end{cases} \tag{6.4}$$

where y_{best} is the present best value. $y_{optimal}$ is the true optimal value. *NFE* is the number of function evaluations and *dim* refers to the dimension.

6.3 COMPARISON EXPERIMENTS

In order to verify the efficiency and robustness of the algorithms, 15 representative benchmark functions are given in this chapter for comparative testing, including ten low-dimensional problems (Ack, GW, Peak, ST, AP, F1, HM, GF, Levy, HN6) and five high-dimensional problems (Schw3, Trid, Sums, Sphere, F16). It is worth noting that most of the benchmark algorithms are highly nonlinear problems.

In contrast, five methods, including EGO, CAND, HAM, MKRG, and MRBF, are employed in this chapter. Among them, EGO is a Kriging-based global optimization algorithm which captures new samples by maximizing the EI function. CAND originally comes from a stochastic RBF algorithm presented by Regis and Shoemaker (2007) and is currently implemented by Müller's surrogate toolbox in this work. HAM is a hybrid

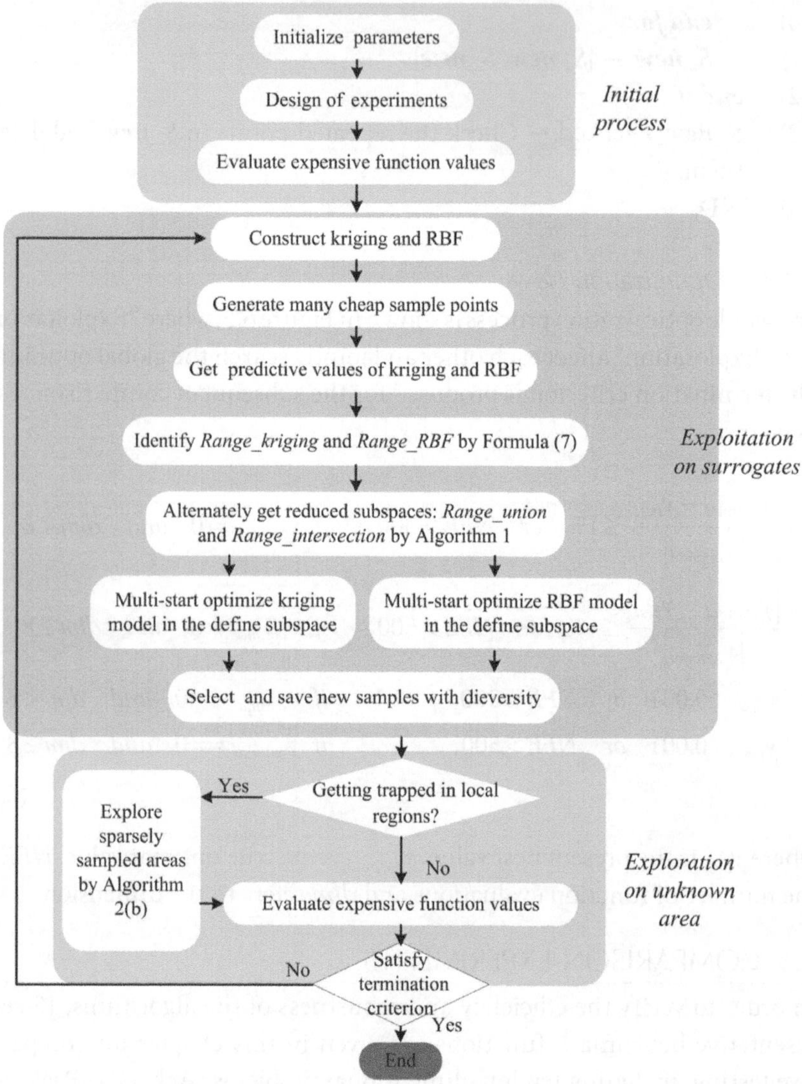

FIGURE 6.5 Flowchart of HSOSR.

meta-model-based method using three surrogates to predict the global optimum, which has a robust performance in most mathematical cases. MKRG and MRBF have the same idea as HSOSR, except that MKRG and MRBF just use their predictive information (One from Kriging and the other from RBF) and explore the global design space. Figure 6.6 shows the iterative results of the six algorithms on all the above-mentioned cases

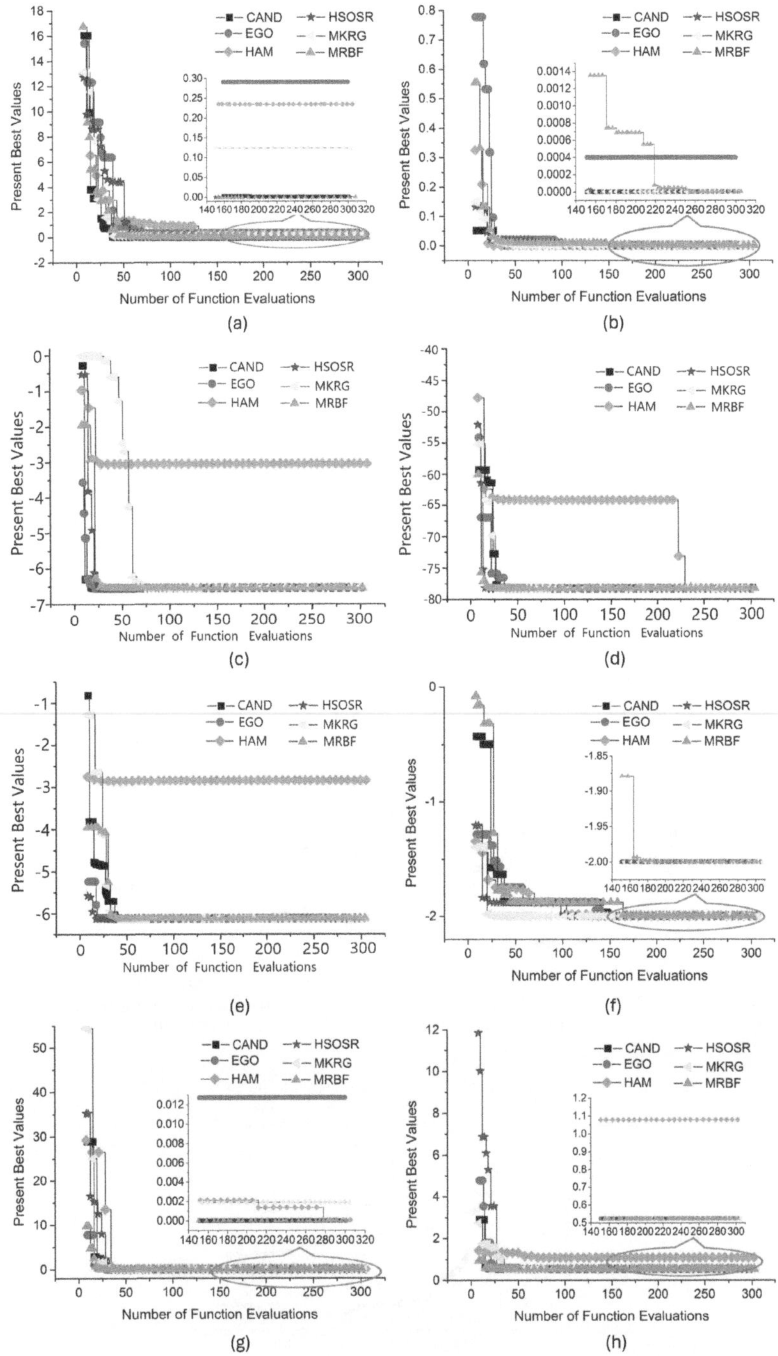

FIGURE 6.6 Iteration diagram of the six algorithms on 15 cases.

(Continued)

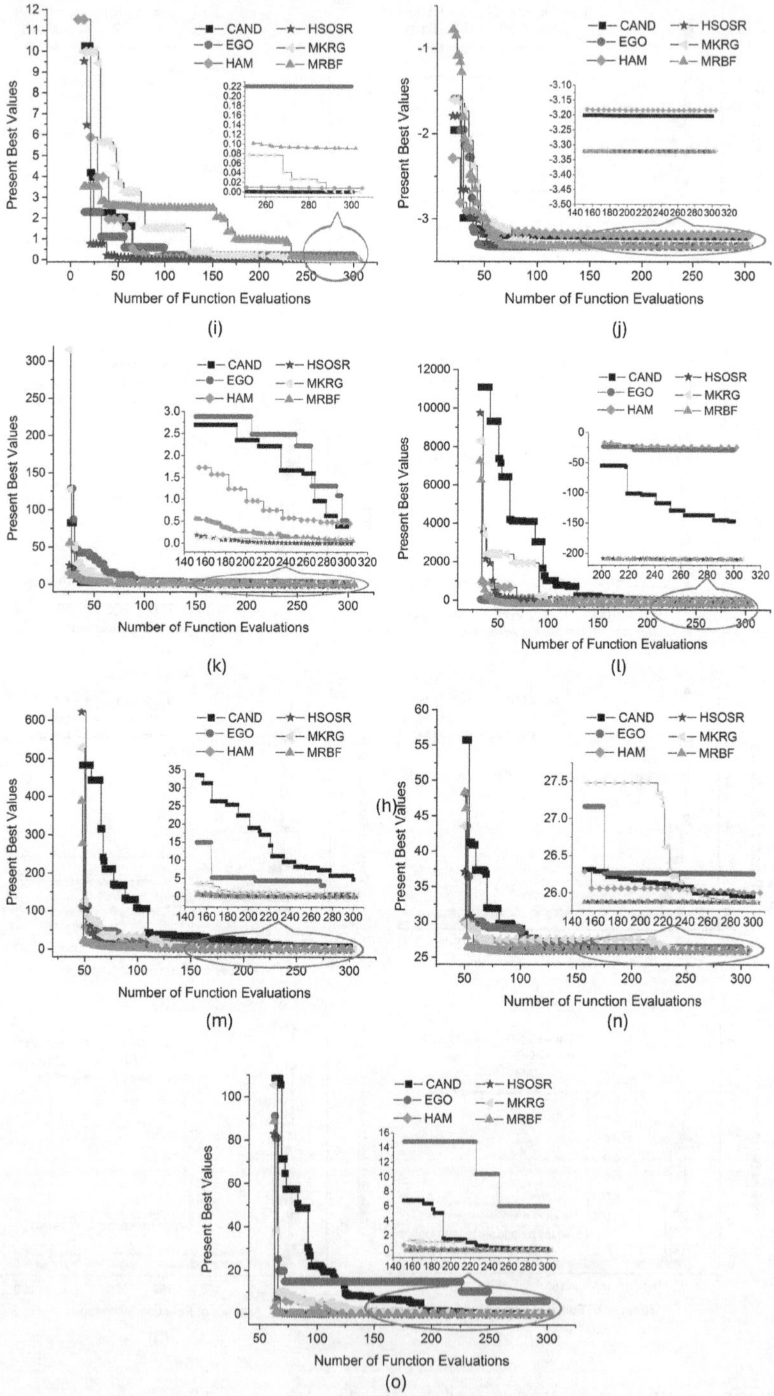

FIGURE 6.6 (*Continued*) Iteration diagram of the six algorithms on 15 cases.

within 300 function evaluations. It is worth mentioning that most cases have clearer subgraphs to describe the convergent parts. Since most of these test cases are multimodal problems, it is easy for an algorithm to get stuck in a local valley. As Figure 6.6a shows, CAND, HSOSR, and MRBF get closer to the true target, but EGO, HAM, and MKRG get stuck in some local optimal regions. It also can be seen from Figure 6.6c, e, h, and j that HAM has the worst performance and cannot jump out of a local region within 300 function evaluations. From Figure 6.6a, g, i, and l, it can be found that EGO has difficulty in dealing with Ackley, Him, Levy, and Trid. Besides, CAND performs worse on higher-dimensional problems (HN6, Schw3, Trid, Sums, F16, and Sphere). Although MKRG decreases slowly at the beginning as Figure 6.6c, i, j, l, and n show, it can go close to the target values at last. Intuitively, HSOSR and MRBF are relatively efficient. In most cases, HSOSR and MRBF can quickly find the global optima. However, MRBF has a slower convergence speed compared with HSOSR in Figure 6.6b, e, f, and i–k. In summary, Figure 6.6 provides a preliminary comparison of the six algorithms on iterative results. On one hand, Figure 6.6 shows the convergence abilities of these different algorithms. On the other hand, it proves that HSOSR is more efficient than others on these benchmark cases.

Since the stochastic nature of the six algorithms, ten tests are repeated on all the cases. Equation (6.4) is employed as the termination criterion in this test. Besides, since EGO spends much execution time on higher-dimensional problems, the allowable NFE in Eq. (6.4) is defined as 300 for the high-dimensional cases. Tables 6.1 and 6.3 show the mean NFE and final best values of the six algorithms. Tables 6.2 and 6.4 list the statistical results of NFE. NFE refers to the number of function evaluations. The best results in Tables 6.1–6.4 are flagged with boldface. In Tables 6.2 and 6.4, Min, Median and Max represent the minimum NFE, median NFE, and maximum NFE, respectively. In the four tables, the results with the symbol ">" indicate that at least one test cannot find target values within a defined NFE. Besides, the numbers in brackets reflect the failure times.

From Tables 6.1–6.4, it can be found that MKRG, MRBF, HAM, EGO, and CAND have several failure times on low-dimensional multimodal problems. Since HAM does not have a strategy that makes the search jump out of a local region, it performs the worst on multimodal problems. As per the previous discussion, Ackley possesses a lot of local optimal solutions. MKRG, EGO, and HAM can hardly find a value below 0.001 on Ackley within 300 function evaluations. Similarly, MKRG, MRBF, EGO, and HAM

TABLE 6.1 Mean NFE and Final Best Values of HSOSR, MKRG and MRBF

	HSOSR		MKRG		MRBF	
Func.	**NFE**	**Best Value**	**NFE**	**Best Value**	**NFE**	**Best Value**
Ackley	**139**	**[1.18e-4, 9.43e-4]**	>300	[0.067, 4.331]	>131	[3.91e-5, 2.580]
GW	**90**	**[1.37e-7, 5.57e-4]**	133.2	[9.42e-7, 4.45e-4]	149.4	[4.65e-5, 7.46e-4]
Peaks	**30.3**	**[−6.551, −6.491]**	36.2	[−6.548, −6.512]	85	[−6.547, −6.487]
ST	38.4	[−78.329, −77.599]	30.1	[−78.332, −77.679]	30.2	[−78.173, −77.585]
Alp	**19.2**	**[−6.128, −6.074]**	35.4	[−6.129, −6.076]	40.3	[−6.119, −6.079]
F1	136.7	[−2.000, −1.993]	>161.5	[−2.000, −1.879]	>186.1	[−1.999, −1.879]
Him	**30.4**	**[5.96e-6, 7.02e-4]**	>142.6	[8.62e-7, 1.23e-2]	50.6	[5.02e-4, 9.13e-4]
GF	52.4	[0.524, 0.527]	30.4	[0.523, 0.528]	>148.4	[0.525, 0.678]
Levy	**190.6**	**[4.91e-4, 9.60e-4]**	>278.7	[3.95e-4, 1.951]	>230.6	[7.97e-4, 3.090]
HN6	**52.6**	**[−3.313, −3.291]**	103.8	[−3.315, −3.289]	92.5	[−3.306, −3.290]
Schw3	**299.8**	**[5.26e-4, 9.84e-4]**	>500	[0.075, 1.823]	>402	[7.67e-4, 2.83e-3]
Trid10	**169.9**	**[−208.785, −207.92]**	171.3	[−208.949, −207.911]	292.6	[−208.336, −207.906]
Sums	**304.6**	**[5.61e-4, 9.96e-4]**	>500	[0.018, 0.059]	315	[5.18e-4, 9.91e-4]
F16	**69**	**[26.016, 26.133]**	174.6	[26.002, 26.128]	71.5	[25.927, 26.129]
Sphere	124.5	[5.32e-4, 9.41e-4]	>500	[8.08e-3, 5.84e-2]	**111.7**	**[5.70e-4, 9.42e-4]**

also have difficulty in dealing with Levy. Intuitively, HSOSR has the most robust performance on all the low-dimensional problems. Furthermore, HSOSR can use fewer function evaluations to get the target values.

For high-dimensional tests, it is difficult for MKRG, HAM, EGO, and CAND to perform well on Schw3, Trid10, Sums, and Sphere. However, MKRG, HAM, and CAND can efficiently get the target value on F16. Both HSOSR and MRBF can solve high-dimensional cases well, but HSOSR uses fewer NFE than MRBF on Schw3 and Trid10. Moreover, HSOSR finds a satisfactory solution on F16 just using about 69 function evaluations.

TABLE 6.2 Statistical NFE of HSOSR, MKRG and MRBF

	HSOSR			MKRG			MRBF		
Func.	**Min**	**Median**	**Max**	**Min**	**Median**	**Max**	**Min**	**Median**	**Max**
Ackley	**88**	**113.5**	**245**	>300	>300	>300(10)	94	112.5	>300(1)
GW	**28**	**93**	**193**	62	150	205	38	162	243
Peaks	**16**	**24**	**59**	20	35	56	20	68	247
ST	16	33	86	16	28	52	12	28	68
Alp	**10**	**20**	**22**	19	34	62	17	35	97
F1	92	131	223	98	159.5	>300(1)	53	196	>300(2)
Him	**27**	**30.5**	**36**	28	98.5	>300(2)	30	46.5	79
GF	27	54	82	16	30	53	25	119.5	>300(2)
Levy	**95**	**187.5**	**299**	203	>300	>300(7)	82	>300	>300(6)
HN6	**37**	**52**	**72**	60	106.5	129	42	75	211
Schw3	**218**	**281**	**464**	>500	>500	>500(10)	283	422.5	>500(3)
Trid10	**115**	**163**	**237**	145	161	246	208	291	397
Sums	**242**	**311**	**336**	>500	>500	>500(10)	234	303.5	426
F16	**60**	**68**	**80**	113	164	299	61	70	85
Sphere	106	119.5	149	>500	>500	>500(10)	**101**	**112.5**	**132**

TABLE 6.3 Mean NFE and Final Best Values of HAM, EGO and CAND

	HAM		EGO		CAND	
Func.	**NFE**	**Best Value**	**NFE**	**Best Value**	**NFE**	**Best Value**
Ackley	>300	[2.78e-3, 1.664]	>300	[0.037, 0.503]	>241.9	[7.49e-4, 2.24e-3]
GW	>164.8	[3.10e-5, 7.40e-3]	97.4	[1.42e-6, 7.73e-4]	>205.3	[1.57e-4, 7.40e-3]
Peaks	>113	[−6.551, −3.050]	31.5	[−6.551, −6.518]	35.3	[−6.550, −6.494]
ST	>71.3	[−78.325, −64.196]	33.5	[−78.332, −77.803]	**27.2**	**[−78.252, −77.555]**
Alp	>53.6	[−6.128, −2.854]	23.5	[−6.126, −6.073]	26.8	[−6.124, −6.080]
F1	85.5	[−2.000, −1.983]	**73.7**	**[−2.000, −1.986]**	>226.3	[−1.999, −1.879]
Him	76.2	[7.70e-6, 7.99e-4]	>112.2	[1.78e-4, 7.39e-3]	82.8	[1.91e-5, 9.94e-4]
GF	>164.9	[0.524, 1.079]	>136.9	[0.523, 0.550]	**25.9**	**[0.523, 0.528]**
Levy	>263	[2.15e-4, 7.10e-2]	>300	[0.016, 0.413]	224.1	[3.55e-4, 9.92e-4]

(Continued)

TABLE 6.3 (*Continued*) Mean NFE and Final Best Values of HAM, EGO and CAND

	HAM		EGO		CAND	
Func.	**NFE**	**Best Value**	**NFE**	**Best Value**	**NFE**	**Best Value**
HN6	>144.4	[−3.317, −3.176]	>82	[−3.310, −3.202]	>157.8	[−3.314, −3.137]
Schw3	>500	[0.144, 2.693]	>300	[0.422, 2.127]	>500	[0.036, 0.447]
Trid10	>500	[−161.737, 26.035]	>300	[−29.129, −18.946]	>493.9	[−207.901, 91.161]
Sums	>500	[1.45e-3, 3.191]	>300	[1.256, 6.426]	>500	[0.027, 0.347]
F16	>351.5	[26.109, 26.651]	>283.2	[25.994, 26.527]	205	[26.042, 26.129]
Sphere	>500	[5.80e-3, 0.414]	>300	[2.495, 9.393]	>500	[3.45e-3, 2.37e-2]

TABLE 6.4 Statistical NFE of HAM, EGO and CAND

	HAM			EGO			CAND		
Func.	**Min**	**Median**	**Max**	**Min**	**Median**	**Max**	**Min**	**Median**	**Max**
Ackley	>300	>300	>300(10)	>300	>300	>300(10)	94	>290.5	>300(5)
GW	27	112.5	>300(4)	32	100	137	117	199.5	>300(2)
Peaks	22	43.5	>300(2)	20	30	45	18	28	60
ST	21	39.5	>300(1)	13	30	83	**17**	**29**	**38**
Alp	14	24	>300(1)	11	23	38	17	19	49
F1	27	66.5	205	**25**	**57.5**	**166**	90	234.5	>300(2)
Him	44	65.5	185	30	32.5	>300(3)	52	79.5	129
GF	63	122	>300(3)	22	58	>300(3)	**17**	**24**	**40**
Levy	115	>300	>300(7)	>300	>300	>300(10)	157	223	278
HN6	48	88	>300(3)	38	46	>300(1)	52	79	>300(4)
Schw3	>500	>500	>500(10)	>300	>300	>300(10)	>500	>500	>500(10)
Trid10	>500	>500	>500(10)	>300	>300	>300(10)	439	>500	>500(9)
Sums	>500	>500	>500(10)	>300	>300	>300(10)	>500	>500	>500(10)
F16	160	343.5	>500(4)	192	>300	>300(8)	182	193	259
Sphere	>500	>500	>500(10)	>300	>300	>300(10)	>500	>500	>500(10)

As a summary of Tables 6.1–6.4, Table 6.5 shows the total NFE mean (TNM) values, failure times, success rates, relative improvements of computational efficiency (RICE) and relative improvements of success rates (RISR) in all the cases. Importantly, RICE and RISR reflect the improved levels of HSOSR than the other five methods. To sum up, HSOSR is a promising global optimization algorithm for EBPs.

TABLE 6.5 Summary of the Final Results

Algorithm	TNM	Failure Times	Success Rate	RICE	RISR
HSOSR	1,747.4	0	100%	—	—
MKRG	>3,097.8	50	66.67%	>77.28%↑	50%↑
MRBF	>2,336.9	14	90.67%	>33.74%↑	10.29%↑
HAM	>3,788.2	75	50%	>116.79%↑	100%↑
EGO	>2,673.9	75	50%	>53.02%↑	100%↑
CAND	>3,452.3	52	65.33%	>97.57%↑	53.07%↑

6.4 CHAPTER SUMMARY

In this chapter, an SBGO algorithm HSOSR is presented, which can solve expensive black-box optimization problems. HSOSR constructs RBF and Kriging models to approximate the true expensive problems, respectively. In each iteration, a group of samples is employed to get the predictive values from RBF and Kriging. Two promising regions from Kriging and RBF are identified by these predictive values. Considering the relations between the two promising regions, two reduced subspaces are created. Furthermore, the optimization search begins to run in the two subspaces alternately. Since RBF and Kriging models can always generate multiple predictive optimal locations, a multi-start optimization algorithm is proposed to find them as supplementary samples. The multi-start optimization search promises that the new samples keep a defined distance from the obtained samples. For the diversity of samples, two different sizes of distance are suggested in this chapter. Once HSOSR gets stuck in a local region, the multi-start optimization algorithm will be run on the estimated mean square error of Kriging to explore the sparsely sampled area.

In order to verify the efficiency and robustness of HSOSR, ten low-dimensional multimodal functions and five high-dimensional functions are tested, and five other algorithms are employed as contrast references. The results show the powerful capacity of HSOSR in dealing with expensive black-box optimization problems. Compared with other classical algorithms, HSOSR can use fewer function evaluations to get close to the true global optimal values.

NOTE

1 Based on "Hybrid Surrogate-based Optimization using Space Reduction (HSOSR) for Expensive Black-box Functions," published in [Applied Soft Computing], [2018]. Permission obtained from [Elsevier].

REFERENCES

Craven, R., Graham, D., & Dalzel-Job, J. (2016). Conceptual Design of a Composite Pressure Hull. *Ocean Engineering, 128*, 153–162. https://doi.org/10.1016/j.oceaneng.2016.10.031

Dyn, N., Levin, D., & Rippa, S. (1986). Numerical Procedures for Surface Fitting of Scattered Data by Radial Functions. *Siam Journal on Scientific and Statistical Computing, 7*(2), 639–659. https://doi.org/10.1137/0907043

Forrester, A. I. J., & Keane, A. J. (2009). Recent Advances in Surrogate-Based Optimization. *Progress in Aerospace Sciences, 45*(1–3), 50–79. https://doi.org/10.1016/j.paerosci.2008.11.001

Gu, J., Li, G. Y., & Dong, Z. (2012). Hybrid and Adaptive Meta-Model-Based Global Optimization. *Engineering Optimization, 44*(1), 87–104. https://doi.org/10.1080/0305215x.2011.564768

Gutmann, H. M. (2001). A Radial Basis Function Method for Global Optimization. *Journal of Global Optimization, 19*(3), 201–227. https://doi.org/10.1023/a:1011255519438

Hardy, R. L. (1971). Multiquadric Equations of Topography and Other Irregular Surfaces. *Journal of Geophysical Research, 76*(8), 1905–+. https://doi.org/10.1029/JB076i008p01905

Kleijnen, J. P. C. (2009). Kriging Metamodeling in Simulation: A Review. *European Journal of Operational Research, 192*(3), 707–716. https://doi.org/10.1016/j.ejor.2007.10.013

Li, Y., Wu, Y., Zhao, J., & Chen, L. (2017). A Kriging-Based Constrained Global Optimization Algorithm for Expensive Black-Box Functions with Infeasible Initial Points. *Journal of Global Optimization, 67*(1–2), 343–366. https://doi.org/10.1007/s10898-016-0455-z

Mirjalili, S., Mirjalili, S. M., & Lewis, A. (2014). Grey Wolf Optimizer. *Advances in Engineering Software* (pp. 46–61).

Myers, R. H., Montgomery, D. C., Vining, G. G., Borror, C. M., & Kowalski, S. M. (2004). Response Surface Methodology: A Retrospective and Literature Survey. *Journal of Quality Technology, 36*(1), 53–77. https://doi.org/10.1080/00224065.2004.11980252

Ong, Y. S., Nair, P. B., & Keane, A. J. (2003). Evolutionary Optimization of Computationally Expensive Problems via Surrogate Modeling. *Aiaa Journal, 41*(4), 687–696. https://doi.org/10.2514/2.1999

Park, J., & Kim, K.-Y. (2017). Meta-Modeling Using Generalized Regression Neural Network and Particle Swarm Optimization. *Applied Soft Computing, 51*, 354–369. https://doi.org/10.1016/j.asoc.2016.11.029

Queipo, N. V., Haftka, R. T., Shyy, W., Goel, T., Vaidyanathan, R., & Tucker, P. K. (2005). Surrogate-Based Analysis and Optimization. *Progress in Aerospace Sciences, 41*(1), 1–28.

Regis, R. G., & Shoemaker, C. A. (2007). A Stochastic Radial Basis Function Method for the Global Optimization of Expensive Functions. *Informs Journal on Computing, 19*(4), 497–509. https://doi.org/10.1287/ijoc.1060.0182

Tang, Y., Chen, J., & Wei, J. (2013). A Surrogate-Based Particle Swarm Optimization Algorithm for Solving Optimization Problems with Expensive Black Box Functions. *Engineering Optimization*, *45*(5), 557–576. https://doi.org/10.1080/0305215x.2012.690759

Viana, F. A. C., Haftka, R. T., & Watson, L. T. (2013). Efficient Global Optimization Algorithm Assisted by Multiple Surrogate Techniques. *Journal of Global Optimization*, *56*(2), 669–689. https://doi.org/10.1007/s10898-012-9892-5

Wang, G. G., & Shan, S. (2007). Review of Metamodeling Techniques in Support of Engineering Design Optimization. *Journal of Mechanical Design*, *129*(4), 370–380. https://doi.org/10.1115/1.2429697

Wild, S. M., Regis, R. G., & Shoemaker, C. A. (2008). ORBIT: Optimization by Radial Basis Function Interpolation in Trust-Regions. *Siam Journal on Scientific Computing*, *30*(6), 3197–3219. https://doi.org/10.1137/070691814

Younis, A., & Dong, Z. (2010). Trends, Features, and Tests of Common and Recently Introduced Global Optimization Methods. *Engineering Optimization*, *42*(8), 691–718. Article Pii 920975372. https://doi.org/10.1080/03052150903386674

Zhou, X. J., Ma, Y. Z., & Li, X. F. (2011). Ensemble of Surrogates with Recursive Arithmetic Average. *Structural and Multidisciplinary Optimization*, *44*(5), 651–671. https://doi.org/10.1007/s00158-011-0655-6

CHAPTER 7

MGOSIC

Multi-Surrogate-Based Global Optimization Using a Score-Based Infill Criterion[1]

7.1 INTRODUCTION

Due to rapid development and continuous progress in modern engineering, optimization design associated with high-fidelity simulation is gaining more attention (Lakshika et al., 2017; Sala et al., 2016; Tyan et al., 2015; H. Wang et al., 2017; G. Zhou et al., 2017). On one hand, advanced simulation techniques provide precise analyses for real-world applications; On the other hand, they also bring enormous computational costs (Gu et al., 2017; Masters et al., 2017; Singh et al., 2017). Many complex simulation models are multimodal, black-box and time-consuming, which is challenging for global optimization.

Commonly, it is difficult for derivative-based optimization methods to solve expensive black-box optimization problems (EBOPs) (Ong et al., 2003). This is because large numbers of operations on expensive models produce a great computational burden, and meanwhile, uncertain error or noise from simulation codes affects the accuracy of approximate derivatives. Additionally, derivative-based methods overly depend on starting points and easily get trapped in a local valley of multimodal problems. Derivative-free optimization algorithms (Jiang et al., 2017; Meng et al., 2016; Pan, 2012; L. Wang et al., 2017) involving evolutionary computation

DOI: 10.1201/9781003636267-7

(EC) or swarm intelligence (SI) have developed for several decades, which can optimize black-box models in parallel. These algorithms like particle swarm optimization (PSO) (Sun et al., 2013), gray wolf optimization (GWO), bat algorithm (BA), differential evolution (DE) (Rocca et al., 2011) and so on, have been widely used in actual applications. Although EC and SI have remarkable advantages in global optimization, they have to utilize large numbers of function evaluations to explore the design space, which is not efficient for EBOPs. An effective approach to address this issue is to build surrogate models in an optimization process.

Surrogate models, namely meta-models or response surfaces generally use obtained expensive samples to construct simple mathematical expressions as approximate models of complex problems (Q. Zhou, Y. Wang, et al., 2017). Commonly used surrogate models such as Kriging, RBF and QRS can predict function values at the to-be-tested locations (Q. Zhou, P. Jiang, et al., 2017). Although prediction error is inevitable, surrogate models can still give useful guidance information for optimization to improve search efficiency. In general, a complete surrogate-based global optimization (SGO) process includes the following steps: (1) Design of experiment (DOE), that is, an initial sampling process; (2) Construct surrogate models dynamically in each cycle; (3) Exploit surrogate models to find promising samples; (4) Explore sparsely sampled regions; (5) Evaluate the exact function values of obtained new samples; (6) Repeat Steps (2) to (5) until the termination criterion is met. A key factor or difficult point to develop an efficient and robust SGO algorithm is how to find a balance between "Exploitation and Exploration." "Exploitation" refers to search based on surrogate models where a local or global optimizer can be employed to find the predictive best sample for subsequent model updating. Although optimization efficiency is improved, pure "Exploitation" may make the above search get stuck in a local valley. "Exploration" denotes search in sparsely sampled areas, which can make an algorithm jump out of local regions and continue looking for the global optimum.

Due to the wide existence of EBOPs in various fields, SGO has attracted a lot of attention. Jones et al. (1998) developed an efficient global optimization (EGO) algorithm, which maximizes an "Expected Improvement" criterion to capture the promising expensive samples. Gutmann (2001) introduced a distinctive SGO strategy that includes two steps: (1) Assume a target value for the true global optimum; (2) Select the next sample (combined with the target value) that will cause the least "bumpiness" of surrogate models. Wang et al. (2004) presented a mode-pursuing sampling method for SGO, which can generate more samples around the function mode and meanwhile detect the regions possibly containing the global minimum based on QRS. Regis and Shoemaker (2007) proposed a stochastic response surface method that can select a supplementary sample from a set of candidate points in each cycle by RBF approximation. Younis and Dong (2010) presented a region elimination algorithm that identifies several key unimodal regions to speed up the local search. Although most of the above-mentioned methods have better global convergence capabilities, they have lower sampling efficiency in each cycle. In other words, these algorithms do not possess strong parallel capabilities.

Therefore, some scholars have begun to pay attention to both the total computation cost and the iterative efficiency (parallelism) in SGO algorithms (Cai et al., 2017). Ong et al. (2003) developed a parallel SGO algorithm that combines a proposed hybrid optimizer with RBF. On the one hand, the hybrid optimizer utilizes an evolutionary algorithm to do global search; On the other hand, it employs the sequential quadratic programming algorithm to realize local search on RBF. Importantly, the parallelism of traditional evolutionary algorithms is retained in their method. Gu et al. (2017) presented a hybrid and adaptive SGO algorithm, HAM, that simultaneously uses Kriging, RBF and QRS to create several sets for parallel sampling. According to the importance of these sets, the number of to-be-selected samples in each set is different. The points that all three surrogate models approve will have a bigger opportunity to be sampled. In order to supplement multiple samples in each cycle, Viana et al. (2013) developed a multi-surrogate EGO (MSEGO) algorithm. Instead of using one single surrogate model Kriging, MSEGO maximizes the "Expected Improvement" criterion over multiple surrogates. Krityakierne et al. (2016) provided a multi-point SGO strategy that draws lessons from the idea of multi-objective optimization. One objective is the expensive function value of a point, and the other one is the minimum distance of the point to other obtained points. Once the Pareto frontier is obtained, multiple sample points can be selected by a candidate search strategy. Li et al. (2016) decomposed the large-scale optimization space into several subspaces for local exploitation and global exploration, which can avoid the difficulties in constructing Kriging with a large size of training data. In addition, a heuristic criterion was proposed to select promising samples from candidate points obtained in these subspaces per iteration.

This chapter introduces a new global optimization algorithm named MGOSIC to solve unconstrained EBOPs. In MGOSIC, three surrogate models, Kriging, radial basis function (RBF) and quadratic response surfaces (QRS) are dynamically constructed, respectively. Additionally, a multi-point infill criterion is proposed to obtain new points in each cycle, where a score-based strategy is presented to mark cheap points generated by Latin hypercube sampling. According to their predictive values from the three surrogate models, the promising cheap points are assigned with different scores. In order to obtain the samples with diversity, a max–min approach is proposed to select promising sample points from the cheap point sets with higher scores. Simultaneously, the best solutions predicted by Kriging, RBF and QRS are also recorded as supplementary samples, respectively. Once MGOSIC gets stuck in a local valley, the estimated mean square error of Kriging will be maximized to explore the sparsely sampled regions. Moreover, the whole optimization algorithm is carried out alternately in the global space and a reduced space. In summary, MGOSIC not only brings a new idea for multi-point sampling but also builds a reasonable balance between exploitation and exploration.

7.2 ALGORITHM FLOW

In this section, the proposed algorithm flow is provided. Before MGOSIC begins, an initialization process is required for the algorithm parameters like design ranges, internal parameters of surrogate models, termination variables, target values and so on. Subsequently, the specific algorithm steps are summarized as follows.

- **Step 1** Utilize optimized Latin hypercube sampling (OLHS) (Jin et al., 2005) to identify initial sample points in the original design range and then evaluate their exact sample values.
- **Step 2** Create a database to save these expensive samples. Besides, sort all the samples by their expensive function values.
- **Step 3** Construct Kriging, RBF and QRS models based on the samples in the database, respectively.

  Figure 7.1 shows a specific example to demonstrate Steps 1 to 3. The employed function is called Himmelblau, which is a multimodal problem. Kriging and RBF can capture the nonlinear feature of Himmelblau, while QRS can just identify a general trend.

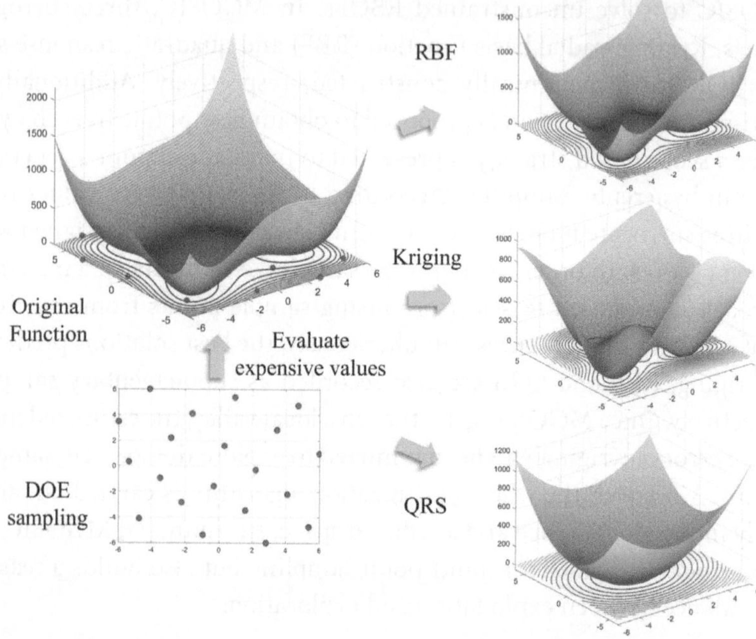

FIGURE 7.1 Construction of surrogate models.

- **Step 4** Create a reduced subspace around the present best solution to speed up the local convergence.

$$
\begin{aligned}
& Lb_i^{Sub} = S_i^{best} - w \cdot \left(Ub_i^{Range} - Lb_i^{Range} \right) \\
& Ub_i^{Sub} = S_i^{best} + w \cdot \left(Ub_i^{Range} - Lb_i^{Range} \right) \\
& if \quad Lb_i^{Sub} < Lb_i^{Range} \\
& then \quad Lb_i^{Sub} \leftarrow Lb_i^{Range} \\
& if \quad Ub_i^{Sub} > Ub_i^{Range} \\
& then \quad Ub_i^{Sub} \leftarrow Ub_i^{Range} \\
& \forall \quad i = 1,2,\ldots,d \\
& Subspace_i = \left[Lb_i^{Sub}, Ub_i^{Sub} \right]
\end{aligned}
\tag{7.1}
$$

where S_i^{best} is the present best solution, Lb_i^{Range} and Ub_i^{Range} are the lower and upper bounds of the original design space, and Lb_i^{Sub} and Ub_i^{Sub} are the bounds of the new subspace. In Eq. (7.1), w is a weight coefficient that determines the size of this subspace. In this chapter, w is set as 0.1.

- **Step 5** Determine which space, the subspace or global space, will be regarded as the search space in accordance with the present number of iterations. Define the numbers of the total cheap points (N_1 and N_2), and the numbers of promising samples (M_1 and M_2) in the subspace and global space, respectively. In the subsequent tests, $N_1 = 10{,}000$ and $N_2 = 1{,}000$, $M_1 = 100$ and $M_2 = 500$.
- **Step 6** Judge whether MGOSIC has got stuck in a local valley. If so, the samples with bigger MSE values of Kriging will be chosen to explore the sparsely sampled area. In the subsequent sections, more details will be provided.
- **Step 7** Evaluate Kriging, RBF and QRS values at all the N cheap sample points and select the top M samples from the three groups of results, respectively. The points that are located in the top M samples of all three surrogate models have a score of 3, and those located in the top M samples of two surrogate models have a score of 2.
- **Step 8** Firstly, save the predictive optimal points from Kriging, RBF and QRS, respectively. Furthermore, select $\boldsymbol{K}_1$ and $\boldsymbol{K}_2$ promising solutions from the point sets with scores 2 and 3, respectively. All these points will be used to update the previous database. More details about **Steps 7** and **8** will be shown in the following section.
- **Step 9** Delete repeated samples to avoid unnecessary computation.
- **Step 10** Evaluate expensive function values at the newly added sample points and sort them. Repeat **Steps 2** to **10** until a termination criterion is satisfied.

For better readability, a flowchart of MGOSIC is shown in Figure 7.2. The termination criterion for the subsequent test is suggested as follows.

$$\begin{cases} if\ dim > 2, & then\quad y_{best} \le target, \quad or \quad NFE > 500 \\ if\ dim \le 2, & then\quad y_{best} \le target, \quad or \quad NFE > 300 \end{cases} \tag{7.2}$$

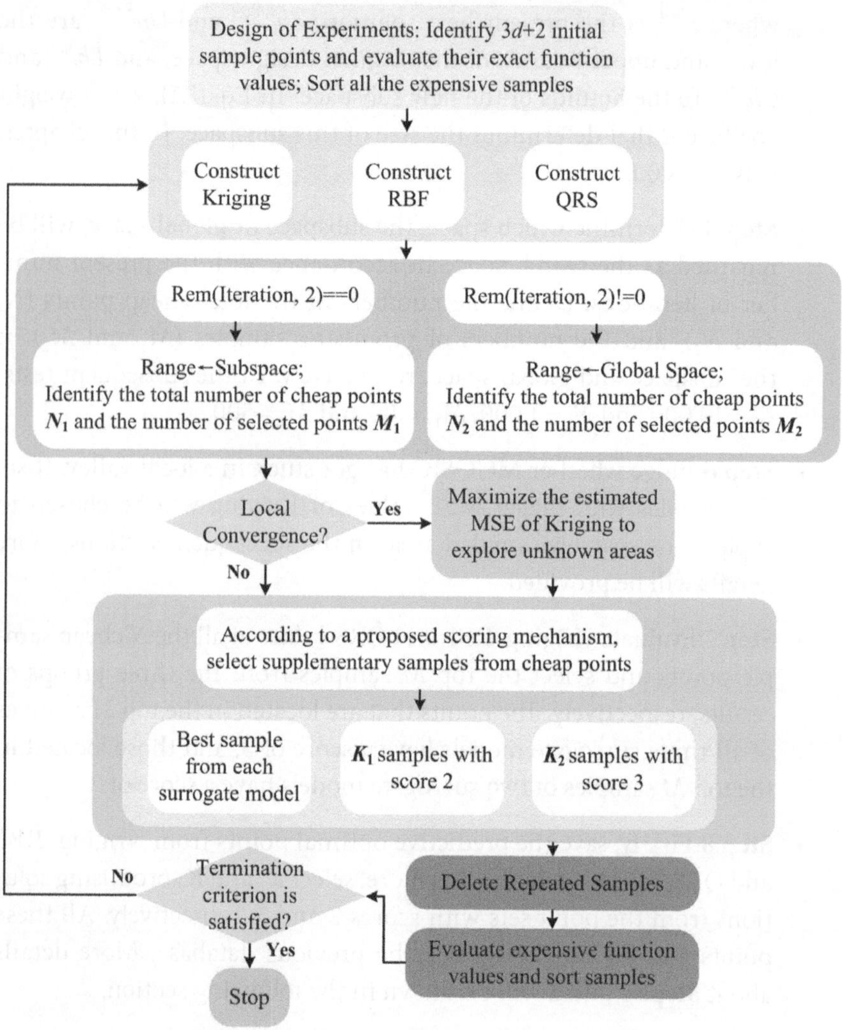

FIGURE 7.2 Flowchart of MGOSIC.

where *NFE* denotes the number of function evaluations, *target* refers to target values of expensive black-box problems, and *dim* represents dimensions.

7.3 MULTI-POINT INFILL CRITERION

Before introducing this proposed infilling criterion, we will give an example to make it easier to understand. Assume that there is a businessman who has no idea about how to choose rabbits but wants to buy ten better ones from 1,000 rabbits. Besides, there are three experienced experts, each of whom can recommend 100 better rabbits for the businessman based on their respective opinions. Under this circumstance, firstly, the businessman should buy the best ones that are recommended by the three experts, respectively; Secondly, the businessman should identify which ones from the 300 rabbits are recommended by all three experts, two experts and one expert. Naturally, the businessman will select more rabbits that are jointly recommended by more experts.

Intuitively, Kriging, RBF and QRS are three experienced experts who can guide an optimization process, MGOSIC is the businessman and sample points are those rabbits. The total number of cheap points is N and the number of recommended sample points is M. In this proposed infilling criterion, three best solutions are first selected based on the three surrogate models. The specific formulas are summarized below.

$$\begin{bmatrix} S_1^1, & S_2^1, & \cdots, & S_d^1, & Y_{krg}^1, & Y_{rbf}^1, & Y_{qrs}^1 \\ S_1^2, & S_2^2, & \cdots, & S_d^2, & Y_{krg}^2, & Y_{rbf}^2, & Y_{qrs}^2 \\ \vdots & \vdots & \ddots & \vdots & \vdots & & \\ S_1^N, & S_2^N, & \cdots, & S_d^N, & Y_{krg}^N, & Y_{rbf}^N, & Y_{qrs}^N \end{bmatrix} \Rightarrow \begin{cases} Matrix_{krg}^{topM} \Rightarrow S_{krg}^{rank1} \\ Matrix_{rbf}^{topM} \Rightarrow S_{rbf}^{rank1} \\ Matrix_{qrs}^{topM} \Rightarrow S_{qrs}^{rank1} \end{cases}$$

$$Matrix_{sm}^{topM} = \begin{bmatrix} S_{sm1}^{rank1}, & S_{sm2}^{rank1}, & \cdots, & S_{smd}^{rank1}, & Y_{sm}^{rank1} \\ S_{sm1}^{rank2}, & S_{sm2}^{rank2}, & \cdots, & S_{smd}^{rank2}, & Y_{sm}^{rank2} \\ \vdots & \vdots & \ddots & & \vdots \\ S_{sm1}^{rankM}, & S_{sm2}^{rankM}, & \cdots, & S_{smd}^{rankM}, & Y_{sm}^{rankM} \end{bmatrix}, \tag{7.3}$$

$$sm \in \{krg, rbf, qrs\}$$

where Y_{krg}^{ranki}, Y_{rbf}^{ranki} and Y_{qrs}^{ranki} are the ith ranked predictive values from Kriging, RBF and QRS, respectively. Besides, S_{krg}^{rank1}, S_{rbf}^{rank1} and S_{qrs}^{rank1} are the best solutions obtained from the three surrogate models. In order to improve the search accuracy, S_{krg}^{rank1}, S_{rbf}^{rank1} and S_{qrs}^{rank1} can be obtained from the three surrogate models by a global optimizer. In this chapter, the GWO is employed to get them.

Subsequently, MGOSIC fuses the sample points from $Matrix_{krg}^{topM}$, $Matrix_{rbf}^{topM}$ and $Matrix_{qrs}^{topM}$ into one big matrix, in which generally there are multiple groups of repeated sample points. A scoring strategy is proposed below.

$$\begin{bmatrix} S_{krg1}^{rank1}, & S_{krg2}^{rank1}, & \cdots, & S_{krgd}^{rank1} \\ \vdots & \vdots & \cdots & \vdots \\ S_{krg1}^{rankM}, & S_{krg2}^{rankM}, & \cdots, & S_{krgd}^{rankM} \\ S_{rbf1}^{rank1}, & S_{rbf2}^{rank1}, & \cdots, & S_{rbfd}^{rank1} \\ \vdots & \vdots & \cdots & \vdots \\ S_{rbf1}^{rankM}, & S_{rbf2}^{rankM}, & \cdots, & S_{rbfd}^{rankM} \\ S_{qrs1}^{rank1}, & S_{qrs2}^{rank1}, & \cdots, & S_{qrsd}^{rank1} \\ \vdots & \vdots & \cdots & \vdots \\ S_{qrs1}^{rankM}, & S_{qrs2}^{rankM}, & \cdots, & S_{qrsd}^{rankM} \end{bmatrix} \tag{7.4}$$

$$\Rightarrow \begin{cases} \{S_1^{Score3}, S_2^{Score3}, \cdots, S_{k1}^{Score3}\} \Leftrightarrow \textit{Appear three times} \\ \{S_1^{Score2}, S_2^{Score2}, \cdots, S_{k2}^{Score2}\} \Leftrightarrow \textit{Appear twice} \\ \{S_1^{Score1}, S_2^{Score1}, \cdots, S_{k3}^{Score1}\} \Leftrightarrow \textit{Appear once} \end{cases}$$

In the big matrix, scores of these points equal the number of their occurrences. k_1, k_2 and k_3 represent the number of sample points in the three sets. The specific pseudo code about the proposed scoring strategy is summarized below.

Algorithm 7.1 Scoring Mechanism

(01) **Begin**
(02) ***S_hybrid*** ← S_{krg}^{topM}, S_{rbf}^{topM} and S_{qrs}^{topM}
(03) ***m*** ← Calculate the total number of sample points in ***S_hybrid***.
(04) ***Score*** ← Define a unit vector with the length ***m***.
(05) ***Z*** ← Define an empty logical variable.
(06) ***for i*** ← 1 to ***m***−1
(07) ***for j*** ← ***i***+1 to ***m***
(08) ***Z*** ←***S_hybrid*** **(*i*, :)==** ***S_hybrid*** **(*j*, :).**
(09) ***Ztemp*** ←True value 1.
(10) ***for k*** ← ***1*** to ***d***
(11) ***Ztemp*** ←***Ztemp*** **&& Z(*k*)**
(12) ***end for***

(13) *if* ***Ztemp*** == 1
(14) ***Score***(***i***) ← ***Score***(***i***)+1;
(15) ***Score***(***j***) ← ***Score***(***j***)+1;
(16) ***end if***
(17) ***end for***
(18) ***end for***
(19) S^{score3} ← Delete repeated points in ***S_hybrid*** ((***Score***==3), :) and save them.
(20) S^{score2} ← Delete repeated points in ***S_hybrid*** ((***Score***==2), :) and save them.
(21) S^{score1} ← Delete repeated points in ***S_hybrid*** ((***Score***==1), :) and save them.
(22) **End**

Figure 7.3 provides an example on Ackley to demonstrate the scoring strategy clearly. Assume that there are ten expensive samples (dots in Figure 7.3a), and Kriging, RBF and QRS are constructed in Figure 7.3b–d,

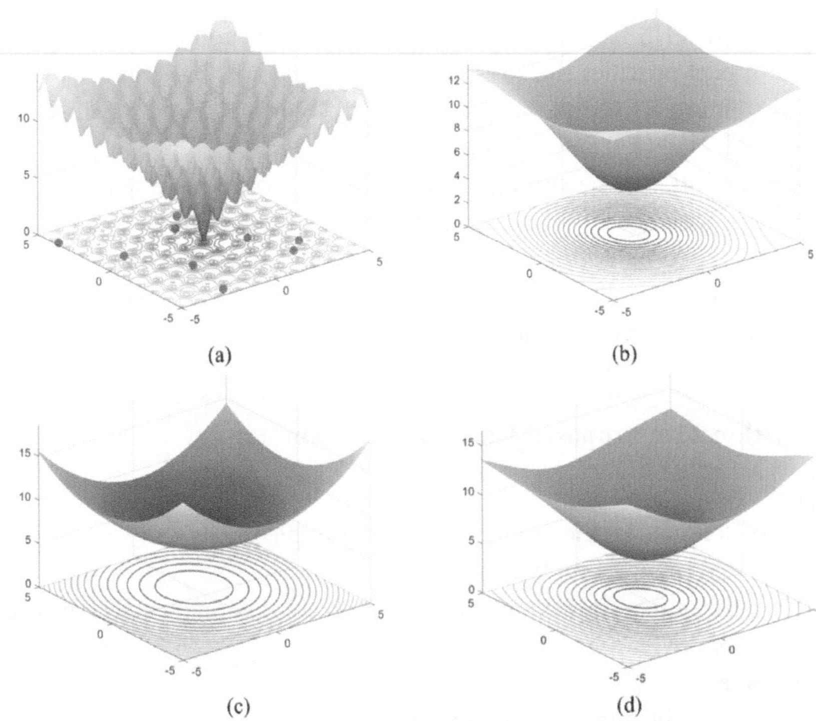

FIGURE 7.3 Ackley and its surrogate models. (a) Original Ackley function. (b) Kriging model of Ackley. (c) RBF model of Ackley. (d) QRS model of Ackley.

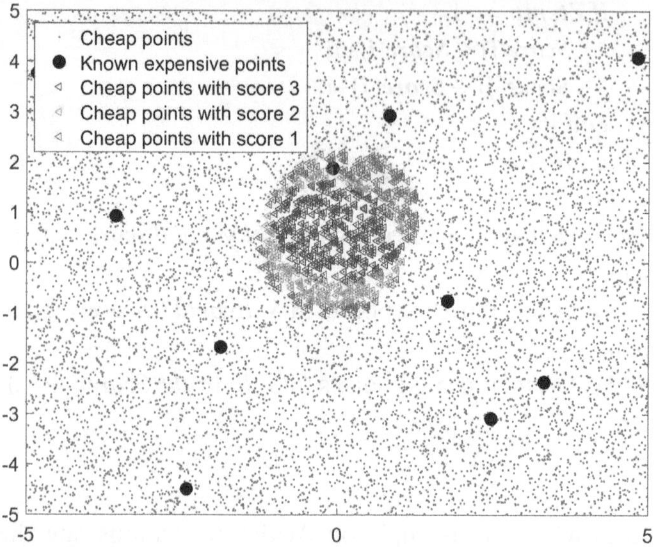

FIGURE 7.4 Illustration of scoring strategy.

respectively. LHS is used to generate 10,000 cheap points and each surrogate model provides their respective top 500 points based on Eq. (7.3). Finally, Figure 7.4 shows the point sets with scores 1, 2 and 3, which are obtained by Eq. (7.4).

In this work, the newly added sample points will be selected from the point sets with scores 3 and 2. Additionally, in order to keep sampling diversity, the to-be-added points need to satisfy a proposed max–min criterion. "max–min" denotes that the minimum distance is maximized, and its pseudocode is shown below.

Algorithm 7.2 A Proposed Max–Min Criterion

Begin
  S_temp ← All the present expensive sample points.
  S^{scoreX}***_new*** ← Empty.
  if the number of points in set $S^{scoreX} > N$ (In this chapter, N is set as 2)
    $K \leftarrow N$.
  else
    $K \leftarrow$ the number of points in S^{scoreX}.
  end if
  for i ← 1 to K

(10) *Dis* ←find the nearest neighbors in *S_temp* for each point in S^{scoreX}, and get the minimum distance vector.
(11) *Max_dis* ←find the maximum distance from *Dis*
(12) *Max_point* ←find the corresponding point in S^{scoreX}.
(13) S^{scoreX}*_new* ← [S^{scoreX}*_new*; *Max_point*]
(14) *S_temp* ← [*S_temp*; *Max_point*]
(15) *end for*
(16) **End**

In Algorithm 7.2, S^{scoreX}*_new* is the selected sample points, which can make MGOSIC have a better space-filling feature in the neighborhood of the present promising regions. Actually, this max–min criterion aims at selecting points that possess the maximum difference with the known expensive samples from the two promising point sets. Figure 7.5 gives an example to explain the max–min approach. Firstly, each new point (dots) needs to find its closest neighbor (squares) and the corresponding minimum distance in the space. As Figure 7.5 shows the four minimum distances are 0.2558, 0.3245, 0.3360 and 0.7933 and Point 3 and Point 4 will be chosen as supplementary sample points.

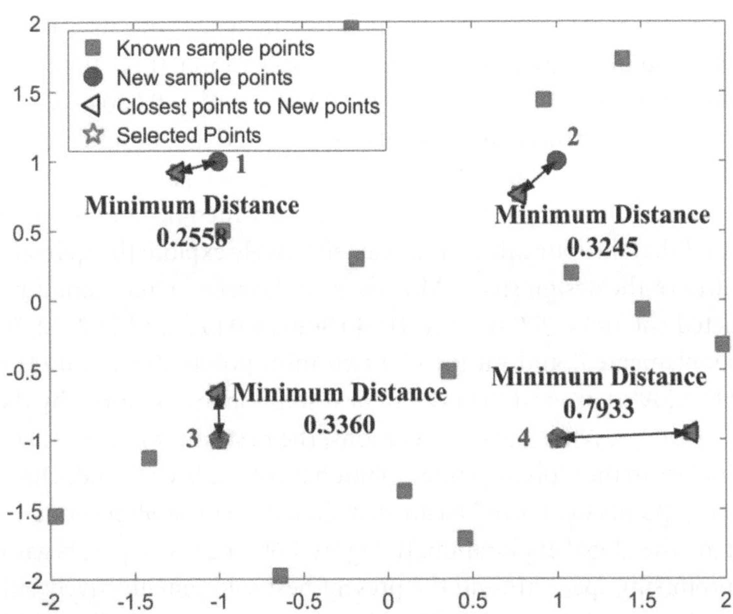

FIGURE 7.5 Illustration of max–min approach.

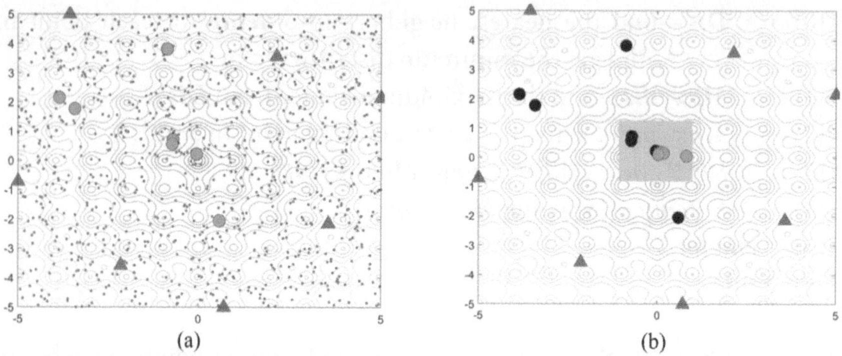

FIGURE 7.6 Search process of MGOSIC. (a) Original Ackley function. (b) Kriging model of Ackley.

As **Step 4** in Section 3.1 has introduced, the best samples from Kriging, QRS and RBF are obtained in the original design space, and the scored samples are alternately selected in a reduced space and the original space. Figure 7.6 shows the search process of MGOSIC on Ackley, where triangles are DOE points, gray small dots are cheap points generated by LHS, big gray dots are updated points in the current iteration, and black dots are supplementary points in the last iteration. In Figure 7.6(a, seven new points are captured from the original design space, and five new points are obtained in Figure 7.6b. During the first iteration, three best predicted solutions [−0.7137, 0.7144], [−0.7346, 0.5641] and [−0.0504, 0.2226] are obtained from the three surrogate modes, respectively. Besides, four points [−0.8786, 3.8146], [−3.8661, 2.1467], [0.5865, −2.0495] and [−3.4302, 1.7663] are selected from the cheap points by the presented infill criterion. It can be found that the four infill points can effectively explore the sparsely sampled area of the design space. Moreover, in the second iteration, three best predicted solutions [0.0412, 0.1741], [0.8092, 0.0452] and [0.0832, 0.1683] are supplemented, and meanwhile two infill points [0.1704, 0.1415] and [0.0243, 0.0981] around the present best solution are acquired by the presented strategy. After the two iterations, the best solution [0.0243, 0.0981] that is close to the global optimal solution [0, 0] has been found. Obviously, the cheap points in Figure 7.6a are distributed over the whole design space, realizing the global exploration. In Figure 7.6b, the cheap points gathering in a promising space around the present best solution effectively enhance the local search. It is worth noting that the coefficient w in Eq. (7.1) determines the size of the reduced space and meanwhile affects the density of cheap points. Essentially, a smaller *w* can bring a high density of cheap points around the present best solution, which promotes the local search. However, when *w* is too small, the search space is overly limited, which may decrease the search efficiency. Hence, the proposed range for *w* is [0.05, 0.15], and *w* is defined as 0.1 in the subsequent tests.

7.4 EXPLORATION OF UNKNOWN AREA

The above-mentioned infilling criterion mainly focuses on the promising locations predicted by Kriging, RBF and QRS. Besides, the proposed max–min criterion can make MGOSIC have a better space-filling performance in a local region, but cannot explore the sparsely sampled regions in the global space. Therefore, the estimated MSE of Kriging is employed to explore unknown areas of the global space. In this work, a local condition is defined to judge whether MGOSIC gets stuck in a local valley or not. In each iteration, the average change of the top ***P*** sample values will be recorded. Furthermore, if they do not change obviously during several successive iterations, the exploration strategy will begin working. The specific pseudo code is listed as follows.

Algorithm 7.3 Exploration Unknown Area

(01) **Begin**
(02) ***Rank_value*** ← Sort all the present best sample values.
(03) ***MeanbestY*(*iteration*)** ← Get mean values of the top ***P*** sample values in each iteration. (In this chapter, ***P*** is set as 3)
(04) ***if*** the number of iterations> ***Q*** (In this chapter, ***Q*** is set as 5)
(05) ***GVI*** ← | ***MeanbestY*(*end*)– *MeanbestY*(*end*-5)**|.
(06) ***end if***
(07) ***if GVI*** < Δ (In this chapter, the default value of Δ is 1e−4)
(08) ***Smse*** ← Get multiple sample points in the original design range by LHS.
(09) ***for i*** ← 1 to ***m*** (In this work, ***m*** equals to 30*d*)
(10) ***MSE*** ← Get the estimated MSE values of Kriging at ***Smse***
(11) ***end for***
(12) ***S_exploration*** ← Sort ***MSE*** and select two samples with the maximal ***MSE*** values.
(13) ***end if***
(14) **End**

7.5 COMPARISON EXPERIMENTS

In order to demonstrate the capability of MGOSIC, two parts of benchmark cases (Gu et al., 2012; Long et al., 2015) including lower-dimensional ($d = 2$–5) and higher-dimensional problems ($d = 6$–20) are used for testing. These representative cases have different characteristics involving multimodal, convex, large-scale and so on. Additionally, the specific target values of all the cases are given in this chapter, and more details are listed in Tables 7.1 and 7.2. It can be found that all the proposed target values are much closer to the true global minima. Finally, an algorithm will stop when the termination criterion in Eq. (7.2) is satisfied.

7.5.1 Preliminary Comparison and Analysis

As our previous introduction, EGO (Jones et al., 1998) is a well-known SGO algorithm and has advantages in low-dimensional multimodal problems. Similarly, CAND presented by Regis and Shoemaker (2007) also has a remarkable performance in low-dimensional problems. Hence, EGO and CAND are tested on two-dimensional cases as a preliminary contrast.

TABLE 7.1 Comparison on Low-Dimensional Problems

	MGOSIC			EGO			CAND		
Func.	Values Range	NFE	NIT	Values Range	NFE	NIT	Values Range	NFE	NIT
Ack	[5.98e−7, 7.19e−4]	75.8	12.4	[5.55e−2, 7.87e−1]	>300(10)	>293	[5.30e−4, 1.93e−3]	>227.8(6)	>220.8
BA	[3.66e−6, 9.67e−4]	35.5	5.9	[1.06e−4, 1.34e−2]	>168.6(4)	>161.6	[6.18e−5, 8.56e−4]	217.8	210.8
Peak	[−6.551, −6.538]	50.9	8.7	[−6.551, −6.505]	27.2	20.2	[−6.550, −6.502]	29.3	22.3
SE	[−1.457, −1.450]	34.5	5.6	[−1.457, −1.450]	45.4	38.4	[−1.456, −1.451]	32.9	25.9
GP	[3.001, 3.005]	93.1	16.8	[3.000, 3.684]	>262(8)	>255	[3.000, 3.009]	111.4	104.4
F1	[−2.000, −1.993]	116.9	19.2	[−2.000, −1.994]	95.4	88.4	[−1.999, −1.992]	202.3	195.3
HM	[3.63e−6, 8.50e−4]	40.5	6.5	[9.71e−5, 6.51e−2]	>225.7(7)	>218.7	[1.09e−5, 9.15e−4]	83.9	76.9
GF	[0.5233, 0.5234]	51	8.9	[0.5233, 0.5459]	>209.8(6)	>202.8	[0.5233, 0.5234]	38.8	31.8
RS	[3.55e−14, 7.77e−4]	52.7	8.7	[5.03e−3, 3.35e−1]	>300(10)	>293	[6.52e−5, 1.990]	>267.5(8)	>260.5

TABLE 7.2 Comparison on Higher-Dimensional Problems

	MGOSIC			DE		
Func.	Values Range	NFE	NIT	Values Range	NFE	NIT
Levy	[3.36e−4, 9.86e−4]	169	27.9	[2.35e−4, 9.67e−4]	2,290	114.5
DP	[2.56e−4, 9.94e−4]	325.8	53.9	[1.80e−4, 9.02e−4]	3,860	193
ST	[−195.77, −195.15]	214.7	34.2	[−195.51, −181.49]	>5,884(2)	>294.2
HN6	[−3.319, −3.301]	77.5	11.7	[−3.312, −3.300]	3,488	174.4
Schw	[7.54e−5, 9.72e−4]	301.9	47.1	[2.24e−4, 9.08e−4]	3,914	195.7
GW	[4.63e−4, 9.94e−4]	332.9	48	[0.426, 0.725]	>1e4(10)	>500
Trid	[−209.99, −209.56]	87.6	13	[−209.73, −209.01]	4672	233.6
Sums	[2.31e−15, 4.27e−13]	145.3	25.1	[7.11e−4, 1.25e−3]	>8,556(1)	>427.8
F16	[25.959, 26.096]	81.5	9.1	[26.021, 26.100]	2,728	136.4
Sphere	[2.93e−4, 9.98e−4]	171	28.5	[9.47e−4, 7.61e−3]	>9,990(9)	>499.5

Considering the randomness of these algorithms, all the following tests are repeated ten times. Table 7.1 shows the comparison results, including the range of obtained best values (Values Range), number of function evaluations (NFE) and number of iterations (NIT). Here, NFE and NIT in Table 7.1 are mean values. The symbol ">" means that target values cannot be found within the maximal NFE, and the numbers in "()" represent the failure times. From Table 7.1, it can be seen that MGOSIC can efficiently find all the target values. Although EGO and CAND have a good performance on Peak, SE and F1, they need more NFE and NIT than MGOSIC to get close to these target values in most cases. Especially, Ack and RS have so many local valleys that EGO and CAND can hardly succeed in most cases. More importantly, EGO and CAND can just add one point in each cycle, which causes larger NIT values than MGOSIC. In addition, the widely used global optimization algorithm DE is also tested for comparison on higher-dimensional cases. For DE, the maximal allowable NFE is 10,000. Like Table 7.1, Table 7.2 gives the similar comparison results. It is obvious that traditional global optimization algorithms need more NFE and NIT than MGOSIC on these higher-dimensional cases.

Additionally, MGOSIC are also compared with two SGO algorithms with multi-point infill criteria. One is called SOCE (Dong et al., 2018) that is a clustering-based global optimization algorithm using Kriging and QRS to build surrogates; the other one named MSEGO was presented by Viana et al. (2013), which extends the original EGO to sample multiple points per cycle by using several surrogates. Tables 7.3 and 7.4 provide the comparison results, where the data of MSEGO and EGO come from the reference Long et al. (2015) and the results of SOCE are obtained from Dong et al. (2018). As Long et al. (2015) mentioned, EGO and MSEGO were tested by Viana's surrogate toolbox (Viana et al., 2013), and MSEGO supplemented three points per cycle in their tests. Besides, due to the adaptive sampling feature, SOCE has an uncertain sampling number, but most of the time it adds three points per cycle. From Tables 7.3 and 7.4, it is clear that all four algorithms can get much closer to the true global optima on SE, Peak, SC and BR which are nonlinear problems with fewer local

TABLE 7.3 Obtained Values of EGO, MSEGO, SOCE and MGOSIC

	EGO		**MSEGO**		**SOCE**		**MGOSIC**	
Func.	**Var. Range**	**Median**	**Var. Range**	**Median**	**Var. Range**	**Median**	**Var. Range**	**Median**
SE	[−1.456, −1.436]	−1.453	[−1.456, −1.454]	**−1.456**	[−1.456, −1.448]	**−1.456**	[−1.457, −1.450]	−1.455
Peak	[−6.550, −6.383]	**−6.550**	[−6.498, −5.979]	−6.498	[−6.551, −6.494]	−6.544	[−6.551, −6.538]	−6.549
SC	[−1.032, −1.031]	−1.031	[−1.024, −0.987]	−1.024	[−1.032, −1.030]	**−1.032**	[−1.032, −1.030]	−1.031
BR	[0.398, 0.400]	**0.398**	[0.398, 0.431]	**0.398**	[0.398, 0.399]	0.399	[0.398, 0.398]	**0.398**
F1	[−1.375, −1.283]	−1.375	[−1.874, −1.636]	−1.874	[−2.000, −1.980]	−1.994	[−2.000, −1.993]	**−1.999**
GF′	[0.966, 3.480]	0.966	[0.001, 0.035]	0.001	[0.003, 0.009]	0.007	[1.17e−4, 9.56e−4]	**5.50e−4**
GP	[7.581, 43.353]	7.581	[3.002, 3.014]	3.002	[3.000, 3.029]	3.008	[3.001, 3.005]	**3.001**
GN	[0.459, 0.459]	0.459	[0.176, 0.627]	0.177	[3.33e−15, 4.81e−3]	7.33e−4	[6.33e−15, 7.18e−4]	**3.61e−4**
HN6	[−3.316, −3.308]	**−3.313**	[−3.208, −3.052]	−3.145	[−3.317, −3.290]	−3.306	[−3.319, −3.301]	−3.311

TABLE 7.4 Mean NFE and NIT of EGO, MSEGO, SOCE and MGOSIC

	EGO		MSEGO		SOCE		MGOSIC	
Func.	Mean NFE	Mean NIT	Mean NFE	Mean NIT	Mean NFE	Mean NIT	Mean NFE	Mean NIT
SE	52	41	109.6	33.5	**33.4**	9.3	34.5	**5.6**
Peak	42.6	31.6	130.4	40.5	**37.3**	11.7	50.9	**8.7**
SC	**32.6**	21.6	131.2	40.7	34.9	10	41	**7**
BR	36.1	25.1	112.6	34.5	**25.9**	7.1	40.2	**6.8**
F1	**52**	41	131.4	40.8	108.5	27.8	116.9	**19.2**
GF'	**52**	41	132.0	41	113.5	35.1	123.6	**22.7**
GP	**52**	41	120.4	37.1	145.9	45.5	93.1	**16.8**
GN	52	41	132.0	41	95.7	27.2	**44.8**	**7.2**
HN6	**68.8**	13.8	176.0	41	89.1	24.7	77.5	**11.7**

minima, but MSEGO requires more NFE. For F1 and GN that possess lots of local minima, EGO and MSEGO have a worse performance. Relatively, MSEGO with the help of multiple surrogate models can find better solutions than EGO on F1 and GN, but NFE also gets larger at the same time. In Tables 7.3 and 7.4, GF′ is the same as GF in Table 7.1, except that the variable range of GF′ is [−5, 5]. Among the four algorithms, EGO has the worst performance on GF′ and GP within 41 iterations, but it is very efficient on HN6. SOCE has an acceptable performance in all nine cases, but it usually uses more function evaluations and iterations than MGOSIC. In summary, compared with others, MGOSIC needs fewer NIT and can always efficiently get the target values on these cases.

7.5.2 Analysis and Discussion

After the preliminary comparisons, MGOSIC has shown its powerful capability in solving expensive black-box problems. In order to further demonstrate its significance, two recently presented SGO algorithms, MSSR (Dong et al., 2016) and HAM (Gu et al., 2012), are tested for comparison. Since the maximal sampling number per iteration (MSNPI) in MGOSIC is 7, and HAM most of the time also adds about seven points per cycle, the MSNPI of MSSR is also defined as seven in this test. Firstly, a group of representative iterative results that can reflect their average performance is listed in Figure 7.7. In order to make it clearer, some sub-graphs like Figure 7.7p and r are locally magnified, and some are improved by the log10 function. Intuitively, MGOSIC can always find the target values more quickly in most cases. Sometimes, MGOSIC may get stuck on multimodal problems like Peak, GP, F1, RS, but it can successfully jump out of the local optimal regions and find the global optima at last. Conversely, HAM lacks an effective exploration strategy, so it frequently misses the global optima. Since MSSR is only guided by Kriging, it overly relies on the predictive capability of Kriging. Therefore, MSSR has worse performance on AK, RS, HN6, Schw and GW. From these iterative figures, it can be found that MGOSIC is more efficient than HAM and MSSR. Moreover,

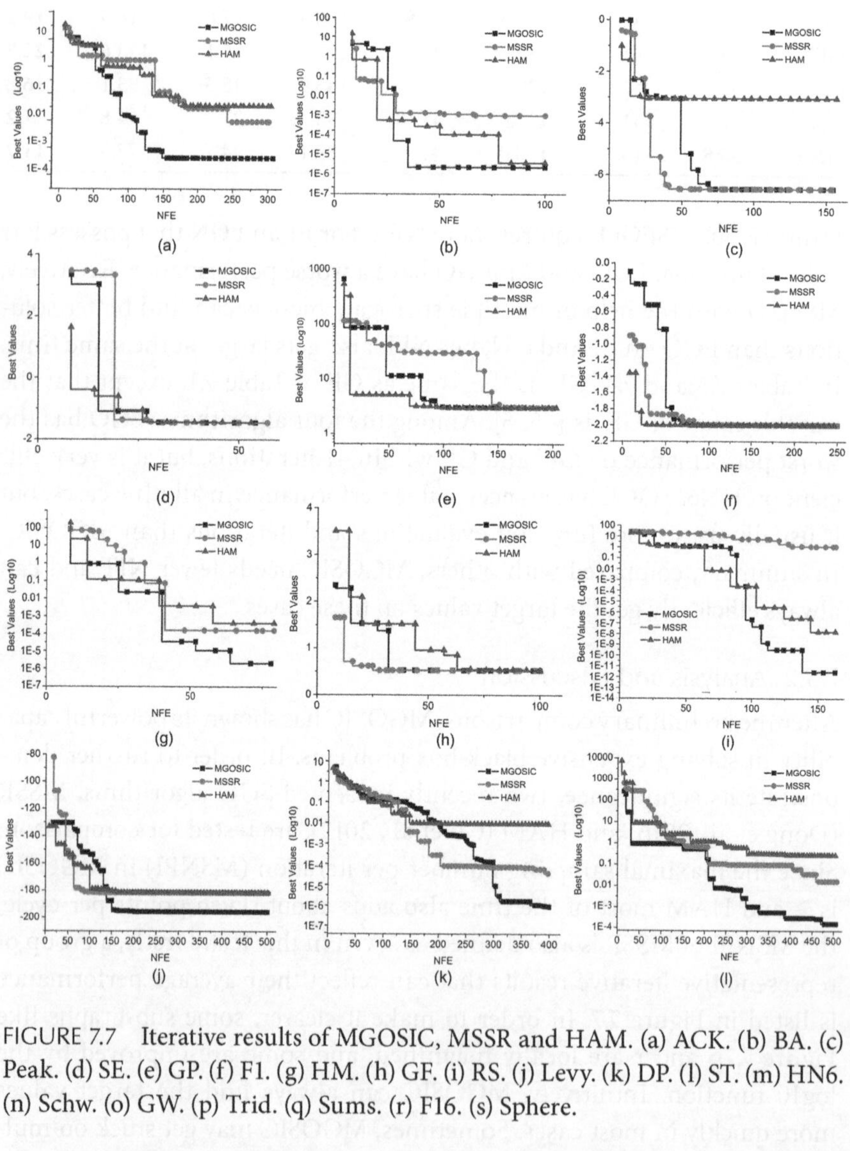

FIGURE 7.7 Iterative results of MGOSIC, MSSR and HAM. (a) ACK. (b) BA. (c) Peak. (d) SE. (e) GP. (f) F1. (g) HM. (h) GF. (i) RS. (j) Levy. (k) DP. (l) ST. (m) HN6. (n) Schw. (o) GW. (p) Trid. (q) Sums. (r) F16. (s) Sphere.

(Continued)

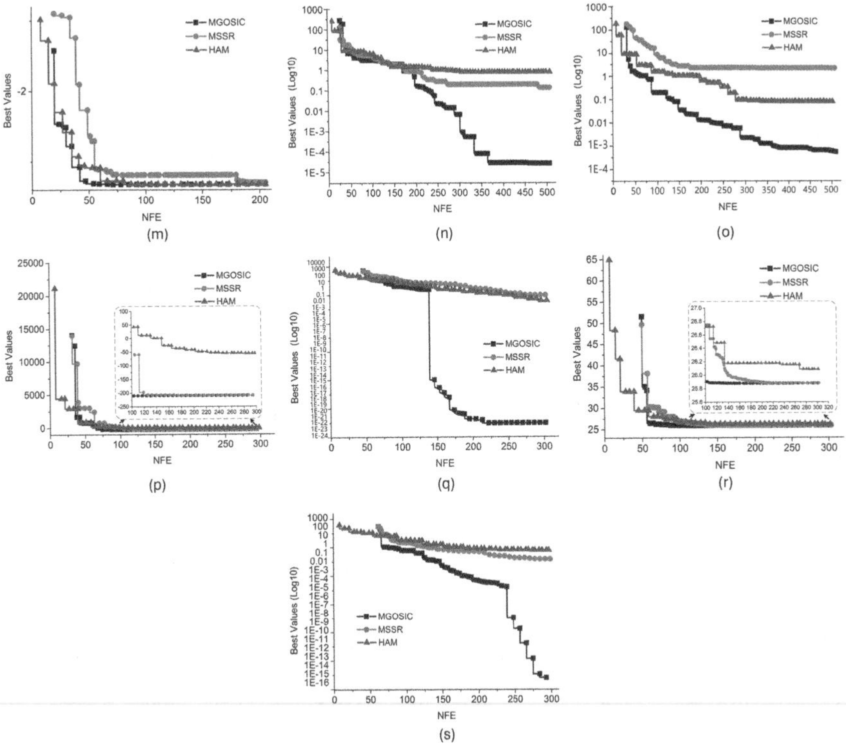

FIGURE 7.7 *(Continued)* Iterative results of MGOSIC, MSSR and HAM. (a) ACK. (b) BA. (c) Peak. (d) SE. (e) GP. (f) F1. (g) HM. (h) GF. (i) RS. (j) Levy. (k) DP. (l) ST. (m) HN6. (n) Schw. (o) GW. (p) Trid. (q) Sums. (r) F16. (s) Sphere.

in order to compare their stability, each test is repeated ten times and the detailed data are shown in Tables 7.5–7.7. The termination criterion in Eq. (7.2) is employed for the three algorithms. Here, "NFE Range," "NIT Range" and "Values Range" denote the ranges of obtained NFE, NIT and best values during the ten tests, respectively. Besides, "R." represents ranks of the three algorithms, which are obtained based on their average performance. In Table 7.7, "SR" is the abbreviation of "Success Rate."

In Tables 7.5–7.7, there is no doubt that MGOSIC has the highest efficiency and strongest stability. MSSR and HAM have a satisfactory performance on low-dimensional problems. MSSR can find the target value on Peak using the fewest NFE, and HAM has the best performance on F1. Although MSSR and HAM can hardly find the target value on ACK, sometimes they can get much closer to 1e−4. Besides, compared with MSSR, HAM has a lower success rate on multimodal problems Peak and SE. Since the target value

TABLE 7.5 Statistical NFE of MGOSIC, MSSR and HAM on All Cases

	MGOSIC			MSSR			HAM		
Func.	NFE Range	Mean	R.	NFE Range	Mean	R.	NFE Range	Mean	R.
Ack	[15, 149]	75.8	1	[>300, >300] (10)	>300	2	[>300, >300] (10)	>300	2
BA	[28, 40]	35.5	1	[51, 141]	89.5	3	[44, 117]	72.8	2
Peak	[18, 99]	50.9	2	[24, 73]	38.4	1	[26, >300] (3)	>114.8	3
SE	[27, 42]	34.5	1	[26, 86]	37.6	2	[22, >300] (2)	>91.5	3
GP	[82, 106]	93.1	1	[76, 165]	122.8	3	[81, 172]	110.8	2
F1	[42, 193]	116.9	2	[27, 242]	158	3	[34, 171]	92.8	1
HM	[30, 49]	40.5	1	[34, 145]	59.2	2	[29, 159]	74.1	3
GF	[28, 69]	51	1	[20, 129]	60.3	2	[>300, >300] (10)	>300	3
RS	[15, 95]	52.7	1	[47, >300] (4)	>200.3	3	[46, 243]	86.8	2
Levy	[90, 285]	169	1	[154, >500] (4)	>337.8	2	[102, >500] (7)	>389.8	3
DP	[241, 461]	325.8	1	[>500, >500](10)	>500	3	[326, >500] (6)	>440.2	2
ST	[62, 389]	214.7	1	[80, >500] (6)	>339.8	2	[124, >500] (5)	>373.5	3
HN6	[52, 228]	77.5	1	[59, 218]	107.4	2	[87, >500] (2)	>181.1	3
Schw	[242, 334]	301.9	1	[>500, >500](10)	>500	2	[>500, >500](10)	>500	2
GW	[263, 428]	332.9	1	[>500, >500](10)	>500	3	[375, >500] (9)	>490.5	2
Trid	[73, 136]	87.6	1	[162, >500] (8)	>438.6	2	[>500, >500](10)	>500	3
Sums	[144, 146]	145.3	1	[>500, >500](10)	>500	3	[368, >500] (7)	>466.7	2
F16	[72, 93]	81.5	1	[103, 197]	161.7	2	[184, >500] (5)	>363.2	3
Sphere	[152, 186]	171	1	[>500, >500](10)	>500	2	[>500, >500](10)	>500	2

of GF is quite strict, HAM cannot find it within 300 function evaluations. Although the case RS has a lot of local optimal solutions, it has an overall downward trend that can be predicted accurately by QRS. Therefore, MGOSIC and HAM that use QRS to construct surrogate models have higher efficiency. With the dimension increasing, the success rate of MSSR and HAM decreases significantly. Especially, MSSR and HAM can hardly

TABLE 7.6 Statistical NIT of MGOSIC, MSSR and HAM on All Cases

	MGOSIC			MSSR			HAM		
Func.	NIT Range	Mean	R.	NIT Range	Mean	R.	NIT Range	Mean	R.
Ack	[2, 24]	12.4	1	[>48, >59]	>55	3	[>49, >55]	>52.4	2
BA	[5, 7]	5.9	1	[12, 25]	17.5	3	[7, 19]	11.3	2
Peak	[3, 19]	8.7	1	[7, 22]	10.6	2	[4, >47]	>17.4	3
SE	[4, 7]	5.6	1	[8, 33]	14.2	3	[3, >48]	>13.9	2
GP	[15, 19]	16.8	1	[15, 28]	21.8	3	[13, 28]	17.8	2
F1	[8, 29]	19.2	2	[10, 96]	55	3	[5, 25]	13.7	1
HM	[5, 8]	6.5	1	[9, 28]	14.6	3	[4, 24]	11.1	2
GF	[5, 12]	8.9	1	[7, 24]	15.4	2	[>45, >49]	>47.3	3
RS	[2, 16]	8.7	1	[13, >84]	>50.5	3	[7, 37]	13	2
Levy	[15, 46]	27.9	1	[35, >85]	>61.6	2	[16, >85]	>64.2	3
DP	[41, 73]	53.9	1	[>74, >93]	>79.8	3	[52, >82]	>70.2	2
ST	[9, 62]	34.2	1	[21, >102]	>61.7	3	[19, >78]	>56.1	2
HN6	[8, 37]	11.7	1	[13, 74]	30.9	3	[13, >78]	>27.4	2
Schw	[37, 52]	47.1	1	[>87, >121]	>98.1	3	[>80, >83]	>82.3	2
GW	[39, 62]	48	1	[>68, >75]	>69.5	2	[70, >99]	>94.9	3
Trid	[11, 20]	13	1	[26, >84]	>68.8	2	[>83, >88]	>85.8	3
Sums	[25, 26]	25.1	1	[>82, >136]	>109.1	3	[61, >88]	>80	2
F16	[7, 12]	9.1	1	[53, 119]	96.3	3	[28, >85]	>59.5	2
Sphere	[24, 32]	28.5	1	[>122, >146]	>132	3	[>94, >98]	>96.1	2

TABLE 7.7 Statistical Best Values of MGOSIC, MSSR and HAM on All Cases

	MGOSIC			MSSR			HAM		
Func.	Values Range	SR	R.	Values Range	SR	R.	Values Range	SR	R.
Ack	[5.98e−7, 7.19e−4]	1	1	[8.96e−3, 2.581]	0	2	[3.33e−3, 5.16e−1]	0	2
BA	[3.66e−6, 9.67e−4]	1	1	[7.73e−5, 7.86e−4]	1	1	[3.98e−6, 7.06e−4]	1	1
Peak	[−6.551, −6.538]	1	1	[−6.551, −6.501]	1	1	[−6.551, −3.050]	0.7	2
SE	[−1.457, −1.450]	1	1	[−1.456, −1.450]	1	1	[−1.457, 2.866]	0.8	2
GP	[3.001, 3.005]	1	1	[3.000, 3.009]	1	1	[3.000, 3.009]	1	1
F1	[−2.000, −1.993]	1	1	[−2.000, −1.992]	1	1	[−2.000, −1.993]	1	1
HM	[3.63e−6, 8.50e−4]	1	1	[2.15e−6, 5.13e−4]	1	1	[8.32e−7, 8.00e−4]	1	1

(Continued)

TABLE 7.7 (*Continued*) Statistical Best Values of MGOSIC, MSSR and HAM on All Cases

	MGOSIC			MSSR			HAM		
Func.	Values Range	SR	R.	Values Range	SR	R.	Values Range	SR	R.
GF	[0.5233, 0.5234]	1	1	[0.5233, 0.5234]	1	1	[0.5235, 0.5276]	0	2
RS	[3.55e−14, 7.77e−4]	1	1	[5.96e−6, 0.995]	0.6	2	[1.46e−6, 8.13e−4]	1	1
Levy	[3.36e−4, 9.86e−4]	1	1	[2.97e−4, 1.04e−2]	0.6	2	[1.51e−4, 5.51e−2]	0.3	3
DP	[2.56e−4, 9.94e−4]	1	1	[1.33e−3, 1.14e−1]	0	3	[1.25e−4, 6.17e−1]	0.4	2
ST	[−195.77, −195.15]	1	1	[−195.62, −167.56]	0.4	3	[−195.52, −181.55]	0.5	2
HN6	[−3.319, −3.301]	1	1	[−3.317, −3.302]	1	1	[−3.316, −3.203]	0.8	2
Schw	[7.54e−5, 9.72e−4]	1	1	[1.38e−2, 1.89e−1]	0	2	[2.51e−2, 1.864]	0	2
GW	[4.63e−4, 9.94e−4]	1	1	[0.691, 2.561]	0	3	[8.07e−4, 0.632]	0.1	2
Trid	[−209.99, −209.56]	1	1	[−209.74, −200.87]	0.2	2	[−207.33, −78.88]	0	3
Sums	[2.31e−15, 4.27e−13]	1	1	[1.01e−2, 2.21e−1]	0	3	[4.87e−4, 1.77e−1]	0.3	2
F16	[25.959, 26.096]	1	1	[26.021, 26.096]	1	1	[25.968, 27.039]	0.5	2
Sphere	[2.93e−4, 9.98e−4]	1	1	[4.90e−3, 9.24e−2]	0	2	[6.30e−3, 4.24e−1]	0	2

find the target values of DP, Schw, GW, Sums and Sphere. Besides, MSSR and HAM have the lower success rate on ST. On the contrary, MGOSIC still has the remarkable performance on high-dimensional problems. It is worth noting that MGOSIC just uses 87.6 and 81.5 function evaluations on Trid and F16, respectively. What is more, for the 20-dimensional problem Sphere, MGOSIC just needs 171 function evaluations. More importantly, MGOSIC uses the fewest NIT to find the target values in most cases, which reflects its outstanding parallel capability. To sum up, MGOSIC is an efficient SGO algorithm that can be applied for EBOPs.

7.5.3 Engineering Applications

In order to demonstrate the engineering applicability of MGOSIC, the optimal shape design of a two-dimensional hydrofoil is used for the test.

The geometric parameterization for the hydrofoil employs the class and shape function transformation (CST) method (Kulfan, 2008) that is originally expressed as follows:

$$\frac{z}{c}\left(\frac{x}{c}\right)=C\left(\frac{x}{c}\right)S\left(\frac{x}{c}\right)+\frac{x}{c}\cdot\frac{z_{TE}}{c} \tag{7.5}$$

$$C\left(\frac{x}{c}\right)=\left(\frac{x}{c}\right)^{N_1}\left[1-\frac{x}{c}\right]^{N_2} \tag{7.6}$$

$$S\left(\frac{x}{c}\right)=\sum_{r=0}^{n}\left[v_r\cdot S_{r,n}\left(\frac{x}{c}\right)\right] \tag{7.7}$$

where $C(\bullet)$ and $S(\bullet)$ are the class and shape functions, respectively. Besides, c refers to the chord length of the hydrofoil, Z_{TE}/c denotes the thickness of the tail flange, and N_1 and N_2 are two coefficients to decide the class of the hydrofoil. It is worth noting that v_r and $S_{r,n}$ come from Bernstein polynomials. In this chapter, we modify the CST formulas to make them appropriate for the proposed optimization problem as below.

$$y_u(x)=y_0(x)+x^{N_1}(1-x)^{N_2}\sum_{i=0}^{n}A_{ui}S_i(x) \tag{7.8}$$

$$y_l(x)=y_0(x)+x^{N_1}(1-x)^{N_2}\sum_{i=0}^{n}A_{li}S_i(x) \tag{7.9}$$

where $y_u(x)$, $y_l(x)$ and $y_0(x)$ refer to the upper bounds, lower bounds and a basic hydrofoil, respectively. Here, x represents the coordinate along the chord of the hydrofoil, and y is the coordinate along the thickness direction. The class coefficients N_1 and N_2 are constants 0.5 and 1, and n is set as 5. Considering that the upper and lower curves have the same radius of the front edge, A_{u0} equals to $-A_{l0}$. Hence, nine Bernstein coefficients A_i are regarded as design variables of this optimization problem, and their design ranges come from two basic airfoils "modified NACA0008" and NACA0016. Figure 7.8 shows the design space of the hydrofoil. Additionally, the length of the chord and the angle of attack are also regarded as design variables. The objective is to minimize the drag coefficient, and meanwhile, the area and lift coefficients are supposed to satisfy inequality constraints.

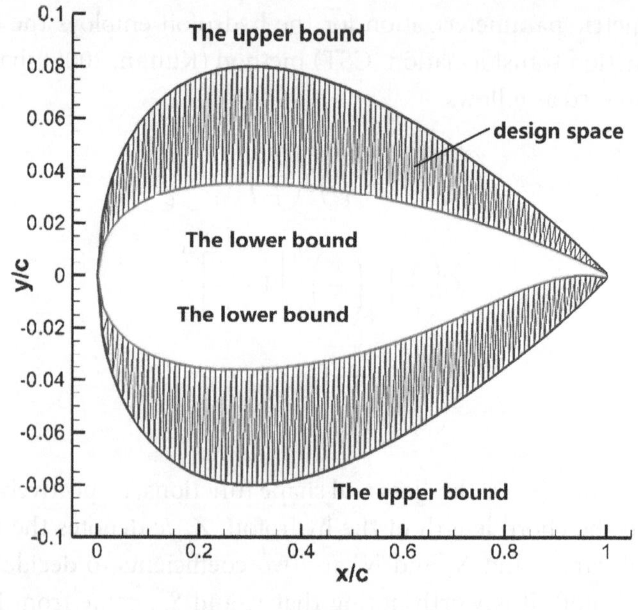

FIGURE 7.8 Design space of the hydrofoil.

The specific optimization formula is summarized as follows.

$$\min \ c_d$$

$$\textit{design range}: X = [\textit{paraments } A, c, aoa]$$

$$0 \le A_1 \le 0.1141; -0.0232 \le A_2 \le 0.1008; -0.010 \le A_3 \le 0.1072;$$

$$-0.0050 \le A_4 \le 0.0815; 0.005 \le A_5 \le 0.111; -0.1008 \le A_6 \le 0.013;$$

$$-0.1072 \le A_7 \le 0.022; -0.0815 \le A_8 \le -0.0258; -0.1112 \le A_9 \le 0.1445;$$

$$0.2 \le c \le 0.3; 3 \le aoa \le 4;$$

$$\begin{aligned} s.t. \quad & c_l \ge 0.3510 \\ & S \ge 0.0051 \\ & thick \ge 0.12 \end{aligned} \tag{7.10}$$

where c is the length of the chord, aoa refers to the angle of attack, c_l is the lift coefficient, S refers to the area and $thick$ represents the maximal thickness. The reference values 0.3510, 0.0051 and 0.12 come from NACA0012, which will be regarded as the reference case. Considering that MGOSIC and other comparison methods are mainly developed for box-constrained problems, Equation (7.10) is modified as follows by a penalty function.

$$\min c_d + P\times\left(\max(0.3510-c_l,0)+\max(0.0051-S,0)+\max(0.12-thick,0)\right)$$
$$designrange: X=[paramentsA,c,aoa] \tag{7.11}$$

where P is the penalty factor that is defined as 10^6. The modified objective function including all the response values like lift coefficient, drag coefficient, area and thickness is directly approximated by surrogates, which is easy to implement in an actual engineering application. The simulation analysis is realized by Computational Fluid Dynamics (CFD), and the maximal iteration number is set as 500 that generally can get satisfactory convergence results. Each analysis process from parametric modeling to CFD simulation will cost about 1.5 minutes. In this chapter, we employ MGOSIC, MSSR and HAM to realize the optimization design of this hydrofoil, and the maximal allowable times of simulation are 300. Figure 7.9 shows the grid partition of the hydrofoil, and Figure 7.10 shows the pressure contour of the reference case NACA0012.

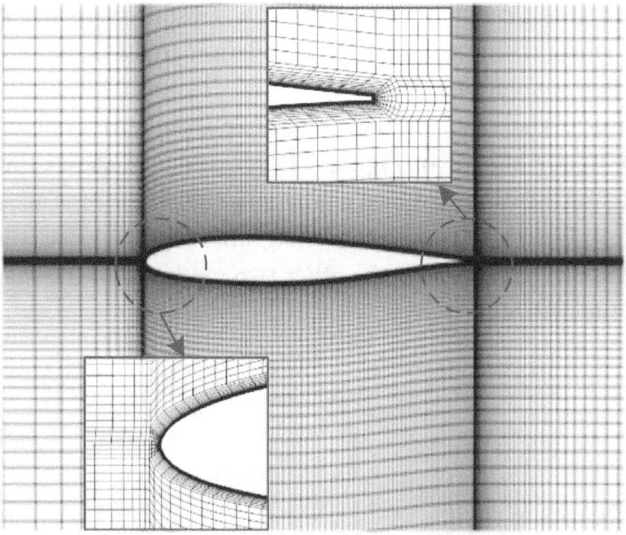

FIGURE 7.9 Grid partition diagram.

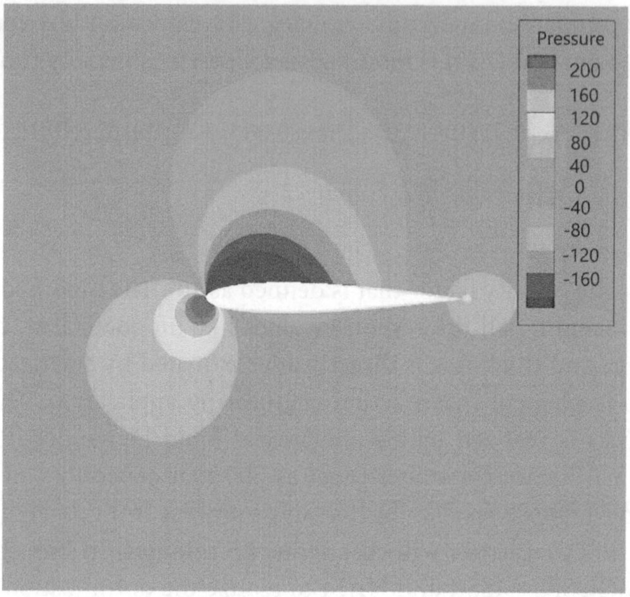

FIGURE 7.10 Pressure contour of NACA0012.

The obtained results from the three methods are listed in Table 7.8. Obviously, MGOSIC gets the minimum drag coefficient, and meanwhile gets larger improvements compared to NACA0012. Additionally, the iterative results of the three global optimization methods are also provided in Figure 7.11. It is clear that MGOSIC has a faster convergence rate. MSSR performs worse within the first 100 simulation analyses, but it can gradually find better solutions. However, HAM can hardly find a better solution after 100 analyses. The best shape and pressure contour obtained by MGOSIC are shown in Figure 7.12. Figure 7.13 shows the comparison results of the obtained best shape and the shape of NACA0012, and Figure 7.14 gives their comparison diagram of pressure curves. In summary, MGOSIC outperforms the other two methods on the shape optimization of the hydrofoil.

TABLE 7.8 Best Results Obtained from MGOSIC, MSSR and HAM

Methods	***cd***	***Cl***	***S***	***Thick***	***Improvement***
NACA0012	0.0153776	0.3510	0.0051	0.12	NA
MGOSIC	0.0148780	0.3678	0.0072	0.1298	3.25%↑
MSSR	0.0149357	0.4300	0.0074	0.1269	2.87%↑
HAM	0.0155002	0.3767	0.0071	0.1309	0.80%↓

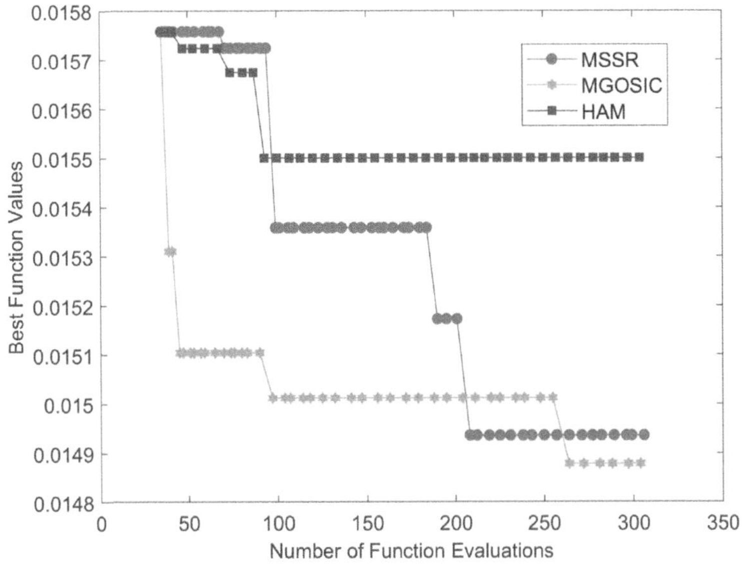

FIGURE 7.11 Comparison of iterative results.

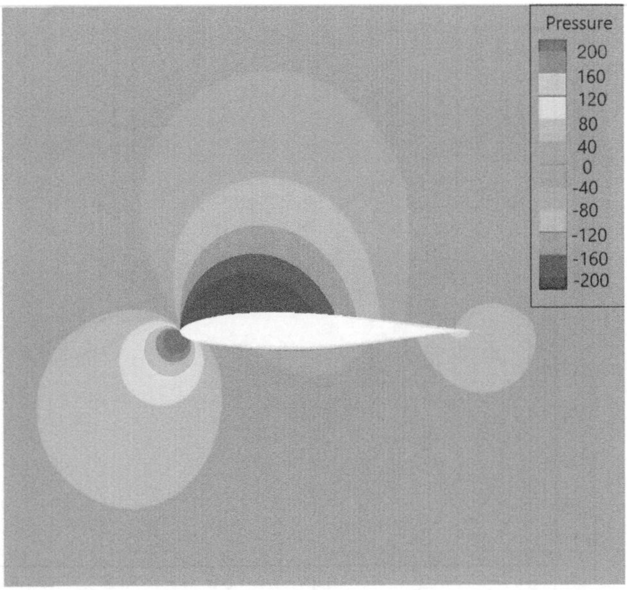

FIGURE 7.12 Pressure contour of the optimal shape.

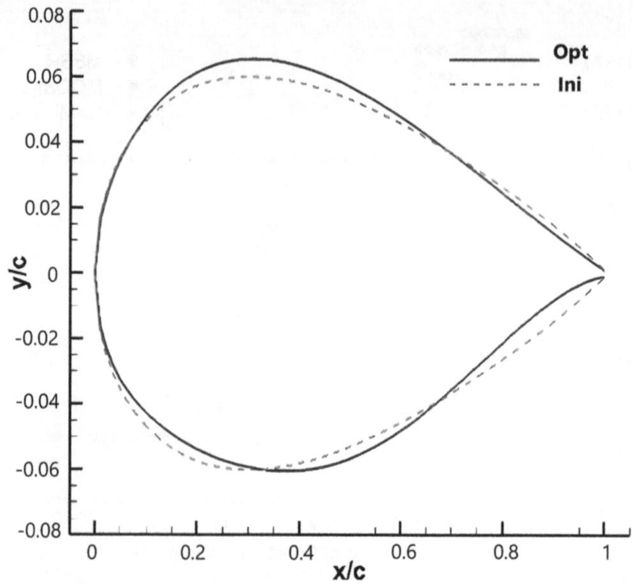

FIGURE 7.13 Comparison diagram of shapes.

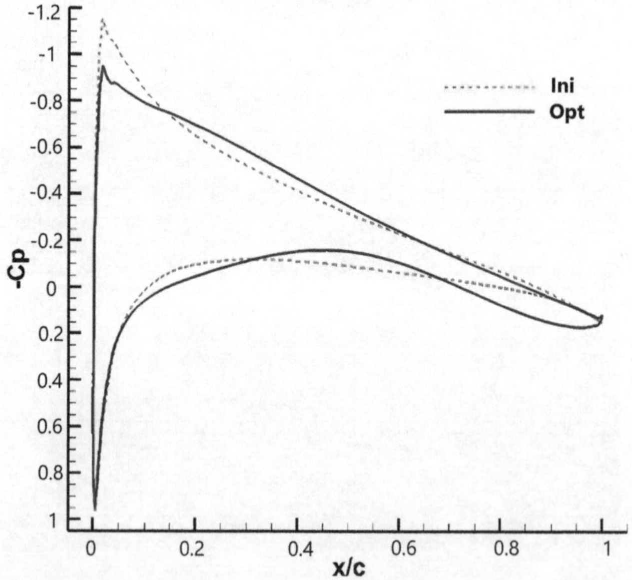

FIGURE 7.14 Comparison diagram of pressure curves.

7.6 CHAPTER SUMMARY

In this chapter, we propose a new SGO algorithm MGOSIC for EBOPs. Traditional multi-surrogate methods mostly utilize weighted sums to construct an ensemble model for optimization, and pay much attention to the choice of these weights. MGOSIC proposes a different strategy that gets multiple sample points in each cycle based on the integrated prediction information from three surrogate models.

In MGOSIC, three approximation methods, Kriging, RBF and QRS, are employed to construct surrogate models, respectively. Besides, a multi-point infilling criterion is presented to capture the new sample points on the three models per iteration. In the proposed infilling criterion, the newly added sample points mainly come from two parts: one is the present best solutions from each surrogate model, and the other one is selected from several promising point sets. These point sets are created by a proposed score-based strategy that marks a lot of cheap sample points based on their predictive values from Kriging, RBF and QRS. The new sample points will be selected from the point sets with higher scores by a proposed max–min approach that maximizes the minimum distance between new points and obtained points. When MGOSIC gets trapped in a local region, the estimated MSE of Kriging will be used to explore the unknown area. Finally, the whole optimization flow is carried out alternately in the global space and a reduced space. Compared with seven existing global optimization algorithms, MGOSIC has the best performance. After the tests on 19 benchmark cases and an engineering application, MGOSIC shows its high efficiency, strong stability and remarkable parallel capability. To sum up, MGOSIC is a promising method to optimize expensive black-box problems.

NOTE

1 Based on "Multi-surrogate-based Global Optimization using a Score-based Infill Criterion," published in [Structural and Multidisciplinary Optimization], [2019]. Permission obtained from [Springer].

REFERENCES

Cai, X., Qiu, H., Gao, L., Yang, P., & Shao, X. (2017). A Multi-Point Sampling Method Based on Kriging for Global Optimization. *Structural and Multidisciplinary Optimization*, *56*(1), 71–88. https://doi.org/10.1007/s00158-017-1648-x

Dong, H., Song, B., Dong, Z., & Wang, P. (2016). Multi-Start Space Reduction (MSSR) Surrogate-Based Global Optimization Method. *Structural and Multidisciplinary Optimization*, *54*(4), 907–926. https://doi.org/10.1007/s00158-016-1450-1

Dong, H., Song, B., Wang, P., & Dong, Z. (2018). Surrogate-Based Optimization with Clustering-Based Space Exploration for Expensive Multimodal Problems. *Structural and Multidisciplinary Optimization*, *57*(4), 1553–1577. https://doi.org/10.1007/s00158-017-1826-x

Gu, J., Li, G., & Gan, N. (2017). Hybrid Metamodel-Based Design Space Management Method for Expensive Problems. *Engineering Optimization*, *49*(9), 1573–1588. https://doi.org/10.1080/0305215x.2016.1261126

Gu, J., Li, G. Y., & Dong, Z. (2012). Hybrid and Adaptive Meta-Model-Based Global Optimization. *Engineering Optimization*, *44*(1), 87–104. https://doi.org/10.1080/0305215x.2011.564768

Gutmann, H. M. (2001). A Radial Basis Function Method for Global Optimization. *Journal of Global Optimization*, *19*(3), 201–227.

Jiang, F., Xia, H., Quang Anh, T., Quang Minh, H., Nhat Quang, T., & Hu, J. (2017). A New Binary Hybrid Particle Swarm Optimization with Wavelet Mutation. *Knowledge-Based Systems*, *130*, 90–101. https://doi.org/10.1016/j.knosys.2017.03.032

Jin, R. C., Chen, W., & Sudjianto, A. (2005). An Efficient Algorithm for Constructing Optimal Design of Computer Experiments. *Journal of Statistical Planning and Inference*, *134*(1), 268–287. https://doi.org/10.1016/j.jspi.2004.02.014

Jones, D. R., Schonlau, M., & Welch, W. J. (1998). Efficient Global Optimization of Expensive Black-Box Functions. *Journal of Global Optimization*, *13*(4), 455–492. https://doi.org/10.1023/A:1008306431147

Krityakierne, T., Akhtar, T., & Shoemaker, C. A. (2016). SOP: Parallel Surrogate Global Optimization with Pareto Center Selection for Computationally Expensive Single Objective Problems. *Journal of Global Optimization*, *66*, 417–437.

Kulfan, B. M. (2008). Universal Parametric Geometry Representation Method. *Journal of Aircraft*, *45*(1), 142–158. https://doi.org/10.2514/1.29958

Lakshika, E., Barlow, M., & Easton, A. (2017). Understanding the Interplay of Model Complexity and Fidelity in Multiagent Systems via an Evolutionary Framework. *IEEE Transactions on Computational Intelligence and AI in Games*, *9*(3), 277–289. https://doi.org/10.1109/tciaig.2016.2560882

Li, Z., Ruan, S., Gu, J., Wang, X., & Shen, C. (2016). Investigation on Parallel Algorithms in Efficient Global Optimization Based on Multiple Points Infill Criterion and Domain Decomposition. *Structural and Multidisciplinary Optimization*, *54*, 747–773.

Long, T., Wu, D., Guo, X., Wang, G. G., & Liu, L. (2015). Efficient Adaptive Response Surface Method Using Intelligent Space Exploration Strategy. *Structural and Multidisciplinary Optimization*, *51*(6), 1335–1362. https://doi.org/10.1007/s00158-014-1219-3

Masters, D. A., Taylor, N. J., Rendall, T. C. S., & Allen, C. B. (2017). Multilevel Subdivision Parameterization Scheme for Aerodynamic Shape Optimization. *AIAA Journal*, *55*(10), 3288–3303. https://doi.org/10.2514/1.J055785

Meng, Z., Pan, J.-S., & Xu, H. (2016). QUasi-Affine TRansformation Evolutionary (QUATRE) Algorithm: A Cooperative Swarm Based Algorithm for Global Optimization. *Knowledge-Based Systems*, *109*, 104–121. https://doi.org/10.1016/j.knosys.2016.06.029

Ong, Y. S., Nair, P. B., & Keane, A. J. (2003). Evolutionary Optimization of Computationally Expensive Problems via Surrogate Modeling. *AIAA Journal, 41*(4), 687–696. https://doi.org/10.2514/2.1999

Pan, W.-T. (2012). A New Fruit Fly Optimization Algorithm: Taking the Financial Distress Model as an Example. *Knowledge-Based Systems, 26*, 69–74. https://doi.org/10.1016/j.knosys.2011.07.001

Regis, R. G., & Shoemaker, C. A. (2007). A Stochastic Radial Basis Function Method for the Global Optimization of Expensive Functions. *Informs Journal on Computing, 19*(4), 497–509.

Rocca, P., Oliveri, G., & Massa, A. (2011). Differential Evolution as Applied to Electromagnetics. *IEEE Antennas and Propagation Magazine, 53*(1), 38–49. https://doi.org/10.1109/map.2011.5773566

Sala, R., Baldanzini, N., & Pierini, M. (2016). Representative Surrogate Problems as Test Functions for Expensive Simulators in Multidisciplinary Design Optimization of Vehicle Structures. *Structural and Multidisciplinary Optimization, 54*(3), 449–468. https://doi.org/10.1007/s00158-016-1410-9

Singh, P., van der Herten, J., Deschrijver, D., Couckuyt, I., & Dhaene, T. (2017). A Sequential Sampling Strategy for Adaptive Classification of Computationally Expensive Data. *Structural and Multidisciplinary Optimization, 55*(4), 1425–1438. https://doi.org/10.1007/s00158-016-1584-1

Sun, C., Zeng, J., Pan, J., Xue, S., & Jin, Y. (2013). A New Fitness Estimation Strategy for Particle Swarm Optimization. *Information Sciences, 221*, 355–370. https://doi.org/10.1016/j.ins.2012.09.030

Tyan, M., Nhu Van, N., & Lee, J.-W. (2015). Improving Variable-Fidelity Modelling by Exploring Global Design Space and Radial Basis Function Networks for Aerofoil Design. *Engineering Optimization, 47*(7), 885–908. https://doi.org/10.1080/0305215x.2014.941290

Viana, F. A. C., Haftka, R. T., & Watson, L. T. (2013). Efficient Global Optimization Algorithm Assisted by Multiple Surrogate Techniques. *Journal of Global Optimization, 56*(2), 669–689. https://doi.org/10.1007/s10898-012-9892-5

Wang, H., Fan, T., & Li, G. (2017). Reanalysis-Based Space Mapping Method, an Alternative Optimization Way for Expensive Simulation-Based Problems. *Structural and Multidisciplinary Optimization, 55*(6), 2143–2157. https://doi.org/10.1007/s00158-016-1633-9

Wang, L., Pei, J., Menhas, M. I., Pi, J., Fei, M., & Pardalos, P. M. (2017). A Hybrid-Coded Human Learning Optimization for Mixed-Variable Optimization Problems. *Knowledge-Based Systems, 127*, 114–125. https://doi.org/10.1016/j.knosys.2017.04.015

Wang, L. Q., Shan, S. Q., & Wang, G. G. (2004). Mode-Pursuing Sampling Method for Global Optimization on Expensive Black-Box Functions. *Engineering Optimization, 36*(4), 419–438. https://doi.org/10.1080/03052150410001686486

Younis, A., & Dong, Z. (2010). Metamodelling and Search Using Space Exploration and Unimodal Region Elimination for Design Optimization. *Engineering Optimization, 42*(6), 517–533.

Zhou, G., Zhao, W., Li, Q., Shen, W., & Wang, C. (2017). Multi-Objective Robust Design Optimization of a Novel NPR Energy Absorption Structure for Vehicles Front Ends to Enhance Pedestrian Lower Leg Protection. *Structural and Multidisciplinary Optimization*, *56*(5), 1215–1224. https://doi.org/10.1007/s00158-017-1754-9

Zhou, Q., Jiang, P., Shao, X., Hu, J., Cao, L., & Wan, L. (2017). A Variable Fidelity Information Fusion Method Based on Radial Basis Function. *Advanced Engineering Informatics*, *32*, 26–39. https://doi.org/10.1016/j.aei.2016.12.005

Zhou, Q., Wang, Y., Choi, S.-K., Jiang, P., Shao, X., & Hu, J. (2017). A Sequential Multi-Fidelity Metamodeling Approach for Data Regression. *Knowledge-Based Systems*, *134*, 199–212. https://doi.org/10.1016/j.knosys.2017.07.033

CHAPTER 8

SCGOSR

Surrogate-Based Constrained Global Optimization Using Space Reduction[1]

8.1 INTRODUCTION

Continuous advancements in modern industry make simulated-based design and optimization imperative (Tolson & Shoemaker, 2007). Although the above-mentioned algorithms have advantages in dealing with expensive black-box optimization problems with boundary constraints, most of them cannot handle nonlinear constrained optimization problems.

When both the objective and constraints are computationally expensive black-box functions, the complexity of optimization gets further increased. Bjorkman and Holmstrom (2000) developed a radial basis function (RBF)-based optimization algorithm that utilized a penalty technique to transform an inequality-constrained problem into a box-constrained problem. Besides, a train design optimization problem was successfully solved with fewer costly function evaluations. Basudhar et al. (2012) presented an efficient global optimization algorithm for constrained problems, where Kriging is used for approximation of the objective function and support vector machines (SVMs) are employed to approximate the boundary of feasible regions. Importantly, one unique SVM can represent several correlated constraints, which considerably simplifies the complexity of constrained optimization. Regis (2011) extended the previous local

DOI: 10.1201/9781003636267-8

metric stochastic RBF (LMSRBF) algorithm to handle costly nonlinear inequality-constrained problems. The constrained LMSRBF algorithm constructs surrogate models for objective and constraint functions, respectively, and identifies candidate points that are predicted to be feasible. Bagheri et al. (2017) presented a "Self-Adjusting Constrained Optimization by RBF Approximation (SACOBRA)" method based on Regis's research. Importantly, SACOBRA can efficiently find feasible solutions without parameter tuning. Parr et al. (2012) presented an enhanced infill sampling criterion that treats objective improvement and constraint satisfaction as two separate functions and uses multi-objective optimization to select update points. Additionally, there is also some literature focusing on multi-objective optimization with expensive objectives and constraints (Audet et al., 2010; Durantin et al., 2016). Muller and Woodbury (2017) also pointed out that algorithms for problems with expensive objectives and constraints are scarce.

Hence, this chapter aims at developing a new global optimization algorithm for computationally expensive black-box-constrained problems. The problem type considered in this chapter can be briefly summed up as follows:

$$\begin{aligned} & \min f(\boldsymbol{x}) \\ S.T.\quad & g_j(\boldsymbol{x}) \le 0, \quad \forall j = 1,\ldots,m \\ & Lb_i \le x_i \le Ub_i, \quad \forall i = 1,\ldots,n \\ & x_i \in R, \quad \forall i = 1,\ldots,n \end{aligned} \tag{8.1}$$

where $\boldsymbol{x}$ is the design variable vector, $f(\boldsymbol{x})$ denotes the costly objective and $\boldsymbol{g}(\boldsymbol{x})$ refers to the costly constraint vector. $\boldsymbol{Lb}$ and $\boldsymbol{Ub}$ are the lower and upper bounds of the design variable $\boldsymbol{x}$, respectively. For the computationally expensive problems described in Eq. (8.1), using a smaller number of function (objective and constraints) evaluations to get the global optimum is important. Since the actual engineering applications involve multimodal or high nonlinear models, both $f(\boldsymbol{x})$ and $\boldsymbol{g}(\boldsymbol{x})$ in Eq. (8.1) may have complex forms.

Actually, this chapter is the extension of our previous work where a Kriging-based global optimization algorithm MSSR was presented. MSSR is mainly developed for unconstrained expensive black-box problems. On constraint handling, MSSR just adds a penalty term to the objective function, and the reduced spaces are created without using penalty-based strategies. Besides, MSSR always constructs the complete surrogate models for costly objective and constraint functions, respectively, which is time-consuming. In this chapter, Kriging is used to approximate the costly objective and constraints. In addition, a proposed multi-start constrained optimization algorithm carries out a search on the Kriging models to get supplementary points in each cycle. In order to find feasible regions, even the global optimum quickly, a penalty-based space reduction strategy is presented. In this strategy, two penalty methods are used respectively to sort the expensive samples and two subspaces are created based on the ranking of the present samples. Considering the difficulty in fitting an accurate large-scale surrogate model, two groups of local surrogate models located in the defined subspaces are dynamically constructed per optimization cycle. Furthermore, once SCGOSR gets stuck in a local valley, the estimated mean square error of Kriging is maximized to explore the sparsely sampled area, guaranteeing the balance between the local and global searches.

8.2 SCGOSR ALGORITHM

In SCGOSR, Kriging is employed to construct surrogate models for costly objective and constraint functions, respectively. In order to add multiple promising samples in each cycle, a multi-start constrained optimization algorithm is proposed to exploit Kriging models. Furthermore, a space reduction strategy is presented to create two subspaces where two groups of local surrogate models are separately constructed. The multi-start optimization is carried out alternately in the two subspaces and the overall design space. Once a local convergence criterion is satisfied, SCGOSR will maximize the estimated MSE of Kriging to explore the sparsely sampled regions. More details will be introduced in the following sections.

8.2.1 Multi-Start Constrained Optimization

Generally, optimization on surrogate models may generate several predictive local optimal solutions, especially when both objective and constraint functions are Kriging models. The true global optimal solution may exist among these potential optimal locations, and thus, it is important to capture these predictive local optimal samples and select more promising ones. In this chapter, a multi-start constrained optimization algorithm is utilized to exploit the Kriging models. Different from MSSR, which is mainly designed for unconstrained problems, this multi-start constrained optimization algorithm utilizes a penalty function to deal with the predicted results and save them in a defined matrix. Besides, in MSSR the distances between points are defined as constants, while in this chapter a distance criterion that relies on the size of the design space is proposed. The specific process is described as follows.

Firstly, several starting points are generated in a defined space by Latin hypercube sampling, and then sequential quadratic programming (SQP) begins to run from these starting points. Samples and the corresponding predictive values obtained by multiple SQP solvers are saved in a matrix ***PLO***. Equation (8.2) gives the specific expression of the multi-start constrained optimization.

$$
\begin{aligned}
&\mathit{Multi\_Start\ Optimization} \\
&\mathit{StartingPoints}: \boldsymbol{x}_i, \quad i = 1,2,\ldots,M \\
&\qquad \begin{array}{ll} SQP & \min \hat{Y}_{krg}(\boldsymbol{x}) \\ S.T. & \hat{g}_{krg}(\boldsymbol{x}) \le 0 \\ & Lb \le \boldsymbol{x} \le Ub \end{array}
\end{aligned} \tag{8.2}
$$

where $\hat{Y}_{krg}(\boldsymbol{x})$ and $\hat{g}_{krg}(\boldsymbol{x})$ are Kriging models of the exact objective and constraint functions, respectively. In Eq. (8.2), M refers to the number of starting points and $\boldsymbol{x}_i$ denotes the ith starting point. Once the predictive local optima are obtained, these samples and predictive values of the objective and constraints will be recorded in matrix ***PLO***. Equation (8.3) describes a penalty method that can transform the Kriging-based objective and constraints into an augmented function. As Eq. (8.4) shows, the new ***PLO*** matrix has M rows and $(n+1)$ columns.

$$
\hat{Y}_{aug}(\boldsymbol{x}) = \hat{Y}_{krg}(\boldsymbol{x}) + P \cdot \sum_{i=1}^{m} \max\left(\hat{g}_{ikrg}(\boldsymbol{x}), 0\right) \tag{8.3}
$$

$$
\boldsymbol{PLO} = \begin{bmatrix} S_1^1, & S_2^1, & \cdots, & S_n^1, & \hat{Y}_{krg}^1 & \hat{g}_{krg1}^1, & \cdots, & \hat{g}_{krgm}^1 \\ S_1^2, & S_2^2, & \cdots, & S_n^2, & \hat{Y}_{krg}^2 & \hat{g}_{krg1}^2, & \cdots, & \hat{g}_{krgm}^2 \\ \vdots & \ddots & \cdots & \vdots & \vdots & \vdots & & \\ S_1^M, & S_2^M, & \cdots, & S_n^M, & \hat{Y}_{krg}^M & \hat{g}_{krg1}^M, & \cdots, & \hat{g}_{krgm}^M \end{bmatrix}
\Rightarrow \begin{bmatrix} S_1^1, & S_2^1, & \cdots, & S_n^1, & \hat{Y}_{aug}^1 \\ S_1^2, & S_2^2, & \cdots, & S_n^2, & \hat{Y}_{aug}^2 \\ \vdots & \ddots & \cdots & \vdots & \vdots \\ S_1^M, & S_2^M, & \cdots, & S_n^M, & \hat{Y}_{aug}^M \end{bmatrix} \tag{8.4}
$$

According to the size of $\hat{Y}_{aug}^i$ in ***PLO***, the matrix is sorted in ascending order. Since multiple SQP solvers may get similar or repeated local optimal solutions, the redundant samples will be deleted from ***PLO***. The samples in Eq. (8.5) need to keep a defined distance as below.

$$\left\|\boldsymbol{S}^i - \boldsymbol{S}^j\right\| > \Delta \cdot \left\|\boldsymbol{Ub} - \boldsymbol{Lb}\right\| \tag{8.5}$$

where ***Ub*** and ***Lb*** are the design bounds and Δ is a weight factor that determines the size of the distance. Generally, a smaller Δ may bring more points that are much closer to each other, but a bigger Δ may make SCGOSR miss some promising points. Hence, the recommended range for Δ is [1e−6, 1e−4]. Besides, the samples that go much closer to the obtained sample will also be eliminated. The final supplementary samples will be chosen from the filtered ***PLO*** and the smaller $\hat{Y}^i_{aug}$ will have the higher priority.

8.2.2 Space Reduction for Constrained Optimization

Space reduction (also called region elimination) can remove the less promising and previously explored regions to decrease the number of costly function evaluations. Mostly, a reduced space is the neighborhood of the present best solution or a small region that encloses several promising solutions. For constrained optimization, the so-called best solution not only possesses the minimum objective value but also has to satisfy all constraints. In order to find these promising samples from the expensive sample set, two penalty functions are utilized, and the specific formulas are shown below.

$$Y_{aug1} = \begin{cases} Y_{obj} & if\ g_i \le 0,\ \ \forall i = 1,\ldots,m \\ Y_{obj} + P & if\ g_i > 0,\ \ \exists i = 1,\ldots,m \end{cases} \tag{8.6}$$

$$Y_{aug2} = Y_{obj} + P \cdot \sum_{i=1}^{m} \max\left(g_i, 0\right) \tag{8.7}$$

$$\begin{aligned} &\begin{bmatrix} S_1^1, & S_2^1, & \cdots, & S_n^1 \\ S_1^2, & S_2^2, & \cdots, & S_n^2 \\ \vdots & \ddots & \cdots & \vdots \\ S_1^K & S_2^K & \cdots, & S_n^K \end{bmatrix} \Rightarrow \begin{bmatrix} Y_{obj}^1 \\ Y_{obj}^2 \\ \vdots \\ Y_{obj}^K \end{bmatrix} \\ + &\begin{bmatrix} g_1^1, & g_2^1, & \cdots, & g_m^1 \\ g_1^2, & g_2^2, & \cdots, & g_m^2 \\ \vdots & \ddots & \cdots & \vdots \\ g_1^K, & g_2^K, & \cdots, & g_m^K \end{bmatrix} \Rightarrow \begin{bmatrix} Y_{aug1}^1 \\ Y_{aug1}^2 \\ \vdots \\ Y_{aug1}^K \end{bmatrix} \cdot \begin{bmatrix} Y_{aug2}^1 \\ Y_{aug2}^2 \\ \vdots \\ Y_{aug2}^K \end{bmatrix} \end{aligned} \tag{8.8}$$

where m is the number of constraints, K is the number of expensive samples and n represents the dimension of design variables. In Eqs. (8.3), (8.6) and (8.7), P is the penalty factor that needs to be noticeably bigger than the objective function value, and Y_{aug1} and Y_{aug2} are two augmented functions. A smaller P will not generate remarkable changes to the augmented objective functions, and thus, the recommended range for P is [1e10, 1e20]. Equation (8.8) shows the sample matrix, expensive objective vector and expensive constraint matrix. Additionally, two augmented objective vectors are obtained by the proposed penalty functions. Intuitively, the first penalty function will punish solutions that just violate any constraint, while the second one can "forgive" solutions that go much closer to the constraint bounds. Relatively speaking, Eq. (8.6) is more rigorous than Eq. (8.7) when dealing with the solutions near bounds. Besides, solutions on both sides of constraint bounds may enhance the approximation accuracy of Kriging models in the vicinity of bounds. In other words, the solutions that violate constraints but are located near constraint bounds are also valuable. Considering this characteristic, $\boldsymbol{Y}_{aug1}$ is minimized to find the present best solution and $\boldsymbol{Y}_{aug2}$ is sorted to obtain the ranks of all the expensive samples. The specific process is described as follows.

$$\min\begin{bmatrix} Y_{aug1}^{1} \\ Y_{aug1}^{2} \\ \vdots \\ Y_{aug1}^{K} \end{bmatrix} \Rightarrow \boldsymbol{S}_{aug1}^{\min} \quad sort\begin{bmatrix} Y_{aug2}^{1} \\ Y_{aug2}^{2} \\ \vdots \\ Y_{aug2}^{K} \end{bmatrix} \Rightarrow \left\{\boldsymbol{S}_{aug2}^{rank1}, \boldsymbol{S}_{aug2}^{rank2}, \ldots, \boldsymbol{S}_{aug2}^{rankK}\right\} \tag{8.9}$$

In Eq. (8.9), $\boldsymbol{S}_{aug1}^{min}$ may not equal to $\boldsymbol{S}_{aug2}^{rank1}$, because they come from two different evaluation criteria. On the basis of the above-obtained better samples, two subspaces are created as follows:

$$\begin{aligned}
&\boldsymbol{Lb}_{sub1} = \boldsymbol{S}_{aug1}^{\min} - w\cdot\left(\boldsymbol{Ub}_{range} - \boldsymbol{Lb}_{range}\right) \\
&if\ Lb_{sub1}(i) < Lb_{range}(i),\ \ then,\ \ Lb_{sub1}(i) = Lb_{range}(i) \\
&\boldsymbol{Ub}_{sub1} = \boldsymbol{S}_{aug1}^{\min} + w\cdot\left(\boldsymbol{Ub}_{range} - \boldsymbol{Lb}_{range}\right) \\
&if\ Ub_{sub1}(i) > Ub_{range}(i),\ \ then,\ \ Ub_{sub1}(i) = Ub_{range}(i) \\
&\forall i = 1,\ldots,n \\
&Subspace1: \left[\boldsymbol{Lb}_{sub1}; \boldsymbol{Ub}_{sub1}\right]
\end{aligned} \tag{8.10}$$

$$
\begin{aligned}
&M = round(r \cdot K) \\
&\boldsymbol{Lb}_{sub2}(i) = \min\left(\left\{\boldsymbol{S}_{aug2}^{rank1}(i);\boldsymbol{S}_{aug2}^{rank2}(i);\cdots\boldsymbol{S}_{aug2}^{rankM}(i)\right\}\right) \\
&\boldsymbol{Ub}_{sub2}(i) = \max\left(\left\{\boldsymbol{S}_{aug2}^{rank1}(i);\boldsymbol{S}_{aug2}^{rank2}(i);\cdots\boldsymbol{S}_{aug2}^{rankM}(i)\right\}\right) \\
&i = 1,2,\ldots,n \\
&Subspace2:\left[\boldsymbol{Lb}_{sub2};\boldsymbol{Ub}_{sub2}\right]
\end{aligned}
\tag{8.11}
$$

In Eq. (8.10), $\boldsymbol{Ub}_{range}$ and $\boldsymbol{Lb}_{range}$ are the upper and lower bounds of the original design space, w is a weight factor, and n refers to the number of dimensions. In Eq. (8.11), r is a ratio coefficient and K represents the number of expensive samples. The two user-defined parameters, "w" and "r" determine the size of the subspaces. If "w" is bigger than 50% or "r" is bigger than 100%, it will lose the significance of the space reduction. On the contrary, if "w" and "r" are too small, the local surrogate models will get inaccurate and SCGOSR may miss some promising solutions. Therefore, the recommended ranges for "w" and "r" are [10%, 20%] and [20%, 40%], respectively. Intuitively, *Subspace*1 is a neighborhood of the present best solution that comes from Eqs. (8.6) and (8.9), while *Subspace*2 encloses several promising samples that are defined by Eqs. (8.7) and (8.9). In SCGOSR, the proposed multi-start constrained optimization algorithm alternately explores the three spaces: *Subspace*1, *Subspace*2 and the global design space. As Ong et al. (2003) suggested, it is difficult to construct an accurate global surrogate model, especially when objective and constraint functions are multimodal problems. Hence, based on the samples in the two subspaces, two groups of local Kriging models for the costly objective and constraints are constructed, respectively.

8.2.3 Exploration on Unknown Area

Generally, a successful global optimization algorithm has the capacity to escape from local optima to explore the unknown area. In SCGOSR, when all the new samples in the matrix *PLO* do not satisfy the diversity requirement, or several successive iterations do not bring better samples, the algorithm will focus on the sparsely sampled regions. Here, the estimated MSE of Kriging is maximized by the multi-start optimization algorithm to search the added sample points. The specific pseudocode is shown as follows:

Algorithm 8.1 Escape from Local Optima

(01) ***Begin***
(02) ***OptSpace*** ← Identify the optimization space (*Subspace*1, *Subspace*2, Design Space) based on the number of iterations
(03) ***Snew*** ← New samples selected from the PLO matrix
(04) $\boldsymbol{Y}_{aug1}^{rank1}, \boldsymbol{Y}_{aug1}^{rank2}, \cdots \boldsymbol{Y}_{aug1}^{rankK}$ ← Sort the augmented function values to get the ranks
(05) $\boldsymbol{Y}_{aug1}^{mean} \leftarrow mean\left(\boldsymbol{Y}_{aug1}^{rank1}, \boldsymbol{Y}_{aug1}^{rank2}, \cdots \boldsymbol{Y}_{aug1}^{rankm}\right)$, Get the mean value of the top m augmented function values based on the ranks
(06) $\boldsymbol{Y}_{mean}$(***iteration***) ← Save and Record $\boldsymbol{Y}_{aug1}^{mean}$ in each iteration.
(07) **if *iteration*>**5
(08) ***GVI*** ← | $\boldsymbol{Y}_{mean}$(**pre_iter**)- $\boldsymbol{Y}_{mean}$(**pre_iter**-5)|. (Here, "**pre_iter**" refers to the present iteration)
(09) **else**
(10) ***GVI*** ← 1e20
(11) **end if**
(12) **if *Snew*** is empty or ***GVI*** <=1e−6
(13) ***Snew_mse*** ← Call multi-start constrained optimization to maximize the estimated MSE of Kriging in ***OptSpace*** based on Eqs. (8.2) to (8.5)
(14) ***Snew*** ← [***Snew***; ***Snew_mse***]
(15) **end if**
(16) ***End***

In Algorithm 8.1, *GVI* is a temporary variable that records the changes of the top m augmented function values. From Lines (04) to (06) of Algorithm 1, it is clear that the top m samples are selected and their mean value is recorded in each cycle. Besides, as Lines (07) to (12) in Algorithm 8.1 show, if the mean value of the top m sample values does not change obviously or ***Snew*** is empty, the multi-start constrained optimization begins to explore the unknown area.

8.2.4 Optimization Flow

In this section, the overall flowchart of SCGOSR is given, and it mainly includes three parts: initialization, exploitation and exploration. For a global optimization algorithm, exploitation refers to the quick search in the vicinity of the present best solution, while exploration denotes supplementing new points in sparsely sampled areas. SCGOSR possesses the capacity of intensive search in a local promising region and meanwhile is also able to jump out from a local valley. The flowchart of SCGOSR is shown in Figure 8.1.

In Figure 8.1, the local convergence criterion is provided in Algorithm 8.1, and the global stopping criterion is defined as below.

$$y_{best} \leq target, \quad or \quad NFE > 500 \tag{8.12}$$

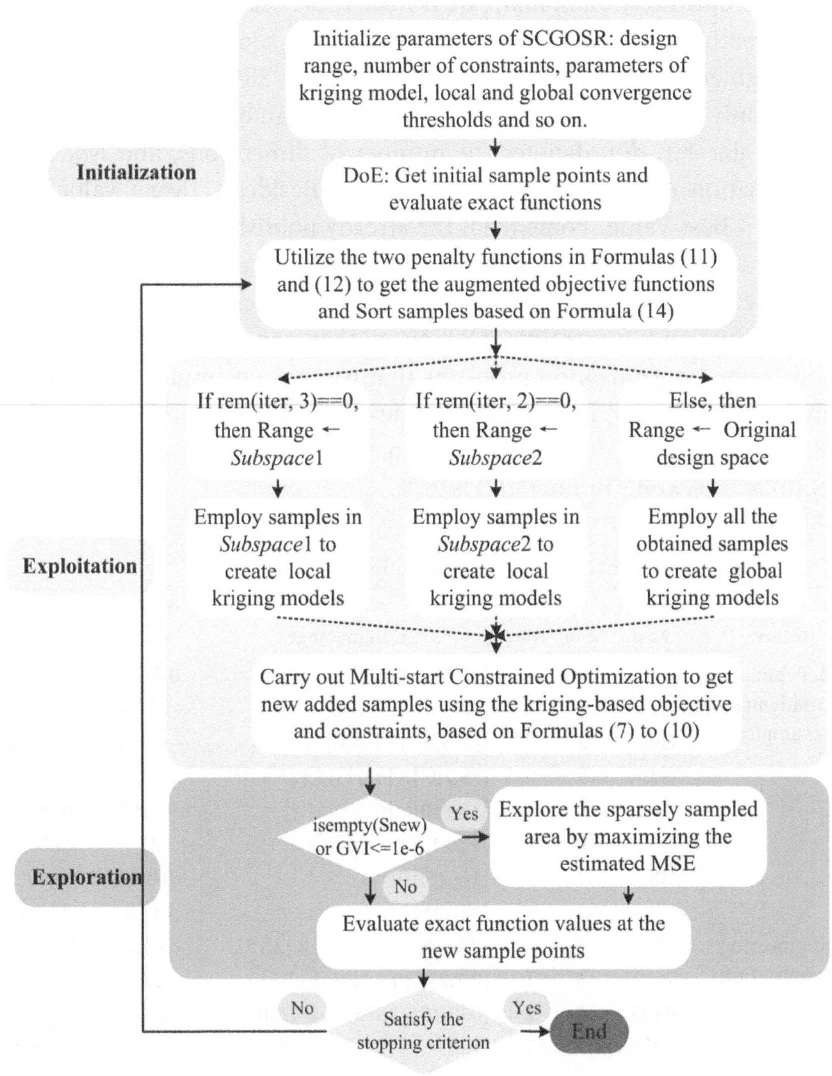

FIGURE 8.1 Flowchart of SCGOSR.

where *target* is a defined target value and *NFE* represents the number of objective or constraint function evaluations.

In Figure 8.1, the function "rem(A, B)" returns the remainder after the division of A by B.

8.3 COMPUTATIONAL EXPERIMENTS

In order to verify the capability and demonstrate the advantage of SCGOSR, different representative benchmark cases in the nonlinear constrained optimization domain are tested. These cases include eight benchmark mathematical examples (BR, SE, GO, G4, G6, G7, G8 and G9) and five engineering applications (TSD, WBD, PVD, SRD and SCBD) that are commonly used. More details of the test cases can be seen in Table 8.1.

In Table 8.1, *dim* denotes the number of dimensions, and *Noc* is the abbreviation of "Number of Constraints." Besides, "Target value" and "Known Best Value" come from the already published papers about constrained problems (Garg, 2014; Thanedar & Vanderplaats, 1995). It is mentionable that these cases have various characteristics and involve different dimensions and constraints. Obviously, they can represent most of the constrained optimization problems that we may encounter in the actual engineering design. In the following tests, the parameter P in Eqs. (8.3), (8.6) and (8.7) equals to 1e10, Δ in Eq. (8.5) is defined as 1e−5, w in Eq. (8.10) is 15%, and r in Eq. (8.11) is 25%.

TABLE 8.1 Nonlinear Constrained Optimization Cases

Category	*Func.*	*dim*	*Noc.*	Design Range	Target Value	Known Best Value
Benchmark mathematical examples	BR	2	1	$[-5,10]\times[0,15]$	0.3980	0.3979
	SE	2	1	$[0,5]^2$	−1.1740	−1.1743
	GO	2	1	$[-0.5,0.5]\times[-1,0]$	−0.970	−0.9711
	G4	5	6	$[78,102]\times[33,45]\times[27,45]^3$	−31,025	−31,025.56
	G6	2	2	$[13,100]\times[0,100]$	−6,960	−6,961.81
	G7	10	8	$[-10,10]^{10}$	25	24.3062
	G8	2	2	$[1e-15,10]^2$	−0.0958	−0.0958
	G9	7	4	$[-10,10]^7$	1,000	680.6301
Engineering applications	TSD	3	4	$[0.05,2]\times[0.25,1.3]\times[2,15]$	0.0128	0.01267
	WBD	4	7	$[0.1,2]\times[0.1,10]^{2\times}[0.1,2]$	1.8	1.7249
	PVD	4	4	$[0.0625,6.1875]^{2\times}[10,200]^2$	6,000	5,885.33
	SRD	7	11	$[2.6,3.6]\times[0.7,0.8]\times[17,28]\times[7.3,8.3]^{2\times}[2.9,3.9]\times[5.0,5.5]$	3,000	2,994.42
	SCBD	10	11	$([2,3.5]\times[35,60])^5$	65,000	62,791

TABLE 8.2 Preliminary Test Results of SCGOSR

Problems	Design Variables	$f(x)$
BR	[9.4248, 2.4750]	0.3979
SE	[2.7450, 2.3523]	−1.1743
GO	[0.1092, −0.6234]	−0.9711
G4	[78, 33, 27.0734, 45, 44.9619]	−31,025.35
G6	[14.0950, 0.8430]	−6,961.80
G7	[2.1640, 2.3825, 8.7750, 5.0870, 0.9753, 1.3864, 1.3067, 9.8169, 8.2413]	24.3187
G8	[1.2315, 4.2450]	−0.0958
G9	[2.0341, 1.9175, −0.6860, 4.4691, −0.2074, 1.7834, 1.6773]	686.8836
TSD	[0.0516, 0.3550, 11.3904]	0.0126653
WBD	[0.2057, 3.4705, 9.0366, 0.2057]	1.7249
PVD	[0.7792, 0.3852, 40.3713, 199.3308]	5,888.66
SRD	[3.5002, 0.7000, 17, 7.3000, 7.7153, 3.3503, 5.2867]	2,994.78
SCBD	[2.9921, 59.8408, 2.7943, 55.3846, 2.5237, 50.4720, 2.2206, 43.9321, 2, 35.0028]	62,874.36

8.3.1 Preliminary Test

Preliminarily, the presented SCGOSR algorithm is tested on the 13 benchmark cases, and the results are listed in Table 8.2. What is more, Figure 8.2 provides the iterative results of SCGOSR on all these cases. It is worth noting that SCGOSR uses "NFE>500" as the global stopping criterion in the preliminary test. It is clear that SCGOSR can easily find the target values of these cases and even get much closer to the global optima shown in Table 8.1. In order to improve the readability of Figure 8.2, clearer results are given in some cases, like BR, G6, G7, G8, G9 and WBD. As Figure 8.2a, e, f, g, i, l, m, o and p shows the initial DoE cannot provide a feasible solution in most cases, but SCGOSR can still capture the feasible solutions with iterations going on.

8.3.2 Comparison and Analyses

Due to the random feature of SCGOSR, ten independent tests were conducted to verify its stability. Additionally, five surrogate-based constrained optimization algorithms (RBFCGOSR, SCGO, MSSR, MS and MSRBF) are tested in contrast. Specifically, RBFCGOSR is the same as SCGOSR except that RBFCGOSR uses cubic RBF to construct the surrogate model; SCGO is the SCGOSR algorithm without space reduction; MSSR is a previously presented global optimization algorithm that can deal with constrained problems; MS is the MSSR algorithm without using space reduction strategies; MSRBF is an RBF-based optimization algorithm using the multi-start

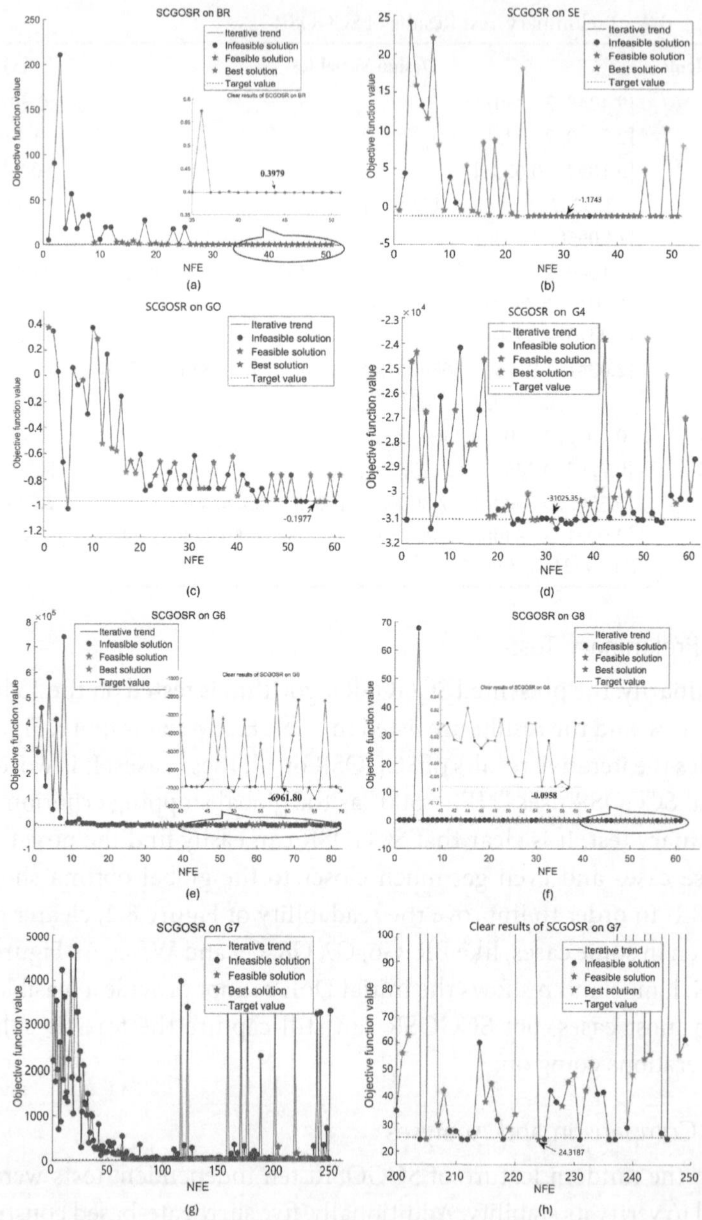

FIGURE 8.2 SCGOSR on benchmark cases. (a) SCGOSR on BR. (b) SCGOSR on SE. (c) SCGOSR on GO. (d) SCGOSR on G4. (e) SCGOSR on G6. (f) SCGOSR on G8. (g) SCGOSR on G7. (h) Clear results of SCGOSR on G7. (i) SCGOSR on G9. (j) Clear results of SCGOSR on G9. (k) SCGOSR on PVD. (l) SCGOSR on SRD. (m) SCGOSR on WBD. (n) Clear results of SCGOSR on WBD. (o) SCGOSR on TSD. (p) SCGOSR on SCBD.

(*Continued*)

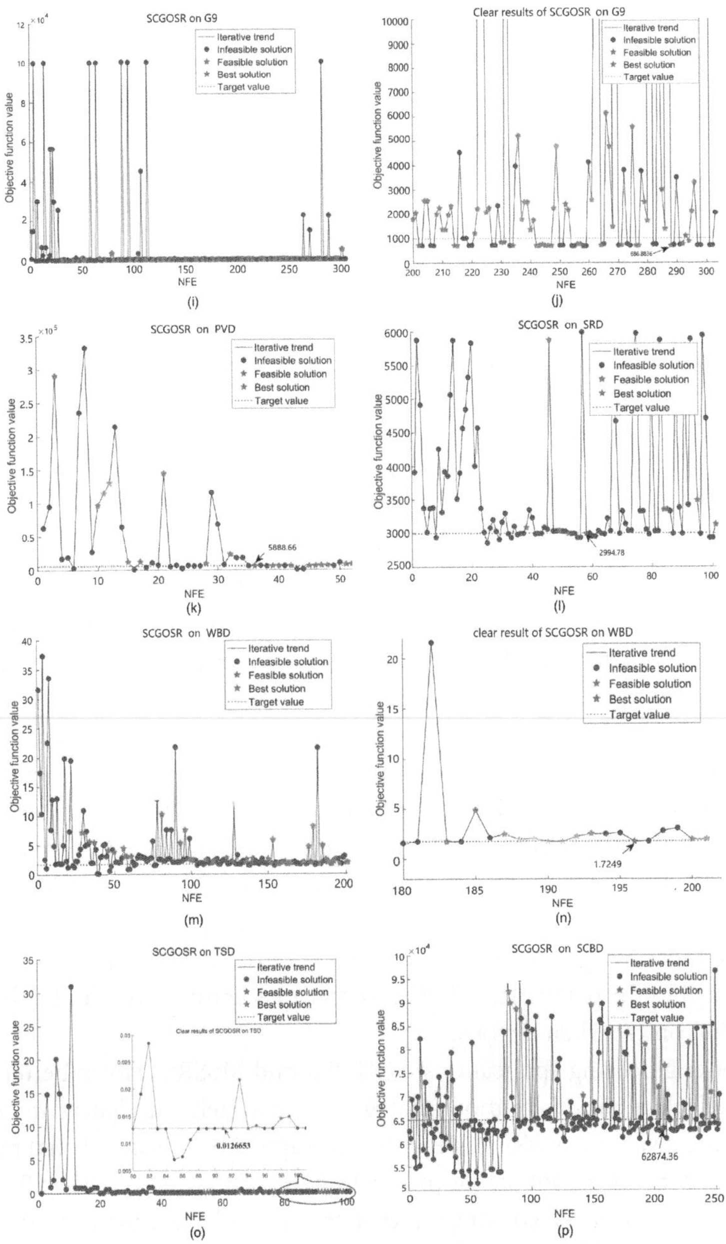

FIGURE 8.2 *(Continued)* SCGOSR on benchmark cases. (a) SCGOSR on BR. (b) SCGOSR on SE. (c) SCGOSR on GO. (d) SCGOSR on G4. (e) SCGOSR on G6. (f) SCGOSR on G8. (g) SCGOSR on G7. (h) Clear results of SCGOSR on G7. (i) SCGOSR on G9. (j) Clear results of SCGOSR on G9. (k) SCGOSR on PVD. (l) SCGOSR on SRD. (m) SCGOSR on WBD. (n) Clear results of SCGOSR on WBD. (o) SCGOSR on TSD. (p) SCGOSR on SCBD.

optimization solvers of SCGOSR. Besides, SCGOSR is also compared with KCGO that is recently published. The ranges of best values are shown in Tables 8.3 and 8.5. The statistical NFE results are shown in Tables 8.3–8.6. Here, the NFE values with the symbol ">" indicate that at least one of the tests cannot find the target value within 500 function evaluations, the numbers in the brackets "()" refer to the number of failures, and the numbers in the brackets "{}" represent how many times the algorithm cannot find feasible solutions.

Intuitively, SCGOSR can find the target values of all the cases within 500 function evaluations, but the other five algorithms have failed cases in varying degrees. In this work, two-dimensional cases like BR, SE, GO, G6 and G8 are nonlinear constrained problems. For these low-dimensional problems, it is clear that MS and MSRBF have a higher possibility of failure, and RBFCGOSR and MSSR can succeed in most cases. BR is a relatively simple case whose global optima can be easily found by all these algorithms. Besides, it is difficult for the two RBF-based algorithms (RBFCGOSR and MSRBF) to quickly find the target value of GO and G8. What is more, MSSR, MS and MSRBF sometimes may go close to the target value of SE but finally, they cannot reach the target.

When the number of dimensions and constraints increases, it will get harder for a surrogate-based optimization algorithm to find target values. For G4, which has five dimensions and six constraints, the proposed optimization flow including SCGOSR, RBFCGOSR and SCGO can successfully find the target value with fewer function evaluations, but MSSR, MS and MSRBF always fail. In the mathematical examples, G7 and G9 seem to be the most complex cases, and thus most of these algorithms have larger NFE values. In particular, MSRBF sometimes cannot even find a feasible solution on G7 when it stops.

For engineering applications, SCGOSR and MSSR perform better. Due to the lack of an exploration strategy that can help an algorithm escape from local valleys, MSRBF is easy to get trapped in a local optimal region. Hence, MSRBF is not stable in most engineering cases. Since SRD and SCBD both have 11 constraints that bring challenges for optimization, RBFCGOSR, MSSR, MS and MSRBF commonly use more function evaluations to search their target values.

Table 8.7 shows the comparison results of SCGOSR and KCGO (Li et al., 2017). In Table 8.7, KCGO provides a group of results that come from the reference. Here, G4′ is a little different from G4 (Garg, 2014) that was previously introduced. The coefficient 0.00026 in G4 is changed to

TABLE 8.3 Best Values and Mean NFE of SCGOSR, RBFCGOSR, and SCGO

	SCGOSR		RBFCGOSR		SCGO	
Func.	**NFE**	**Best Value**	**NFE**	**Best Value**	**NFE**	**Best Value**
BR	25.1	[0.3979, 0.3980]	69	[0.3979, 0.3980]	26.7	[0.3979, 0.3980]
SE	25.9	[−1.1743, −1.1740]	43.2	[−1.1743, −1.1741]	**24.3**	**[−1.1743, −1.1741]**
GO	51.1	[−0.9711, −0.9706]	>137.6	[−0.9711, −0.7653]	85.7	[−0.9711, −0.9708]
G4	**53.9**	**[−31,026, −31,025]**	252.6	[−31,026, −31,025]	55.7	[−31,026, −31,025]
G6	78.5	[−6,961.8, −6,961.4]	**46.4**	**[−6,961.8, −6,961.2]**	>464	[−6,961.6, −6,937.3]
G7	**178.2**	**[24.3149, 24.9969]**	247.2	[24.3062, 24.8145]	>290.6	[24.4436, 27.824]
G8	**51.8**	**[−0.0958, −0.0958]**	>178.1	[−0.0958, −0.0936]	115.5	[−0.0958, −0.0958]
G9	115.6	[826.30, 981.86]	124.2	[845.75, 974.07]	165.2	[730.19, 990.14]
TSD	**75.7**	**[1.267e−2, 1.278e−2]**	>293.3	[1.273e−2, 1.287e−2]	110.2	[1.267e−2, 1.279e−2]
WBD	**101.9**	**[1.7249, 1.7888]**	194	[1.7449, 1.7983]	>392.4	[1.7745, 2.7610]
PVD	42.9	[5,885.3, 5,982.1]	174.2	[5,885.4, 5,972.7]	31.9	[5,885.3, 5,981.7]
SRD	**88.1**	**[2,994.5, 2,997.8]**	232.6	[2,994.5, 2,994.5]	147.8	[2,994.5, 2,999.3]
SCBD	**152.5**	**[62,861, 64,895]**	>256.9	[62,791, 70,594]	216.3	[63,079, 64,699]

TABLE 8.4 Statistical NFE of SCGOSR, RBFCGOSR and SCGO

	SCGOSR			RBFCGOSR			SCGO		
Func.	**Min**	**Median**	**Max**	**Min**	**Median**	**Max**	**Min**	**Median**	**Max**
BR	21	25.5	29	28	60	160	17	24	48
SE	18	24.5	41	25	33	128	**20**	**25**	**30**
GO	17	28	156	22	60.5	>500(1)	18	21	297
G4	**32**	**35.5**	**174**	188	253	307	21	25.5	181
G6	33	65.5	171	**27**	**44.5**	**71**	123	>500	>500(9)
G7	**102**	**199.5**	**239**	107	200.5	459	64	>293.5	>500(5)
G8	**24**	**47.5**	**80**	32	104.5	>500(2)	44	97	297
G9	54	112	198	58	121	213	31	187.5	235
TSD	**43**	**69**	**114**	113	244	>500(1)	62	96.5	249
WBD	**72**	**97**	**153**	112	180.5	372	186	408	>500(3)
PVD	27	41	63	92	159	285	26	31.5	36
SRD	**35**	**60.5**	**272**	143	219.5	345	35	127	331
SCBD	**62**	**119.5**	**297**	134	222.5	>500(1)	42	209	470

TABLE 8.5 Best Values and Mean NFE of MSSR, MS and MSRBF

	MSSR		MS		MSRBF	
Func.	**NFE**	**Best Value**	**NFE**	**Best Value**	**NFE**	**Best Value**
BR	22.6	[0.3979, 0.3980]	**21.8**	**[0.3979, 0.3979]**	>145.9	[0.3979, 0.3981]
SE	>162.8	[−1.1743, −1.1739]	>233.7	[−1.1743, −1.1735]	>74.5	[−1.1743, −1.1729]
GO	**33.6**	**[−0.9711, −0.9701]**	41.8	[−0.9711, −0.9703]	>357.7	[−0.9708, −0.1664]
G4	>272.3	[−31,026, −31,020]	>310.5	[−31,026, −30,742]	>232.2	[−31,026, −31,024]
G6	>253.5	[−6,961.4, −6,958.3]	>454.6	[−6,960.9, −6,918.7]	147.2	[−6,961.8, −6,961.8]
G7	>147.8	[24.3540, 25.1828]	>213.4	[24.3342, 27.8559]	>500	[25.0043, 1e10] {*1}
G8	68.9	[−0.0958, −0.0958]	94.8	[−0.0958, −0.0958]	>427.8	[−0.0958, −2.89e−5]
G9	**109.1**	**[828.79, 999.87]**	160.8	[822.56, 999.99]	>438.1	[946.63, 11,690]
TSD	95.4	[1.267e−2, 1.279e−2]	100.7	[1.268e−2, 1.279e−2]	179.2	[1.267e−2, 1.278e−2]
WBD	156	[1.7354, 1.7976]	>348.4	[1.7643, 2.8849]	>311.6	[1.7333, 2.7608]
PVD	30.4	[5,891.2, 5,951.5]	**29.7**	**[5,885.4, 5,965.3]**	>150.2	[5,885.4, 6,025.5]
SRD	>209.6	[2,994.5, 3,019.2]	>322.3	[2,994.5, 3,018.7]	>328.3	[2,994.5, 5,448.7]
SCBD	284.4	[62,858, 64,648]	307	[62,791, 64,731]	>387.8	[62,791, 1e10]

TABLE 8.6 Statistical NFE of MSSR, MS and MSRBF

Func.	MSSR			MS			MSRBF		
	Min	Median	Max	Min	Median	Max	Min	Median	Max
BR	17	23	28	**19**	**20.5**	**30**	45	79	>500(1)
SE	25	124.5	>500(1)	22	116	>500(4)	20	26	>500(1)
GO	**13**	**33**	**54**	14	42.5	74	17	>500	>500(7)
G4	21	>288.5	>500(5)	22	>500	>500(6)	161	198	>500(1)
G6	15	190	>500(4)	230	>500	>500(8)	53	75	408
G7	72	104	>500(1)	73	98	>500(3)	>500	>500	>500(10)
G8	39	64.5	109	42	95.5	152	31	>500	>500(8)
G9	**60**	**102.5**	**189**	83	145	295	139	>500	>500(8)
TSD	52	101	153	53	84	249	75	167	399
WBD	98	133	411	178	358	>500(2)	60	299	500(4)
PVD	26	31	37	**23**	**28**	**49**	46	63	>500(1)
SRD	31	201.5	>500(1)	34	364	>500(2)	120	305	>500(3)
SCBD	52	310.5	384	146	317.5	466	147	>500	>500(6)

TABLE 8.7 Comparison of SCGOSR and KCGO

	SCGOSR			KCGO	
Cases	**Mean NFE**	**Range of NFE**	**Range of Best Value**	**NFE**	**Best Value**
G4′	60.3	[33, 163]	[**−30,665.54**, −30,665.20]	**24**	−30,665.51
G6	78.5	[33, 171]	[**−6,961.8**, −6,961.4]	**31**	−6,677.68
G7	178.2	[**102**, 239]	[24.3149, 24.9969]	107	**24.3093**
G8	51.8	[**24**, 80]	[**−0.0958**, −0.0958]	39	−0.0956
G9	115.6	[**54**, 198]	[**826.30**, 981.86]	163	860.9243
TSD	75.7	[43, 114]	[**1.267e−2**, 1.278e−2]	**38**	0.0135
WBD	101.9	[**72**, 153]	[**1.7249**, 1.7888]	115	2.3230
SRD	88.1	[**35**, 272]	[**2,994.5**, 2,997.8]	43	2,999.76

0.0006262 in G4′. Now, the known global optima of G4′ is −30,665.54. Obviously, KCGO has impressive performance on these cases. KCGO can find an approximate optimum on G4′ only using 24 function evaluations. SCGOSR can always get satisfactory values on G4′ but needs at least 33 function evaluations. However, SCGOSR sometimes can find the true global optimum −30,665.54. Similarly, SCGOSR can find much better values than KCGO on G6, but KCGO uses fewer function evaluations. For G7, KCGO can get more accurate results than SCGOSR, while SCGOSR sometimes can be more efficient. Intuitively, SCGOSR mostly outperforms KCGO on G8, because the best value of KCGO is outside of the SCGOSR's value range and SCGOSR can use fewer function evaluations. For G9, SCGOSR is able to find a better value of 826.30 than KCGO with a smaller NFE. Besides, the mean NFE of SCGOSR (115.6) is also much smaller than 163. For the three engineering applications TSD, WBD and SRD, there is no doubt that SCGOSR gets more accurate results than KCGO. What is more, SCGOSR also performs efficiently on the three applications.

No matter the mathematical examples or engineering applications, SCGOSR always has impressive performance. More importantly, SCGOSR shows advantages in stability and efficiency compared with other algorithms. In summary, SCGOSR is a promising constrained optimization algorithm for expensive black-box problems.

8.3.3 Further Comparison and Analyses

Additionally, in order to demonstrate the extensive applicability of SCGOSR, further experiments are set up. The constrained optimization algorithm "superEGO" is used as contrast. "superEGO" (Sasena et al., 2002) was developed to solve computationally expensive problems with disconnected feasible regions. In this chapter, two benchmark examples, Gomez and newBranin suggested by Sasena et al. (2002) are tested. newBranin has three feasible regions that only cover about 3% of the design space, and the disconnected feasible regions of Gomez cover approximately 19% of the design space. According to the reference, SCGOSR also utilizes LHS to generate ten initial sample points and runs ten times on the two examples. Besides, SCGOSR will stop and NFE will be recorded when a feasible point is obtained within a box (±1% of the design space range) around the true global optimal solution. The main data of Table 8.8 comes from the reference, and it is clear that superEGO2 performs the best in the two examples. It is worth noting that SCGOSR also has impressive performance. For newBranin, SCGOSR is much closer to the superEGO's results. However, SCGOSR needs more function evaluations than superEGO2 on Gomez. What is more, SCGOSR outperforms the deterministic optimization algorithm DIRECT (Jones, 2001), the gradient-based algorithm SQP, and the nature-inspired algorithm SA (Kirkpatrick et al., 1983).

In order to demonstrate how SCGOSR works on the problems with disconnected feasible regions, two graphical examples are shown in Figures 8.3 and 8.4. In the two figures, the stars are the global optimal solutions, the squares are DoE sample points, the black circles are previously added points and the blue circles are the currently added points. Besides, the dashed lines refer to the constraint bounds.

We advisedly provide the two cases with "worse initial sample points." In other words, the initial sample points cannot offer positive guidance for SCGOSR to find the global optimum at the beginning. From Figure 8.3, it can be seen that the search at first focuses on the "wrong" feasible regions. Figure 8.3b and c shows that the search gradually goes close to the most important feasible region. Additionally, since SCGOSR can capture

TABLE 8.8 Comparison on newBranin and Gomez

	Average Number of Function Evaluations	
Algorithm	**newBranin**	**Gomez**
SCGOSR	24.6	47.5
superEGO1	22.2	66.3
superEGO2	22.0	36.5
DIRECT	76	93
SQP	363	831
SA	5,371	7,150

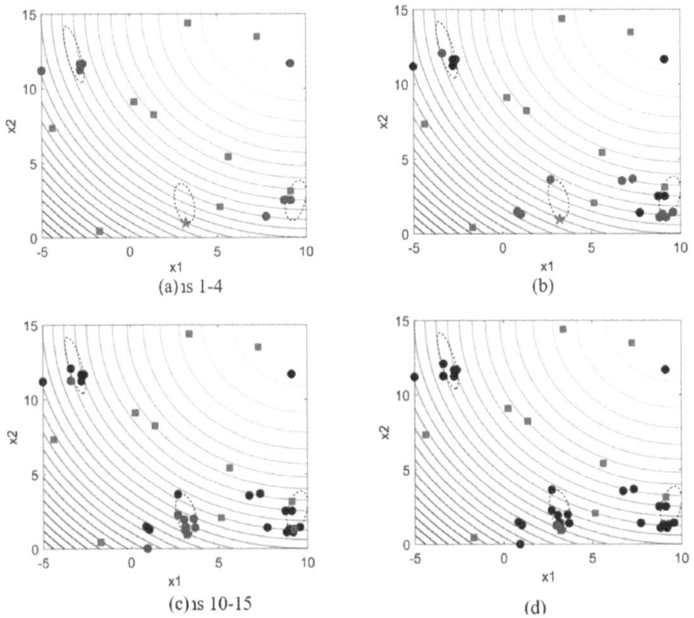

FIGURE 8.3 Iterative results of SCGOSR on newBranin. (a) Iterations 1–4. (b) Iterations 5–9. (c) Iterations 10–15. (d) Iterations 16–17.

multiple local optimal points in each cycle, the three feasible regions are sufficiently explored. Finally, 44 function evaluations are used to find the best solution [3.2340, 0.9547].

Intuitively, Gomez is more complex than newBranin as illustrated in Figure 8.4. It is clear that the search begins from the left feasible region that includes an initial sample point. During the first 20 iterations, SCGOSR is busy exploring "wrong" feasible regions. After SCGOSR explores five feasible regions, it begins to pay attention to the neighborhood of the global optimal point. Finally, SCGOSR finds the satisfactory feasible solution [0.1110, −0.6233] by 35 iterations and 79 function evaluations. In summary, SCGOSR can also solve complex problems with disconnected feasible regions.

8.3.4 Specific Analyses on Space Reduction

After the previous comparison with other methods, SCGOSR has shown its remarkable capability. According to the comparison results of SCGOSR and SCGO in Table 8.3, it is clear that the two subspaces speed up the search process of SCGOSR. In order to analyze the contribution of using *Subspace*1 and *Subspace*2, separately, two algorithms SCGOSR_S1 and

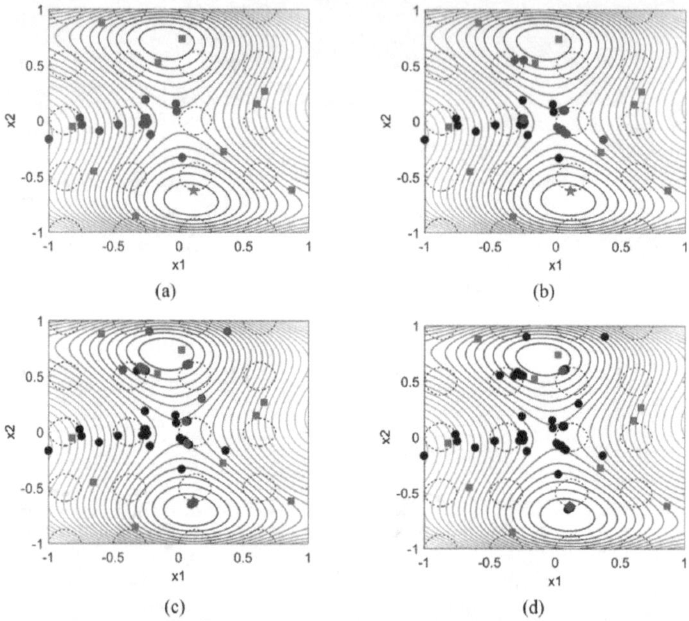

FIGURE 8.4 Iterative results of SCGOSR on Gomez. (a) Iterations 1–4. (b) Iterations 5–9. (c) Iterations 10–15. (d) Iterations 16–17.

TABLE 8.9 Best Values and Mean NFE of SCGOSR_S1 and SCGOSR_S2

	SCGOSR_S1		SCGOSR_S2	
Func.	**Mean NFE**	**Best Values**	**Mean NFE**	**Best Values**
BR	23.7	[0.3979, 0.3980]	28.3	[0.3979, 0.3980]
SE	27.7	[−1.1743, −1.1742]	26.5	[−1.1743, −1.1741]
GO	64.2	[−0.9711, −0.9704]	26.7	[−0.9711, −0.9701]
G4	96.2	[−31026, −31025]	37	[−31,026, −31,025]
G6	80.3	[−6,961.8, −6,960.0]	>362.8 (7)	[−6,961.8, −6,914.1]
G7	152.6	[24.3083, 24.9307]	220.3	[24.3086, 24.8412]
G8	79.6	[−0.0958, −0.0958]	46.2	[−0.0958, −0.0958]
G9	96.9	[782.31, 988.15]	118.5	[720.83, 982.69]
TSD	120.8	[1.267e−2, 1.276e−2]	77.7	[1.267e−2, 1.272e−2]
WBD	89.5	[1.7249, 1.7998]	>150.7(1)	[1.7286, 2.5605]
PVD	40.9	[5,907.3, 5,995.1]	36.8	[5,885.4, 5,959.5]
SRD	168.6	[2,994.5, 2,999.5]	60.9	[2,994.5, 2,999.9]
SCBD	223.4	[62,792, 64,846]	157.7	[62,791, 64,318]

SCGOSR_S2 are tested. The two algorithms are the same as SCGOSR, except that the two subspaces are used, respectively. Table 8.9 shows the statistical results of the two algorithms. Combining the results of SCGOSR

TABLE 8.10 Ranking of SCGOSR, SCGOSR_S1 and SCGOSR_S2

Ranking	SCGOSR	SCGOSR_S1	SCGOSR_S2
BR	2	1	3
SE	1	3	2
GO	2	3	1
G4	2	3	1
G6	1	2	3
G7	2	1	3
G8	2	3	1
G9	2	1	3
TSD	1	3	2
WBD	2	1	3
PVD	3	2	1
SRD	2	3	1
SCBD	1	3	2
Total ranking	23	29	26

in Table 8.3 with the results in Table 8.9, we give the specific ranking of the three algorithms in Table 8.10. Intuitively, SCGOSR_S2 fails several times on G6 and WBD, but it has the best performance on GO, G4, G8, PVD and SRD. Differently, SCGOSR_S1 uses the fewest function evaluations on BR, G7, G9 and WBD, but performs badly on SRD and SCBD. Relatively speaking, SCGOSR has the most stable performance and its total ranking is the best. Although the combinative utilization of the two subspaces may increase NFE on some problems, it makes the space reduction strategy more robust.

8.4 CHAPTER SUMMARY

In this work, a surrogate-based global optimization algorithm for computationally expensive black-box problems (SCGOSR) is presented. It is worth mentioning that SCGOSR can handle problems with costly objectives and constraints, which frequently appear in actual engineering design. In SCGOSR, Kriging is used to construct surrogate models that will be updated with iterations going on. Besides, a multi-start optimization method is proposed to exploit surrogate models, and newly added samples are selected from predictive local optimal solutions. In order to speed up the search on Kriging, two subspaces are created based on two penalty functions. Among them, *Subspace*1 is the vicinity of the present best solution, and *Subspace*2 is a region that encloses several promising solutions.

Furthermore, two groups of local surrogate models are constructed by the samples in the two subspaces, respectively. On the one hand, local surrogates can improve the local convergence efficiency. On the other hand, local surrogates make SCGOSR spend less time in constructing Kriging models for objective and constraint functions. The proposed multi-start optimization is carried out alternately on *Subspace*1, *Subspace*2 and the overall design space. Once SCGOSR gets trapped in a local optimal region and a proposed local convergence criterion is satisfied, SCGOSR begins to explore the sparsely sampled area.

Finally, through the comparison tests on eight mathematical examples and five engineering applications, SCGOSR shows the powerful capacity in dealing with expensive black-box-constrained optimization problems.

NOTE

1 Based on "SCGOSR: Surrogate-based Constrained Global Optimization using Space Reduction," published in [Applied Soft Computing], [2018]. Permission obtained from [Elsevier].

REFERENCES

Audet, C., Savard, G., & Zghal, W. (2010). A Mesh Adaptive Direct Search Algorithm for Multiobjective Optimization. *European Journal of Operational Research, 204*(3), 545–556. https://doi.org/10.1016/j.ejor.2009.11.010

Bagheri, S., Konen, W., Emmerich, M., & Baeck, T. (2017). Self-Adjusting Parameter Control for Surrogate-Assisted Constrained Optimization under Limited Budgets. *Applied Soft Computing, 61*, 377–393. https://doi.org/10.1016/j.asoc.2017.07.060

Basudhar, A., Dribusch, C., Lacaze, S., & Missoum, S. (2012). Constrained Efficient Global Optimization with Support Vector Machines. *Structural and Multidisciplinary Optimization, 46*(2), 201–221. https://doi.org/10.1007/s00158-011-0745-5

Bjorkman, M., & Holmstrom, K. (2000). Global Optimization of Costly Nonconvex Functions Using Radial Basis Functions. *Optimization and Engineering, 1*(4), 373–397. https://doi.org/10.1023/a:1011584207202

Durantin, C., Marzat, J., & Balesdent, M. (2016). Analysis of Multi-Objective Kriging-Based Methods for Constrained Global Optimization. *Computational Optimization and Applications, 63*(3), 903–926. https://doi.org/10.1007/s10589-015-9789-6

Garg, H. (2014). Solving Structural Engineering Design Optimization Problems Using an Artificial Bee Colony Algorithm. *Journal of Industrial and Management Optimization, 10*(3), 777–794. https://doi.org/10.3934/jimo.2014.10.777

Jones, D. R. (2001). *Direct global optimization algorithm*. Springer US.

Kirkpatrick, S., Gelatt, C. D., Vecchi, & P., M. (1983). Optimization by Simulated Annealing. *Science, 220,* 671–680.

Li, Y., Wu, Y., Zhao, J., & Chen, L. (2017). A Kriging-Based Constrained Global Optimization Algorithm for Expensive Black-Box Functions with Infeasible Initial Points. *Journal of Global Optimization, 67,* 343–366.

Muller, J., & Woodbury, J. D. (2017). GOSAC: Global Optimization with Surrogate Approximation of Constraints. *Journal of Global Optimization, 69*(1), 117–136. https://doi.org/10.1007/s10898-017-0496-y

Ong, Y. S., Nair, P. B., & Keane, A. J. (2003). Evolutionary Optimization of Computationally Expensive Problems via Surrogate Modeling. *AIAA Journal, 41*(4), 687–696.

Parr, J. M., Forrester, A. I. J., Keane, A. J., & Holden, C. M. E. (2012). Enhancing Infill Sampling Criteria for Surrogate-Based Constrained Optimization. *Journal of Computational Methods in Sciences & Engineering, 12*(1–2), 25–45.

Regis, R. G. (2011). Stochastic Radial Basis Function Algorithms for Large-Scale Optimization Involving Expensive Black-Box Objective and Constraint Functions. *Computers & Operations Research, 38*(5), 837–853. https://doi.org/10.1016/j.cor.2010.09.013

Sasena, M. J., Papalambros, P. Y., & Goovaerts, P. (2002). Global optimization of problems with disconnected feasible regions via surrogate modeling. *AIAA-2002-5573, 9th AIAA/ISSMO Symposium on Multidisciplinary Analysis and Optimization*, Atlanta, Georgia, September 4–6, 2002.

Thanedar, P. B., & Vanderplaats, G. N. (1995). Survey of Discrete Variable Optimization for Structural Design. *Journal of Structural Engineering-ASCE, 121*(2), 301–306. https://doi.org/10.1061/(ASCE)0733-9445(1995)121:2(301)

Tolson, B. A., & Shoemaker, C. A. (2007). Dynamically Dimensioned Search Algorithm for Computationally Efficient Watershed Model Calibration. *Water Resources Research, 43*(1), Article W01413. https://doi.org/10.1029/2005wr004723

CHAPTER 9

KTLBO

Kriging-Assisted Teaching–Learning-Based Optimization to Solve Computationally Expensive Constrained Problems[1]

9.1 INTRODUCTION

As high-fidelity simulation techniques are leaping forward and being applied extensively, computationally expensive black-box global optimization has turned out to be one of the most challenging problems in engineering optimization (Dong, Song, Wang, et al., 2018; Li et al., 2020; Ororbia et al., 2020). Overall, the more accurate the simulation analysis, the more computation budget it will bring. Thus, engineers are required to take some time to achieve satisfactory accuracy. Besides, costly black-box constraints may further complicate optimization and impose greater challenges (Bagheri et al., 2017; Li, 2019; Miranda-Varela & Mezura-Montes, 2018; Muller & Woodbury, 2017). Sometimes, feasible solutions are difficult to find in actual simulation-based engineering applications with acceptable computational budgets (Akbari & Kazerooni, 2020; Wu et al., 2018). Specifically, the expensive black-box constrained problems (EBCPs) that the chapter focuses on can be described as follows:

$$\begin{aligned} &\min f(\boldsymbol{x}) \qquad x \in [lb, ub] \\ &s.t.\ C_i(\boldsymbol{x}) \le 0, \quad i = 1, \ldots, m. \end{aligned} \tag{9.1}$$

DOI: 10.1201/9781003636267-9

Where $[lb, ub]$ denotes the search space; $f(\boldsymbol{x})$ represents the objective function; $C_i(\boldsymbol{x})$ denotes the ith inequality constraint; and m is the total number of inequality constraints. It is assumed that both $f(\boldsymbol{x})$ and $C_i(\boldsymbol{x})$ are time-consuming black boxes. If the objective function $f(\boldsymbol{x})$ is calculated at an unknown point $\boldsymbol{x}$, the corresponding constraints $C_i(\boldsymbol{x})$ can be attained concurrently. In other words, $f(\boldsymbol{x})$ and $C_i(\boldsymbol{x})$ are different response values from one simulation model.

Since most of the mentioned time-depending and black-box simulation models are unable to present an explicit mathematical expression, the conventional gradient-based mathematical programming methods become inferior. In the existing literature, swarm intelligence (SI) and evolutionary computation (EC) combined with some constraint-handling techniques are widely employed to address black-box constrained problems (Mezura-Montes & Coello Coello, 2011). Farmani and Wright (2003) presented a self-adaptive fitness formulation for constrained optimization by referencing their previous work (Wright & Farmani, 2001), where a penalty-function method was proposed for genetic algorithm (GA). In the optimized version, constraint violations were represented by a single infeasibility measure function involving a two-stage penalty strategy, which could decrease the dimensionality of the problem and make the method more dynamic and self-adaptive. Daneshyari and Yen (2012) developed a constrained multi-swarm particle swarm optimization (CPSO) method in a cultural framework (cultural CPSO), where numerous concepts from the cultural algorithm were employed to optimize PSO's updating mechanism and swarm-communication capability. In cultural CPSO, objective and constraint violation values are normalized, and a V–F space is established to form a modified fitness formulation for comparisons of particles. Wang and Cai (2012) developed an algorithm combining multiobjective optimization with differential evolution (CMODE) to solve constrained optimization by using their previous Cai-Wang (CW) algorithm. In CMODE (Wang & Cai, 2012), objective and constraint violation functions emerged into a biobjective optimization formulation as an attempt to minimize objective values and degree of constraint violations. Unlike CW (Wang & Cai, 2012), CMODE employed DE as the search engine to decrease the number of tuning parameters and proposed a more efficient replacement mechanism for infeasible solutions. Furthermore, Wang et al. (2016) introduced a novel constrained optimization method: integrating feasibility rules with objective function information (FROFI). A novel replacement mechanism and a mutation strategy have been cooperatively adopted to generate promising offspring and achieve global exploration.

Though SI and EC algorithms (Chen et al., 2018; Kar, 2016; Mavrovouniotis et al., 2017) can effectively solve complex black-box optimization, they are overly determined by the number of function assessments, which is inappropriate for computationally expensive problems. In most cases, one simulation may require several minutes or hours, while thousands of calls to the simulation models will cause unbearable computation costs, thereby enormously extending the design cycle. When the time-to-market requirement is tight, an efficient optimization method that requires fewer calls to the expensive model is indispensable (Liu et al., 2014).

Dong, Li, et al. (2018) developed a multi-surrogate-based global optimization method using a score-based infill criterion (MGOSIC), where Kriging, radial basis function (RBF), and polynomial response surface (PRS) are separately employed to build dynamically updated surrogate models. In addition, a score-based infilling criterion is presented to find the candidate sample sets. The points that can perform better on most of the surrogate models will have higher scores. Furthermore, high-score points farther from the known expensive samples will be first introduced into the expensive sample set. Most of the existing surrogate-based optimization (SBO) methods are developed to solve expensive black-box unconstrained problems, and they cannot directly apply to EBCPs. As proposed by Haftka et al. (2016), "When it comes to adaptive sampling algorithms for constrained optimization, the state of the art is less advanced." Regis (2011) extended his previous work and developed a constrained local metric stochastic RBF (ConstrLMSRBF) method that separately builds RBF models for objective and constraint functions. Among the candidate points, the ones predicted to be feasible are first collected. If none of the candidate points are feasible on surrogate models, the point with the least number of constraint violations will be selected. Though ConstrLMSRBF can effectively process some EBCPs, it requires at least one feasible point as the initial sample to drive the subsequent optimization loop. Sometimes, it is difficult to identify the feasible solutions of an actual EBCP at the beginning. Liu et al. (2017) presented an improved constrained optimization algorithm, termed eDIRECT-C, for EBCPs, where a DIRECT-type (Jones et al., 1993) constraint-handling technique using the Voronoi diagram (Liu et al., 2015) was proposed to separately deal with feasible and infeasible cells. Though eDIRECT-C does not contain user-defined parameters and can effectively explore the unknown feasible area, it requires more running time and thus is not appropriate for large-scale and multi-constraint problems. Dong, Song, Dong, et al. (2018) developed a surrogate-based constrained global optimization method using space reduction (SCGOSR), where a multi-start optimization strategy was proposed to capture the promising points from the local and global spaces of Kriging. SCGOSR outperforms other algorithms on most of the benchmark cases, while it overly relies on Kriging's predicting accuracy. If Kriging has a larger prediction error on some problems, SCGOSR will be mistakenly guided by Kriging and exhibit poor performance. Wang et al. (2019) proposed a global and local surrogate-assisted DE (GLoSADE) algorithm for EBCPs, consisting of two phases. At the global phase, DE acts as the search engine to generate potential samples and the generalized regression neural network is adopted to classify these points, achieving the global exploration; at the local phase, the interior point method coupled with RBF is employed to improve the individuals of the population, eventually accelerating the convergence. Surrogate-assisted evolutionary algorithms (e.g., GLoSADE) (Yu et al., 2019) comply with the stochastically sampling mechanism of metaheuristic algorithms, while they exploit the potential information from surrogates, which have aroused considerable attention recently (Dong et al., 2019).

In this chapter, an efficient surrogate-assisted SI method is developed by exploiting the unique optimization framework of teaching–learning-based optimization (TLBO) (Rao et al., 2011) and Kriging's prediction mechanism. Since TLBO was originally developed to process constrained mechanical design optimization, a novel method based on TLBO is expected to efficiently process computationally expensive inequality-constrained optimization. Since TLBO consists of two phases to generate new points, two Kriging-guided sampling strategies that can effectively balance the local search and global exploration are correspondingly proposed. In the Kriging-assisted teaching phase (KATP), the neighborhoods around the present best solution are sufficiently exploited to accelerate the convergence, and a constrained expectation of improvement (EI) function considering the probability of feasibility is set as a prescreening tool to select the potential individuals from the learners. However, in the Kriging-assisted learning phase (KALP), a constrained mean square error (MSE) function more concerned with Kriging's prediction uncertainty is proposed to select the learners located at the sparsely sampled feasible region for global exploration. Through the joint search of the proposed two phases, the new KTLBO algorithm can efficiently solve EBCPs. For this, KTLBO uses Kriging to construct a dynamically updated surrogate model for the objective and constraint functions and establishes a constraint-optimization–oriented data management strategy for archiving, sorting and updating valuable samples.

9.2 TEACHING–LEARNING-BASED OPTIMIZATION

TLBO first presented by Rao et al. (2011) refers to a phenomenon-inspired method that exploits a population to iteratively search the global optimal solution. Uniquely, TLBO imitates how knowledge spreads in a class (population), where the individuals consist of several learners and one teacher. The teacher possessing the highest-level knowledge can guide the learners to get improved, so the overall knowledge level of this class will ultimately shift to the teacher. Moreover, one learner can be inspired by other learners: if you are better, I can follow you; otherwise, I can try the opposite direction. To sum up, TLBO involves two search phases: teaching and learning. To be more specific, Figure 9.1 illustrates the detailed formulas and algorithm steps.

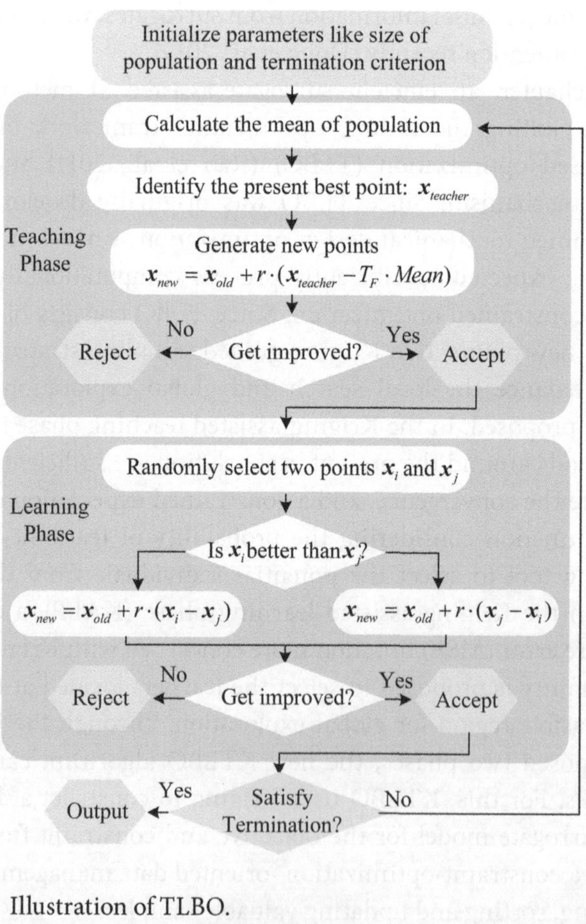

FIGURE 9.1 Illustration of TLBO.

9.3 THE PROPOSED KTLBO

In the presented KTLBO, the design of experiments (DoE) is first employed to yield a group of well-distributed points that should be assessed by real objective and constraint functions. Thereafter, these expensive samples are organized to build surrogate models of objectives and constraints, respectively. In this chapter, a novel sampling method combining metaheuristic search mechanism and the prediction capability of Kriging is presented, to achieve a reasonable balance of global exploration and local exploitation. For each cycle of KTLBO, real data are required to undergo assessment, preprocessing, classifying, surrogate modeling and updating, while the potential candidate points determined from Kriging-assisted teaching and learning phases should go through prescreening and repeatability detecting. After several iterations, the predicting performance of the mentioned Kriging models is gradually enhanced, and more potential points around the true feasible area or global optimum will be captured. Figure 9.2 presents the overall flow of KTLBO, and more details will be explained in the following sections.

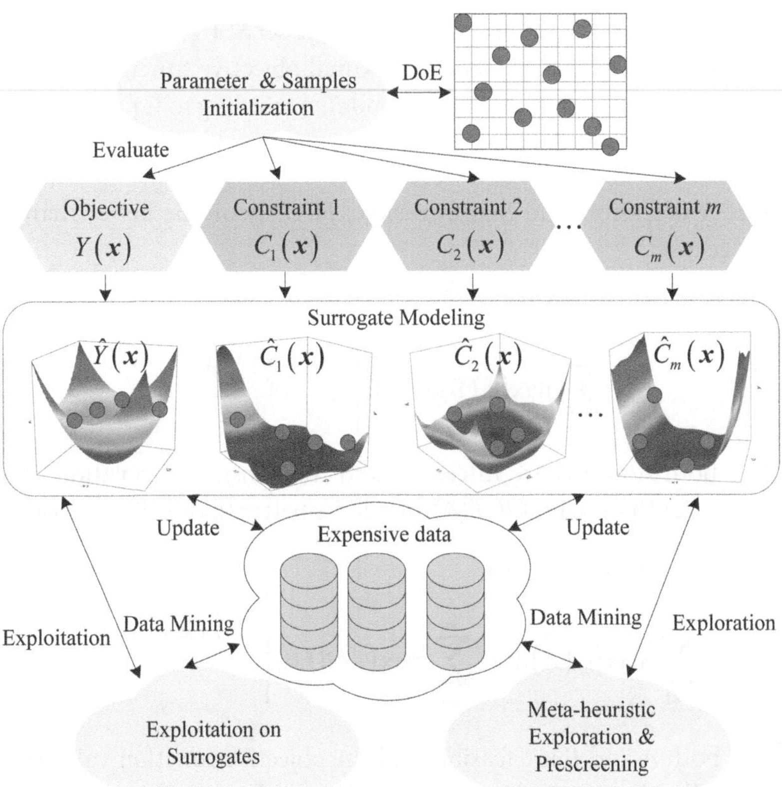

FIGURE 9.2 Data flow of KTLBO.

9.3.1 Initialization of KTLBO

At the initial phase of KTLBO, some basic parameters (e.g., design range, numbers of variables and constraints, population size and number of initial sample points) are initialized and defined, respectively. Next, optimized Latin hypercube sampling (OLHS) is employed to obtain the initial point set $\boldsymbol{S} = \{\boldsymbol{x}^{(1)}, \boldsymbol{x}^{(2)}, \ldots, \boldsymbol{x}^{(N)}\}$ and its corresponding objective and constraints values $\boldsymbol{Y}=\left\{y^{(1)},y^{(2)},\ldots,y^{(N)}\right\}$ and $\boldsymbol{C}=\left\{\boldsymbol{c}^{(1)},\boldsymbol{c}^{(2)},\ldots,\boldsymbol{c}^{(N)}\right\}$, where $\boldsymbol{S}$ and $\boldsymbol{Y}$ denote two vectors, and $\boldsymbol{C}$ is a matrix. To efficiently compare the expensive samples in a constrained problem, a penalty-function method is written:

$$F\left(\boldsymbol{x}^{(j)},\boldsymbol{Y},\boldsymbol{C}\right)=\begin{cases}\max(\boldsymbol{Y})+\sum_{i=1}^{m}\max\left(c_i\left(\boldsymbol{x}^{(j)}\right),0\right), & \text{if } \max\limits_{1\le i\le m}\left(\max\left(c_i\left(\boldsymbol{x}^{(j)}\right),0\right)\right)>0\\ \boldsymbol{Y}\left(\boldsymbol{x}^{(j)}\right), & \text{if } \max\limits_{1\le i\le m}\left(\max\left(c_i\left(\boldsymbol{x}^{(j)}\right),0\right)\right)=0\end{cases}$$
$$\forall j=1,2,\ldots,N \tag{9.2}$$

where $\boldsymbol{x}^{(j)}$ denotes the jth point in the sample set $\boldsymbol{S}$; $\boldsymbol{Y}$ represents the objective values set; and max($\boldsymbol{Y}$) is the maximal objective function value. It is assumed that there are two points A and B. According to Eq. (9.2), it is easy to draw three conclusions.

1. If A is feasible and B is infeasible, $F(A)$ should be better than $F(B)$ because:

$$\left.\begin{array}{l}\boldsymbol{Y}(A)\le\max(\boldsymbol{Y})\\ \sum_{i=1}^{m}\max\left(c_i(B),0\right)>0\end{array}\right\}\Rightarrow F(A)<F(B) \tag{9.3}$$

2. If both A and B are infeasible and the constraint violation of A is smaller than that of B, $F(A)$ should be better than $F(B)$ because:

$$\left.\begin{array}{l}\max(\boldsymbol{Y})=\max(\boldsymbol{Y})\\ \sum_{i=1}^{m}\max\left(c_i(A),0\right)<\sum_{i=1}^{m}\max\left(c_i(B),0\right)\end{array}\right\}\Rightarrow F(A)<F(B) \tag{9.4}$$

3. If both A and B are feasible and the objective function value of A is smaller than that of B, $F(A)$ should outperform $F(B)$ because:

$$\boldsymbol{Y}(A)<\boldsymbol{Y}(B)\Rightarrow F(A)<F(B) \tag{9.5}$$

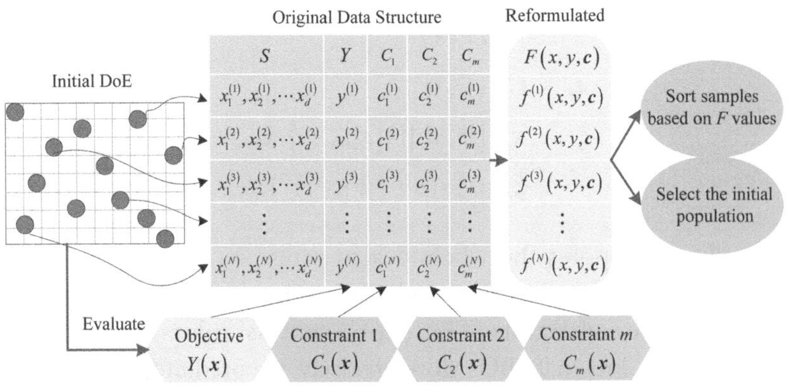

FIGURE 9.3 Data flow of initial phase.

Obviously, if both A and B are infeasible, Eq. (9.2) considers more about how seriously they violate the constraints, which will promote the algorithm to find feasible solutions efficiently. Besides, the penalty function F is employed to classify all samples in $\boldsymbol{S}$, $\boldsymbol{Y}$ and $\boldsymbol{C}$, and select the promising individuals as the population members ***Pop***. Figure 9.3 illustrates the data structure and flow of the initial phase, underpinning the subsequent sampling loop. Moreover, since the initial samples are extensively distributed over the whole design space, the first population ***Pop*** has a better space-filling performance. With the loop continuing and more promising samples added, the ***Pop*** in teaching phase will concentrate on the present best solution to accelerate convergence, and the ***Pop*** in learning phase may continuously exhibit a wide distribution to achieve global exploration. More details can be found in the following sections.

9.3.2 Kriging-Assisted Teaching Phase

The optimization loop includes two phases: one is the KATP that sufficiently exploits the local area around the present best solution; the other one is KALP that can effectively search the sparsely sampled area. In KATP, the predicted local optimal solution $\boldsymbol{x}_{plo}$ should be first captured in a local area enclosing the present best solution $\boldsymbol{x}_{best}$. Since the Kriging models for objective and constraint functions have been built, TLBO directly acts as an optimizer to search the surrogate models. Equation (9.6) expresses the pure exploitation of surrogates.

$$\begin{aligned} & \min \hat{Y}(\boldsymbol{x}) \\ s.t. \quad & \hat{C}_i(\boldsymbol{x}) \le 0 \\ & \boldsymbol{x} \in [lb, ub] \cap [\boldsymbol{x}_{best} - \xi, \boldsymbol{x}_{best} + \xi] \\ & \xi = 0.1 \cdot (ub - lb), \quad i = 1, 2, \ldots, m \end{aligned} \tag{9.6}$$

where $[lb, ub]$ denotes the whole design range; $\hat{Y}(\boldsymbol{x})$ and $\hat{C}_i(\boldsymbol{x})$ are the Kriging models of objective and constraint functions, respectively. Considering the constraints of Eq. (9.6), a contrast rule is adopted to compare any two points in TLBO. More precisely, if a predicted point is better, it is assumed as feasible or at least has a lower constraint violation value. Equation (9.7) accounts for the details about the contrast rule, and it will be used for selecting teachers and smarter learners in TLBO.

$$\begin{cases} p_i \prec p_j & \text{if } \hat{v}(p_i) = 0 \wedge \hat{v}(p_j) > 0 \\ p_i \prec p_j & \text{if } \hat{v}(p_i) = \hat{v}(p_j) = 0 \wedge \hat{Y}(p_i) < \hat{Y}(p_j) \\ p_i \prec p_j & \text{if } \hat{v}(p_i) > 0 \wedge \hat{v}(p_j) > \hat{v}(p_i) \end{cases}$$

$$i \neq j, \quad \forall i, \quad j = 1, 2, \ldots, P \tag{9.7}$$

$$\hat{v}(x) = \sum_{i=1}^{m} \max\left(\hat{C}_i(x), 0\right)^2$$

where p_i and p_j denote two predicted points in one population; $\hat{v}$ represents the constraint violation; P is the population size in TLBO; $\hat{Y}$ is the predicted objective function; and $\hat{C}_i$ is the ith predicted constraint. After considerable generations, the predicted best point can be found.

On the other hand, the current ***Pop*** whose individuals originate from the expensive samples ***S Y C*** begins to generate the new individuals. Figure 9.4 presents the data structure and flow of KATP. In each cycle, ***Pop*** will generate M groups of new positions based on the metaheuristic teaching mechanism and archive these newcomers into a candidate sample pool, from which a proposed prescreening strategy is used to select the most promising points that keep balanced exploitation and exploration.

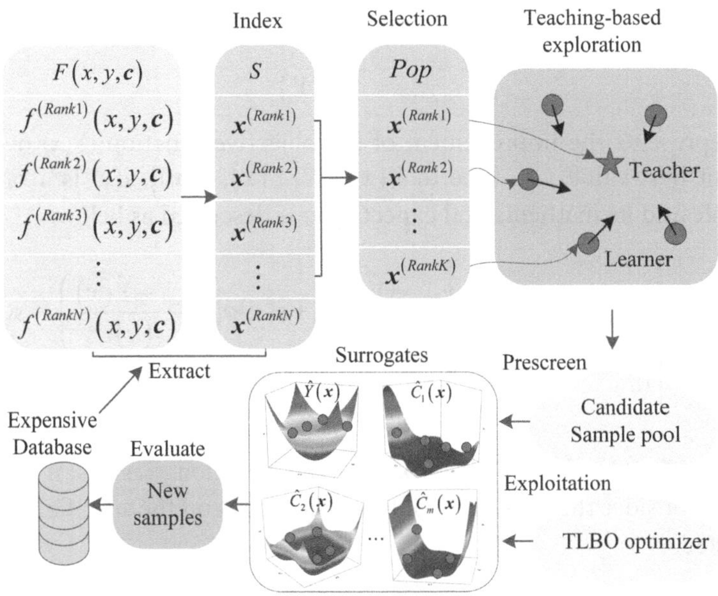

FIGURE 9.4 Data flow of Kriging-assisted teaching phase.

As Figure 9.4 shows, $Pop=\{\boldsymbol{x}^{(Rank1)},\boldsymbol{x}^{(Rank1)},\ldots,\boldsymbol{x}^{(RankK)}\}$ is classified by Eq. (9.2) and then selected from S. Based on the TLBO's search mechanism, the new positions are estimated by the following equations.

$$\boldsymbol{x}_i^j=\boldsymbol{x}^{(Ranki)}+r^j\left(\boldsymbol{x}^{(Rank1)}-T_F{}^j\cdot Q\right)$$
$$Q=\frac{1}{K}\cdot\sum_{i=1}^{K}\boldsymbol{x}^{(Ranki)} \tag{9.8}$$
$$i=1,2,\ldots,K,\quad j=1,2,\ldots,M$$

where $\boldsymbol{x}_i$ denotes the new position generated by $\boldsymbol{x}^{(Ranki)}$; j denotes the jth group; T_F is a random integer between {1, 2}; r is a random number in the range [0, 1]; K refers to the size of ***Pop***; M is the number of groups. Lastly, $K\times M$ new points are archived into a temporary sample pool for prescreening.

Since the EI strategy can identify potential points that balance Kriging's prediction values and space-filling performance, KTLBO utilizes the EI function to select promising points from the sample pool. To be more specific, the EI equations are written as:

$$I(\boldsymbol{x}) = \max\left(y_{best} - Y(\boldsymbol{x}), 0\right)$$
$$y_{best} = \min(\boldsymbol{Y}) \tag{9.9}$$

$I(\boldsymbol{x})$ represents the improvement of the objective function; $Y(\boldsymbol{x})$ over the current best value y_{best}. Due to $Y(\boldsymbol{x}) \sim N\left(\hat{Y}(\boldsymbol{x}), s^2(\boldsymbol{x})\right)$, $I(\boldsymbol{x})$ is a random variable and its mathematical expectation is described as below:

$$E\left[I(\boldsymbol{x})\right] = \begin{cases} \left(y_{best} - \hat{Y}(\boldsymbol{x})\right)\Phi\left(\dfrac{y_{best} - \hat{Y}(\boldsymbol{x})}{s(\boldsymbol{x})}\right) + \hat{s}(\boldsymbol{x})\phi\left(\dfrac{y_{best} - \hat{Y}(\boldsymbol{x})}{s(\boldsymbol{x})}\right), & s(\boldsymbol{x}) \neq 0 \\ 0, & s(\boldsymbol{x}) = 0 \end{cases} \tag{9.10}$$

Equation (9.10) refers to the EI of objective function. Besides, it is necessary to consider the possibility of feasibility at $\boldsymbol{x}$, and the specific formulas are listed.

$$PF = P\left[C \leq 0\right] = P\left[\frac{C - \hat{C}(\boldsymbol{x})}{s_c(\boldsymbol{x})} \leq -\frac{\hat{C}(\boldsymbol{x})}{s_c(\boldsymbol{x})}\right] = \Phi\left(-\frac{\hat{C}(\boldsymbol{x})}{s_c(\boldsymbol{x})}\right) \tag{9.11}$$

where $C(\boldsymbol{x})$ also complies with the normal distribution $N\left(\hat{C}(\boldsymbol{x}), s_c^2(\boldsymbol{x})\right)$. To improve the readability, Figure 9.5 illustrates the EI strategy. For a problem with m constraints, the final EI expression can be formulated as:

$$E_c\left[I(\boldsymbol{x})\right] = E\left[I(\boldsymbol{x})\right] \times \prod_{i=1}^{m} P_i\left[c_i \leq 0\right] \tag{9.12}$$

It is clear that Eq. (9.12) considers the potential contribution of a new point to the objective function, as well as its feasibility. Thus, Eq. (9.12) is regarded as a sorting criterion to find the maximal EI value of each group.

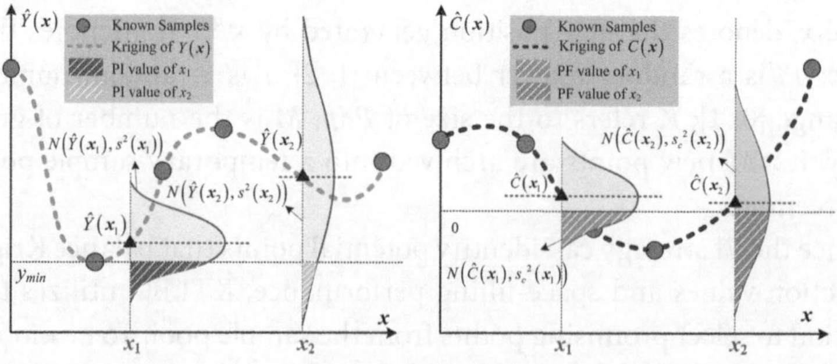

FIGURE 9.5 Illustration of teaching-based prescreening theory.

$$\boldsymbol{x}_i^* = \underset{\boldsymbol{x} \in \boldsymbol{x}_i}{\operatorname{argmax}}\left(E_c\left[I(\boldsymbol{x})\right]\right)$$

$$\boldsymbol{x}_i = \left\{\boldsymbol{x}_i^1, \boldsymbol{x}_i^2, \ldots, \boldsymbol{x}_i^M\right\} \tag{9.13}$$

$$i = 1, 2, \ldots, K$$

Lastly, a group of new points $\{\boldsymbol{x}_1^*, \boldsymbol{x}_2^*, \ldots, \boldsymbol{x}_K^*\}$ that balance the exploration of unexplored areas and exploitation of Kriging are attained. Furthermore, the predicted best solution $\boldsymbol{x}_{plo}$ and these selected new points will be assessed and saved into the expensive sample set for the next cycle. The detailed pseudo code of KATP is provided in Algorithm 9.1.

Algorithm 9.1 Kriging-assisted Teaching Phase

Input: Sample sets **S, Y, C, F**; The number of constraints m; the number of design variables d; the population ***Pop***; the ***Pop***'s size K; the number of sampling groups M;

Output: Updated sample sets **S, Y, C, F**

```
Begin
  KRG ← { KRG_obj, KRG_c1, KRG_cm } /* Build Kriging models based on
    (S, Y), (S, C)*/;
  x_plo ← Get the predicted best solution based on Eqs. (9.11) and
    (9.12) using TLBO;
  Flag ← Check the repeatability to the samples set S /* Use K-nearest
    Neighbors */;
  If Flag = True /* True implies that x_pbest is not repeated*/
    y_pbest, c_pbest ← Calculate the objective and constraint function val-
      ues of x_pbest;
    S, Y, C ← S∪x_pbest, Y∪y_pbest, C∪c_pbest;
  End If
  Teacher ← Identify the most promising sample from S by Eq. (9.7);
  Q ← Evaluate mean positions of Pop;
  For i from 1 to K
    For j from 1 to M
      x_i^j ← Get the new point by Eq. (9.13)
    End For
```

(15) $x_i^* \leftarrow$ Select the most promising individual by Eqs. (9.14)–(9.18)
(16) $Flag \leftarrow$ Check the repeatability to the samples set $\boldsymbol{S}$ /* K-nearest Neighbors */;
(17) **If** $Flag$ = True /* True implies that x_i^* is not repeated*/
(18) $y_i^*, c_i^* \leftarrow$ Calculate the objective and constraint function values of x_i^*;
(19) $\boldsymbol{S}, \boldsymbol{Y}, \boldsymbol{C} \leftarrow \boldsymbol{S} \cup x_i^*, \boldsymbol{Y} \cup y_i^*, \boldsymbol{C} \cup c_i^*$;
(20) **End If**
(21) **End For**
(22) $\boldsymbol{F} \leftarrow$ Update the penalty function values set based on Eq. (9.7).
(23) **Return** Updated sample sets $\boldsymbol{S}, \boldsymbol{Y}, \boldsymbol{C}, \boldsymbol{F}$
(24) End

9.3.3 Kriging-Assisted Learning Phase

In KALP, TLBO is first employed to get the predicted global optimal solution $\boldsymbol{x}_{pgo}$ from Kriging models, where the search range has been changed to global design space [lb, ub]. Besides, in KALP, the manner to form the current population $\boldsymbol{Pop}$ is also inconsistent with KATP. Figure 9.6 illustrates the corresponding data flow. The point with the best F value is first chosen, and then $K - 1$ points are randomly selected from the remaining $N - 1$ points. This selection manner makes the samples in $\boldsymbol{Pop}$ more diverse and distribute more extensively, which promotes the search for unknown areas. According to Figure 9.6, $\boldsymbol{Pop}$ will generate M groups of new points by following TLBO's learning mechanism.

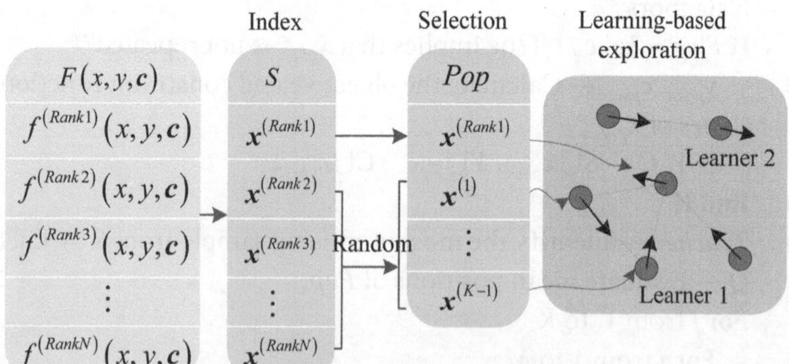

FIGURE 9.6 Data flow of Kriging-assisted learning phase.

$$\begin{cases} \boldsymbol{x}_t^j = \boldsymbol{x}^{(t)} + \boldsymbol{r}^j\left(\boldsymbol{x}^{(t)} - \boldsymbol{x}^{(s)}\right), & \mathit{if}\ \boldsymbol{x}^{(t)} \prec \boldsymbol{x}^{(s)} \\ \boldsymbol{x}_t^j = \boldsymbol{x}^{(t)} + \boldsymbol{r}^j\left(\boldsymbol{x}^{(s)} - \boldsymbol{x}^{(t)}\right), & \mathit{if}\ \boldsymbol{x}^{(s)} \prec \boldsymbol{x}^{(t)} \end{cases}$$
$$\boldsymbol{x}^{(t)}, \boldsymbol{x}^{(s)} \in Pop = \left\{\boldsymbol{x}^{(1)}, \boldsymbol{x}^{(2)}, \ldots, \boldsymbol{x}^{(K)}\right\} \tag{9.14}$$
$$j = 1, 2, \ldots, M$$

where r denotes a random number in the range [0, 1]; K is the size of ***Pop***; M represents the number of groups. Likewise, $K \times M$ new points are generated, which are saved into a temporary sample pool for prescreening.

As discussed above, the estimated MSE $s^2(\boldsymbol{x})$ can indicate the sample density of the design space. A point with larger MSE value implies that it is located in a sparsely sampled area. Figure 9.7 gives a more intuitive explanation, where the MSE values of generated points are 0, whereas $\boldsymbol{x}_1$ and $\boldsymbol{x}_2$ are relatively larger. In fact, points with larger MSE values should be added to enhance the global exploration capability.

For constrained problems, the feasibility of points should be considered. Accordingly, a prescreening method combining MSE and possibility of feasibility expressed in Eq. (9.11) is proposed:

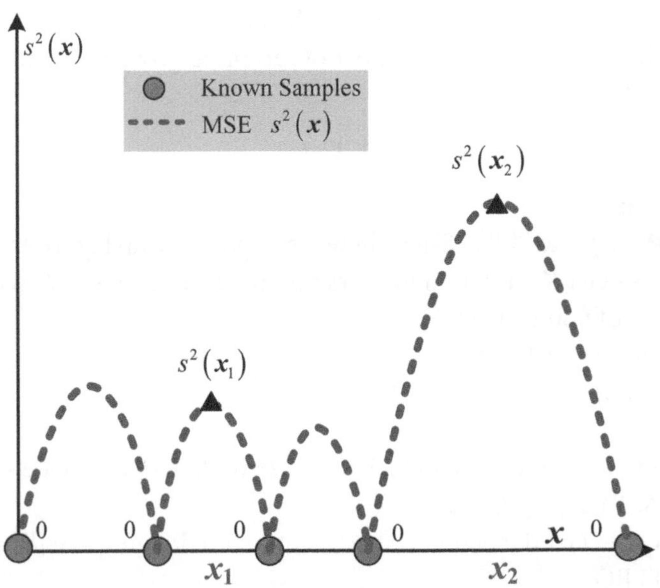

FIGURE 9.7 Illustration of learning-based prescreening theory.

$$SP_c(\boldsymbol{x}) = s(\boldsymbol{x}) \times \prod_{i=1}^{m} P_i[c_i \le 0] \tag{9.15}$$

Equation (9.15) reveals that a point x with a larger $s(x)$ value and higher probability of feasibility will be more attractive. Thus, Eq. (9.15) is regarded as a sorting criterion to determine the maximal SP_c value of each group.

$$\begin{gathered} \boldsymbol{x}_t^* = \underset{x \in \boldsymbol{x}_i}{\operatorname{argmax}}\left(SP_c(\boldsymbol{x})\right) \\ \boldsymbol{x}_t = \left\{\boldsymbol{x}_t^1, \boldsymbol{x}_t^2, \ldots, \boldsymbol{x}_t^M\right\} \\ t = 1, 2, \ldots, K \end{gathered} \tag{9.16}$$

Similar to KATP, a group of new points $\{\boldsymbol{x}_1^*, \boldsymbol{x}_2^*, \ldots, \boldsymbol{x}_K^*\}$ located in the sparsely sampled area is generated. The mentioned new points and the predicted best solution $\boldsymbol{x}_{pgo}$ from the current Kriging models will be saved into the expensive sample set. Specifically, the pseudo-code regarding KALP is summarized in Algorithm 9.2.

Algorithm 9.2 Kriging-assisted Learning Phase

Input: The number of constraints m; the number of design variables d; the ***Pop***'s size K; the number of sampling groups M; sample sets **S**, ***Y***, ***C***, ***F***

Output: Updated sample sets **S**, ***Y***, ***C***, ***F***

(01) **Begin**
(02) $\boldsymbol{Pop}(1) \leftarrow \boldsymbol{x}^{(Rank1)}$ /* Select the best sample from ***S*** as Figure 9.6 shows*/
(03) $T \leftarrow$ Get K-1 random integers ranging from 2 to N /* N is the number of points in **S***/
(04) **For** i from 1 to K-1
(05) $\boldsymbol{Pop}(i) \leftarrow \boldsymbol{x}^{(RankT(i))}$
(06) **End For**
(07) **KRG** $\leftarrow$ { KRG_{obj}, KRG_{c1}, KRG_{cm} } /* Build Kriging models based on (**S**, ***Y***), (**S**, ***C***)*/;
(08) $\boldsymbol{x}_{pgo} \leftarrow$ Get the predicted best solution on design space [lb, ub] using TLBO;
(09) *Flag* $\leftarrow$ Check the repeatability to the samples set **S** /* K-nearest neighbors */;

If *Flag* = True /* True implies that $\boldsymbol{x}_{pbest}$ is not repeated*/
    $\boldsymbol{y}_{pbest}, \boldsymbol{c}_{pbest} \leftarrow$ Calculate the objective and constraint function values of $\boldsymbol{x}_{pbest}$;
    $\boldsymbol{S}, \boldsymbol{Y}, \boldsymbol{C} \leftarrow \boldsymbol{S} \cup \boldsymbol{x}_{pbest}, \boldsymbol{Y} \cup \boldsymbol{y}_{pbest}, \boldsymbol{C} \cup \boldsymbol{c}_{pbest}$;
End If
For t from 1 to K
    $s \leftarrow$ Identify an index $s \in \{1,2,\cdots,K\}, s \neq t$;
    For j from 1 to M
        $\boldsymbol{x}_t^j$ Get the new point by Eq. (9.14)
    End For
    $\leftarrow \boldsymbol{x}_t^*$ Select the most promising individual by Eqs. (9.15) and (9.16)
    Flag $\leftarrow$ Check the repeatability to the samples set $\boldsymbol{S}$ /* K-nearest neighbors */;
    If *Flag* = True /* True implies that $\leftarrow$ is not repeated*/
        $\boldsymbol{y}_i^* \ \boldsymbol{c}_i^* \leftarrow$ Calculate the objective and constraint function values of $\boldsymbol{x}_t^*$;
        $\boldsymbol{S}, \boldsymbol{Y}, \boldsymbol{C} \leftarrow \boldsymbol{S} \cup \boldsymbol{x}_t^*, \boldsymbol{Y} \cup \boldsymbol{y}_t^*, \boldsymbol{C} \cup \boldsymbol{c}_t^*$;
    End If
End For
$\boldsymbol{F} \leftarrow$ Update the penalty function values set based on Eq. (7)
Return Updated sample sets **S**, **Y**, **C**, **F**
End

9.3.4 Overall Optimization Framework of KTLBO

To clearly demonstrate the whole optimization flow, an illustration with specific algorithm steps is given in Figure 9.8. Three areas displaying different colors separately represent initial phase, KATP and KALP, and the logic of the three phases is clearly presented. After the initial phase, KATP and KALP are conducted alternately to realize efficient global optimization. It is clear that KTLBO will continue to work until reaching the maximum allowable number of function evaluations.

9.4 COMPARISON EXPERIMENTS

In this chapter, KTLBO is compared with six well-known and recently published algorithms MSSR (refer to Chapter 4) (Dong et al., 2016), SCGOSR (Dong, Song, Dong, et al., 2018), ConstrLMSRBF (Regis, 2011), TLBO (Rao et al., 2011), CMODE (Wang & Cai, 2012) and FROFI (Wang et al., 2016). Specifically, MSSR, SCGOSR and ConstrLMSRBF refer to SBO algorithms

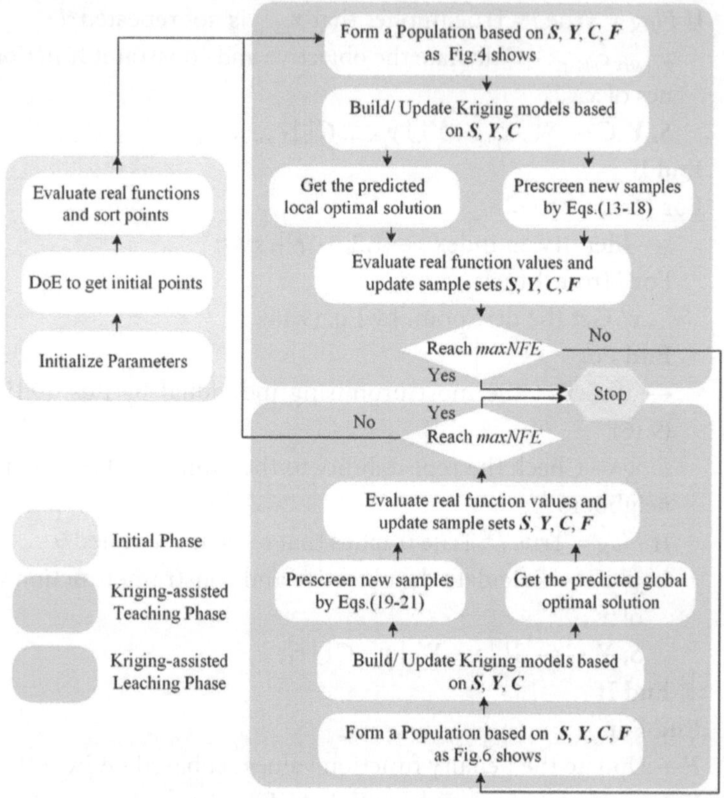

FIGURE 9.8 Overall optimization flow of KTLBO.

and have proved handling black-box optimization problems with costly objectives and constraints on various mathematical cases. In contrast, TLBO, CMODE and FROFI act as three efficient metaheuristic algorithms for constrained optimization and have shown superior performance on black-box constrained optimization problems. To verify KTLBO's capability, 18 benchmark cases exhibiting a range of characteristics are collected as test cases, whose specific information is listed in Table 9.1.

In the 18 benchmark cases, there are 15 extensively used mathematical cases, consisting of 13 CEC2006 cases (Yang et al., 2020), two famous multimodal cases GO and SE (Dong, Song, Dong, et al., 2018), as well as three classical engineering applications TSD, SRD and SCBD (Dong, Song, Dong, et al., 2018). Their design dimension *dim* ranges from 2 to 20, and the number of constraints (*Noc*) falls in the range (1–38). Moreover, *LI* denotes "linear inequality constraint" and *NI* represents "Nonlinear inequality constraint."

TABLE 9.1 Specific Characteristics of 18 Test Cases

Category	Func.	*dim*	*Noc*	*LI*	*NI*	Known Best Value	Type of Obj.
Mathematical cases (13 CEC2006 cases and two widely used cases)	g01	13	9	9	0	−15.0000	Quadratic
	g02	20	2	0	2	−0.8036	Nonlinear
	g04	5	6	0	6	−30,665.5387	Quadratic
	g06	2	2	0	2	−6,961.8139	Cubic
	g07	10	8	3	5	24.3062	Quadratic
	g08	2	2	0	2	−0.0958	Nonlinear
	g09	7	4	0	4	680.6301	Polynomial
	g10	8	6	3	3	7,049.2480	Linear
	g12	3	1	0	1	−1.0000	Quadratic
	g16	5	38	4	34	−1.9052	Nonlinear
	g18	9	13	0	13	−0.8660	Quadratic
	g19	15	5	0	5	32.6556	Nonlinear
	g24	2	2	0	2	−5.5080	Linear
	GO	2	1	0	1	−0.9711	Polynomial
	SE	2	1	0	1	−1.1743	Nonlinear
Engineering application cases	TSD	3	4	1	3	0.01267	Polynomial
	SRD	7	11	4	7	2,994.4711	Polynomial
	SCBD	10	11	5	6	62,791	Polynomial

Since SBO algorithms are generally use fewer function evaluations (FEs) to yield satisfactory solutions, and metaheuristic algorithms require more FEs, two groups of experiments are set. In the first experiment, KTLBO is compared with SCGOSR, MSSR and ConstrLMSRBF, and the maximal number of function evaluations (NFE) is set to 200. In the second, the maximal NFE (maxNFE) is defined as 500, and KTLBO is compared with TLBO, CMODE and FROFI. For all the parameters of SCGOSR, MSSR and ConstrLMSRBF, their default values [42, 25, 38] are used for test, whereas their maximal NFE is defined as 200. For CMODE, FROFI and TLBO, the maxNFE reaches 500, the population size is defined as 10, and all the other parameters remain at their default values [16, 19, 46]. For KTLBO, the size of Pop K is 3, the number of sampling groups M is 10, and the number of DoE samples reaches $2d + 1$, where d denotes the number of dimensions. Besides, KTLBO adopts OLHS [48] to yield its initial DoE samples.

Tables 9.2 and 9.3 list the statistical results on the 13 CEC2006 cases, where SR denotes the successful ratio to find the feasible solutions after the maximal NFE, W-t refers to the Wilcoxon rank sum test, and all the best results are marked in bold. Intuitively, KTLBO can find feasible solutions in all these cases, since its SR is always 100%. SCGOSR exhibits unstable

TABLE 9.2 Statistical Results on CEC2006 Cases (NFE = 200)—Part 1

Problem	Criteria	KTLBO	SCGOSR	MSSR	ConstrLMSRBF
g01	Best	**−15.000**	−14.959	−3.000	−13.596
	Median	**−15.000**	−11.944	−2.085	−9.439
	Worst	**−15.000**	−7.828	−1.169	−3.725
	Mean	**−15.000**	−11.687	−2.085	−9.326
	Std	**0.000**	1.880	1.294	3.148
	SR	**100%**	**100%**	10%	**100%**
	W-t		(+)	(+)	(+)
g02	Best	−0.398	−0.258	−0.335	**−0.400**
	Median	−0.273	−0.203	−0.180	**−0.286**
	Worst	**−0.159**	−0.155	−0.151	−0.142
	Mean	−0.273	−0.209	−0.185	**−0.288**
	Std	0.062	0.062	**0.038**	0.068
	SR	**100%**	**100%**	**100%**	**100%**
	W-t		(+)	(≈)	(+)
g04	Best	**−30,665.539**	**−30,665.539**	−30,665.537	—
	Median	**−30,665.539**	−30,665.520	−30,663.910	—
	Worst	**−30,665.538**	−30,562.619	−30,617.768	—
	Mean	**−30,665.539**	−30,658.768	−30,659.544	—
	Std	**0.000**	23.235	11.123	—
	SR	**100%**	**100%**	**100%**	0%
	W-t		(+)	(+)	(+)
g06	Best	−6,961.803	**−6,961.814**	−6,961.776	—
	Median	−6,961.784	**−6,961.804**	−6,957.937	—
	Worst	**−6,961.762**	−6,961.730	−6,952.356	—
	Mean	−6,961.784	**−6,961.795**	−6,958.198	—
	Std	**0.014**	0.023	2.922	—
	SR	**100%**	**100%**	**100%**	0%
	W-t		(−)	(+)	(+)
g07	Best	24.376	**24.309**	31.539	32.402
	Median	24.419	**24.405**	112.481	38.662
	Worst	**24.507**	30.139	217.320	42.044
	Mean	**24.436**	26.060	126.714	38.598
	Std	**0.044**	2.372	66.111	2.389
	SR	**100%**	**100%**	50%	**100%**
	W-t		(≈)	(+)	(+)
g08	Best	**−0.096**	**−0.096**	**−0.096**	**−0.096**

(Continued)

TABLE 9.2 (*Continued*) Statistical Results on CEC2006 Cases (NFE = 200)—Part 1

Problem	Criteria	KTLBO	SCGOSR	MSSR	ConstrLMSRBF
	Median	**−0.096**	**−0.096**	**−0.096**	−0.094
	Worst	−0.090	**−0.096**	**−0.096**	−0.088
	Mean	−0.095	**−0.096**	**−0.096**	−0.093
	Std	0.002	**0.000**	**0.000**	0.002
	SR	**100%**	**100%**	**100%**	**100%**
	W-t		(−)	(+)	(+)
g09	Best	**682.635**	683.524	830.918	736.743
	Median	736.662	**703.344**	1,313.413	908.523
	Worst	891.725	**818.397**	1,903.922	1,183.825
	Mean	744.492	**714.327**	1,309.897	923.902
	Std	52.052	**34.229**	297.661	122.212
	SR	**100%**	**100%**	**100%**	**100%**
	W-t		(−)	(+)	(+)

TABLE 9.3 Statistical Results on CEC2006 Cases (NFE = 200)—Part 2

Problem	Criteria	KTLBO	SCGOSR	MSSR	ConstrLMSRBF
g10	Best	7,051.015	7,176.968	**7,050.922**	—
	Median	7,061.990	11,125.224	**7,051.715**	—
	Worst	7,108.000	14,567.211	**7,052.509**	—
	Mean	7,064.030	10,667.396	**7,051.715**	—
	Std	12.679	3,106.373	**1.123**	—
	SR	**100%**	25%	10%	0%
	W-t		(+)	(+)	(+)
g12	Best	**−1.000**	**−1.000**	**−1.000**	**−1.000**
	Median	**−1.000**	−0.997	−0.965	**−1.000**
	Worst	**−1.000**	−0.924	−0.822	−0.960
	Mean	**−1.000**	−0.991	−0.944	−0.994
	Std	**0.000**	0.017	0.058	0.011
	SR	**100%**	**100%**	**100%**	**100%**
	W-t		(+)	(+)	(+)
g16	Best	**−1.905**	**−1.905**	**−1.905**	—
	Median	**−1.905**	**−1.905**	**−1.905**	—
	Worst	−1.459	**−1.820**	−1.650	—
	Mean	−1.813	**−1.895**	−1.860	—
	Std	0.156	**0.024**	0.075	—

(*Continued*)

TABLE 9.3 (*Continued*) Statistical Results on CEC2006 Cases (NFE = 200)—Part 2

Problem	Criteria	KTLBO	SCGOSR	MSSR	ConstrLMSRBF
	SR	**100%**	**100%**	**100%**	0%
	W-t		(≈)	(≈)	(+)
g18	Best	**−0.866**	**−0.866**	−0.859	−0.447
	Median	**−0.866**	−0.608	−0.616	−0.355
	Worst	**−0.864**	−0.209	−0.239	−0.217
	Mean	**−0.865**	−0.584	−0.603	−0.343
	Std	**0.000**	0.212	0.174	0.064
	SR	**100%**	90%	75%	95%
	W-t		(+)	(+)	(+)
g19	Best	**37.951**	297.193	301.434	232.529
	Median	**44.020**	518.120	722.746	490.591
	Worst	**73.471**	986.840	1143.817	749.958
	Mean	**45.731**	592.086	710.173	514.584
	Std	**7.596**	212.263	214.210	152.950
	SR	**100%**	**100%**	**100%**	**100%**
	W-t		(+)	(+)	(+)
g24	Best	**−5.508**	**−5.508**	**−5.508**	−4.054
	Median	**−5.508**	**−5.508**	−5.507	−4.053
	Worst	**−5.508**	−5.507	−5.452	−4.049
	Mean	**−5.508**	**−5.508**	−5.499	−4.053
	Std	**0.000**	0.000	0.017	0.001
	SR	**100%**	**100%**	**100%**	**100%**
	W-t		(+)	(+)	(+)

performance on g10 and g18, since it may fail to find feasible solutions during 20 runs. It is easy to observe that MSSR has difficulties on g01, g07, g10 and g18. For instance, MSSR can only succeed twice on g01 during the 20 runs. Compared with others, ConstrLMSRBF is a special algorithm, because it requires at least one feasible solution in the initial sample set to drive the subsequent loop. Hence, a feasible solution of KTLBO is substituted into the initial samples of ConstrLMSRBF, to make it work. However, ConstrLMSRBF exhibits the worst performance on g04, g06, g10, g16 and g18.

In addition, KTLBO can get solutions closer to the true global optima in most cases. Intuitively, KTLBO stays ahead on g01, g04, g07, g10, g12, g18, g19 and g24, while SCGOSR gets first ranks on g06, g08, g09 and g16.

For g06 and g08, SCGOSR outperforms KTLBO, whereas their results are significantly close. MSSR exhibits acceptable performance, and it can approach the true global optima in most cases. However, compared with KTLBO and SCGOSR, MSSR exhibits relatively weaker convergence ability. For instance, the values of MSSR on g04, g06, g07, g09, g18 and g19 are obviously lower than those of SCGOSR and KTLBO. Among the four algorithms, ConstrLMSRBF is indicated to encounter more difficulties in these cases. Tables 9.2 and 9.3 clearly show that ConstrLMSRBF can hardly achieve convergence during 200 FEs or even find feasible solutions in some cases. Though ConstrLMSRBF exhibits unstable performance, it sometimes achieves higher efficiency than MSSR. For instance, it can find more effective mean and median results on g01, g02, g07, g09, g12 and g19 than MSSR. For g02, ConstrLMSRBF outperforms the other three algorithms for the RBF's superior ability to solve high-dimensional problems. In summary, among the four algorithms, KTLBO has more significant advantages in the 13 CEC2006 cases. Table 9.4 lists the comparison results of the four algorithms in the low-dimensional cases and engineering applications. KTLBO still performs efficiently and stably. KTLBO can reach the true global optima of SRD and SE for all the runs. Besides, it can easily approach the true global optima of SCBD, TSD and GO. In contrast, ConstrLMSRBF exhibits worse performance in the five benchmark cases. SCGOSR and MSSR achieve similar performance, while SCGOSR is indicated to be more robust. In summary, Tables 9.2–9.4 draw the same conclusion that KTLBO solves computationally expensive and black-box-constrained optimization problems efficiently. Figure 9.9 illustrates the iterative results of KTLBO, which can reflect KTLBO's average performance during the 20 runs. Figure 9.9 plots the KTLBO history data generated during a completed search. For g01, g04, g07, g09, g12, g16, g18, g19, g24, SCBD, SE and SRD, clearer figures are also added. Intuitively, most of these figures show that the sample values generated by DoE fluctuate more significantly, while the points generated by the iterative process mainly focus on the feasible or global optimal area. For instance, no feasible samples are found on G7, G8 and G10 at first, whereas many pink feasible points are captured with iteration continuing. Besides, as impacted by KTLBO's global exploration mechanism, the algorithm may still have some opportunities to search the unknown infeasible area. As indicated by many cases in Figure 9.9, though the global optimal area has been identified, KTLBO still samples some infeasible points far away from the present best point.

TABLE 9.4 Statistical Results on GO, SE and Engineering Cases (NFE = 200)

Problem	Criteria	KTLBO	SCGOSR	MSSR	ConstrLMSRBF
SRD	Best	**2,994.471**	**2,994.471**	2,994.473	—
	Median	**2,994.471**	2,994.536	2,996.901	—
	Worst	**2,994.471**	3,009.420	3,019.273	—
	Mean	**2,994.471**	2,995.991	3,000.715	—
	Std	**0.000**	3.805	7.896	—
	SR	**100%**	**100%**	**100%**	0%
	W-t		(+)	(+)	(+)
SCBD	Best	62,791.528	**62,791.491**	65,798.689	—
	Median	**62,791.688**	67,703.486	72,331.571	—
	Worst	**62,792.069**	77,506.995	78,398.626	—
	Mean	**62,791.734**	68,242.677	72,793.390	—
	Std	**0.151**	4,160.120	5,563.921	—
	SR	**100%**	**100%**	25%	0%
	W-t		(+)	(+)	(+)
TSD	Best	0.012666	0.012666	**0.012665**	—
	Median	0.012681	0.012697	**0.012665**	—
	Worst	**0.012792**	0.012788	0.013306	—
	Mean	**0.012691**	0.012705	0.012697	—
	Std	**0.000030**	0.000035	0.000143	—
	SR	**100%**	**100%**	**100%**	0%
	W-t		(≈)	(−)	(+)
GO	Best	**−0.971**	**−0.971**	**−0.971**	−0.743
	Median	**−0.971**	**−0.971**	−0.969	0.042
	Worst	−0.744	−0.871	−0.034	0.465
	Mean	**−0.960**	−0.938	−0.877	−0.076
	Std	0.051	**0.047**	0.208	0.489
	SR	**100%**	**100%**	**100%**	**100%**
	W-t		(+)	(+)	(+)
SE	Best	**−1.174**	**−1.174**	**−1.174**	−1.172
	Median	**−1.174**	**−1.174**	**−1.174**	−1.158
	Worst	**−1.174**	**−1.174**	−1.171	62.187
	Mean	**−1.174**	**−1.174**	**−1.174**	10.430
	Std	**0.000**	**0.000**	0.001	22.570
	SR	**100%**	**100%**	**100%**	**100%**
	W-t		(≈)	(+)	(+)

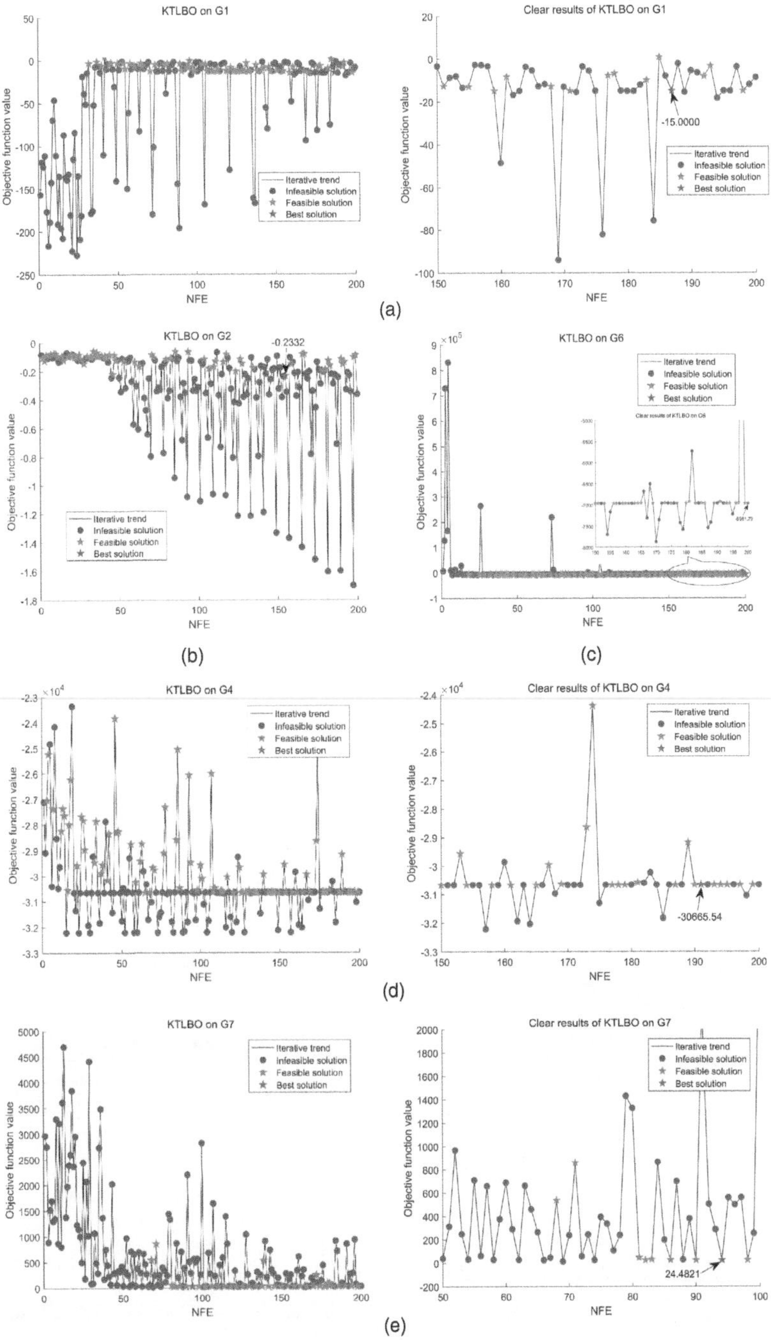

FIGURE 9.9 Iterative results of KTLBO on the 18 cases.

(Continued)

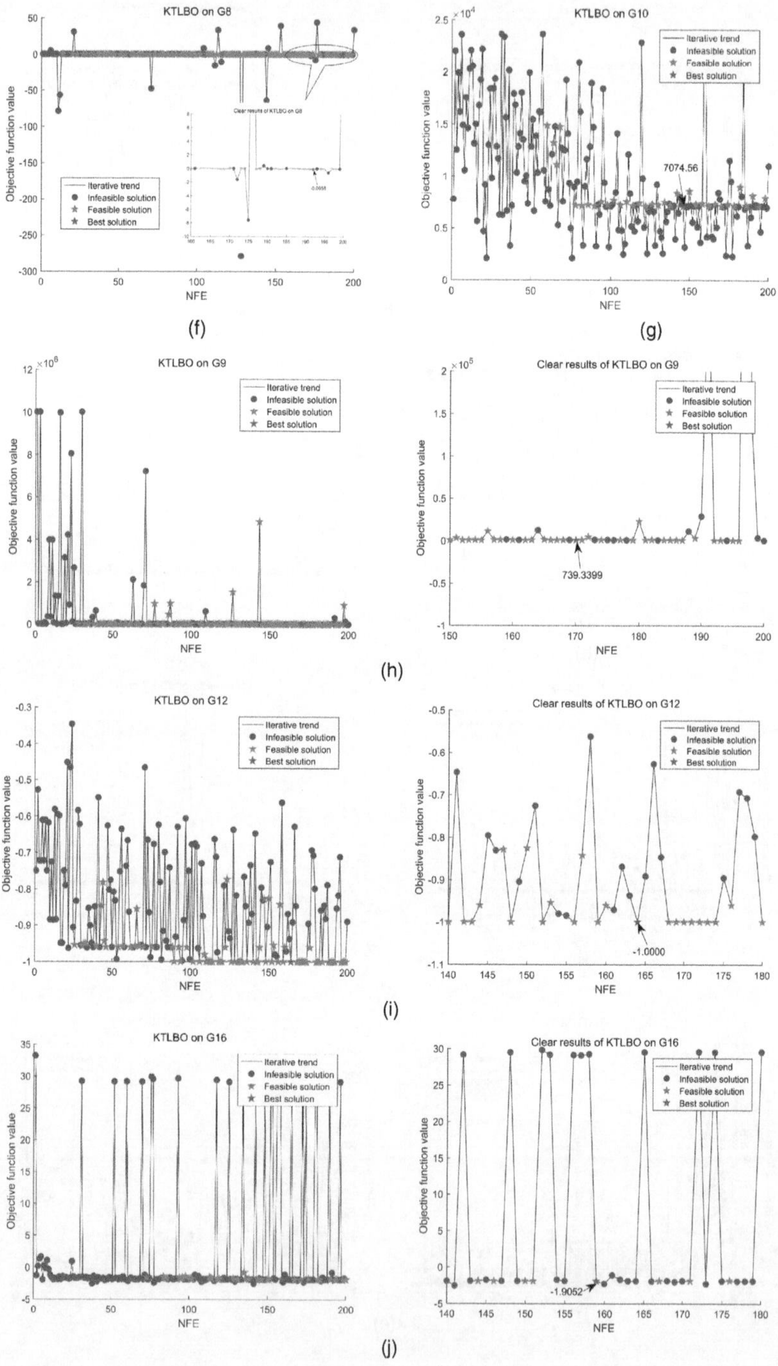

FIGURE 9.9 (*Continued*) Iterative results of KTLBO on the 18 cases.

(*Continued*)

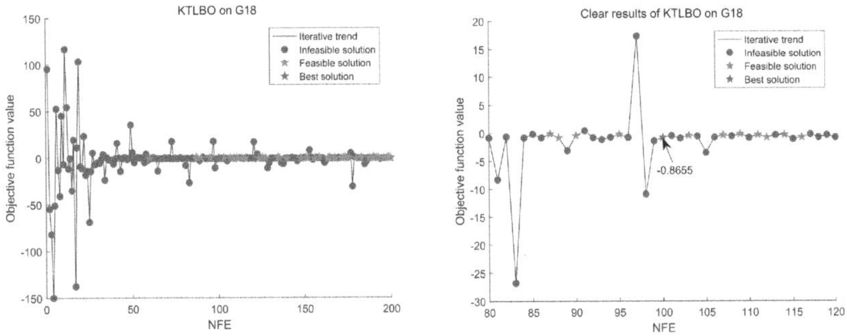

(k)

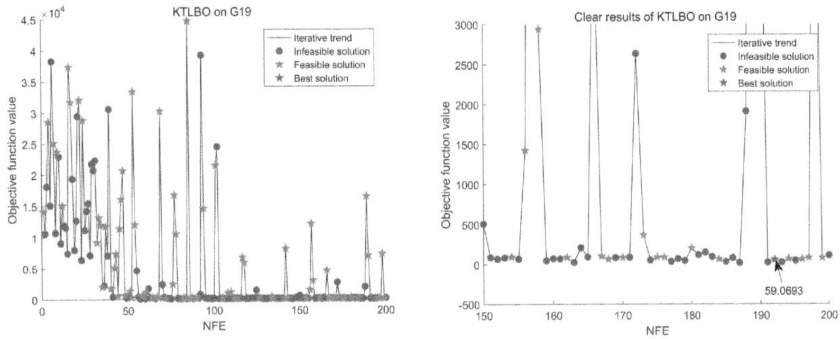

(l)

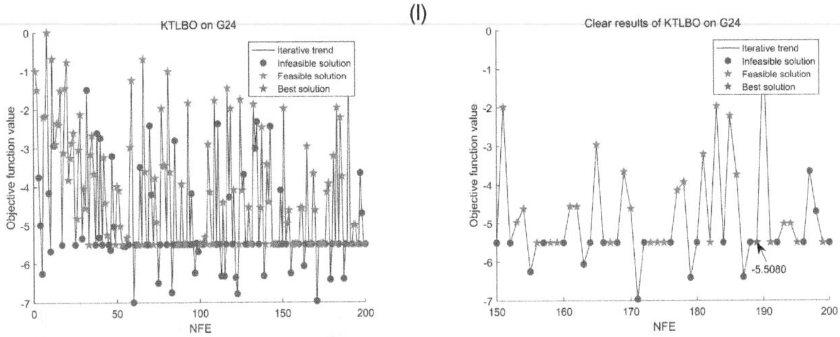

(m)

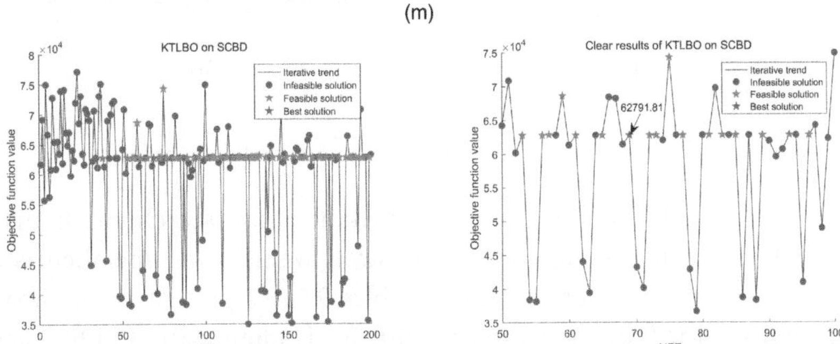

(n)

FIGURE 9.9 *(Continued)* Iterative results of KTLBO on the 18 cases.

(Continued)

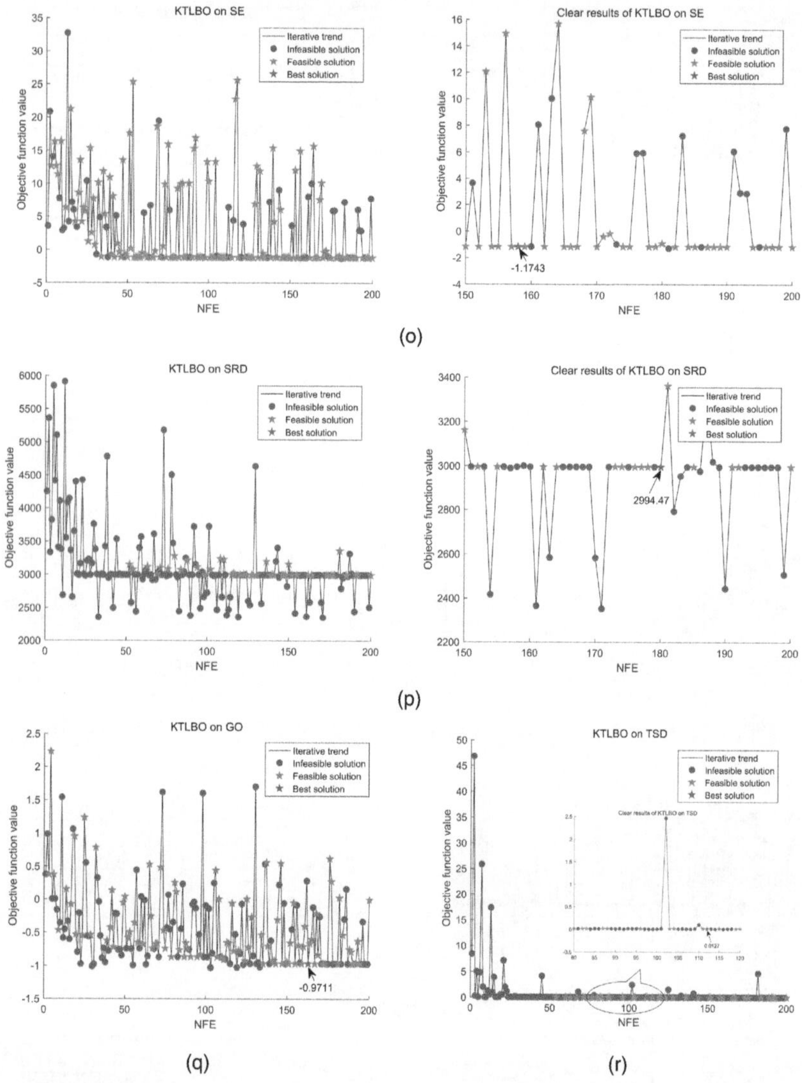

FIGURE 9.9 (*Continued*) Iterative results of KTLBO on the 18 cases.

Since KTLBO complies with the metaheuristic search mechanism, KTLBO is further compared with three well-known metaheuristic-constrained optimization methods. Tables 9.5–9.7 present the comparison results of the four algorithms in 500 FEs. Undoubtedly, KTLBO using 500 FEs can yield more accurate results than that in Tables 9.2–9.4. For many cases (e.g., g01, g04, g06, g07, g08, g09, g12, g18, g24, SE, GO, TSD, SRD and SCBD), KTLBO basically has reached the true global optima.

TLBO performs more robustly among the other three comparison algorithms because it is more likely to find feasible solutions during 20 runs. CMODE and FROFI may always fail in some cases. For instance, FROFI can hardly process g01, g10 and g18, while CMODE cannot process g18. Moreover, CMODE achieves lower SR values on g01, g07 and g10. More function calls are required for CMODE and FROFI to identify the feasible area. Relatively, TLBO, CMODE and FROFI achieve better performance

TABLE 9.5 Statistical Results on CEC2006 Cases (NFE = 500)—Part 1

Problem	Criteria	KTLBO	TLBO	CMODE	FROFI
g01	Best	**−15.000**	−9.041	−6.780	—
	Median	**−15.000**	−6.668	−5.526	—
	Worst	**−15.000**	−3.124	−3.431	—
	Mean	**−15.000**	−6.409	−5.298	—
	Std	**0.000**	1.842	1.231	—
	SR	**100%**	**100%**	30%	0%
	W-t		(+)	(+)	(+)
g02	Best	**−0.443**	−0.344	−0.346	−0.356
	Median	**−0.355**	−0.267	−0.246	−0.245
	Worst	**−0.289**	−0.177	−0.182	−0.181
	Mean	**−0.356**	−0.268	−0.253	−0.257
	Std	0.046	**0.040**	**0.040**	0.052
	SR	**100%**	**100%**	**100%**	**100%**
	W-t		(+)	(+)	(+)
g04	Best	**−30,665.539**	−30,657.709	−30,577.162	−30,422.543
	Median	**−30,665.539**	−30,527.893	−30,246.187	−30,207.457
	Worst	**−30,665.539**	−29,624.622	−29,630.264	−29,922.280
	Mean	**−30,665.539**	−30,376.660	−30,201.293	−30,219.223
	Std	**0.000**	322.562	255.884	116.434
	SR	**100%**	**100%**	**100%**	**100%**
	W-t		(+)	(+)	(+)
g06	Best	**−6,961.812**	−6,616.246	−6,936.174	−6,809.207
	Median	**−6,961.799**	−6,123.592	−6,598.398	−6,439.650
	Worst	**−6,961.778**	−2,080.231	−1,767.477	−4,006.968
	Mean	**−6,961.798**	−5,283.219	−5,660.160	−6,145.102
	Std	**0.009**	1,614.756	1,689.016	759.007
	SR	**100%**	60%	85%	95%
	W-t		(+)	(+)	(+)
g07	Best	**24.335**	147.721	158.689	48.267

(Continued)

TABLE 9.5 (*Continued*) Statistical Results on CEC2006 Cases (NFE = 500)—Part 1

Problem	Criteria	KTLBO	TLBO	CMODE	FROFI
	Median	**24.362**	1,000.305	315.775	107.470
	Worst	**24.447**	1,549.875	707.693	325.735
	Mean	**24.370**	869.814	374.483	137.090
	Std	**0.030**	473.553	237.994	77.036
	SR	**100%**	45%	20%	90%
	W-t		(+)	(+)	(+)
g08	Best	**−0.096**	**−0.096**	**−0.096**	**−0.096**
	Median	**−0.096**	**−0.096**	−0.095	**−0.096**
	Worst	**−0.096**	−0.026	−0.029	−0.029
	Mean	**−0.096**	−0.092	−0.081	−0.092
	Std	**0.000**	0.016	0.026	0.015
	SR	**100%**	**100%**	90%	**100%**
	W-t		(+)	(+)	(+)
g09	Best	**680.646**	692.552	700.421	702.259
	Median	**680.736**	737.429	818.555	744.267
	Worst	**681.581**	829.155	1,345.010	882.689
	Mean	**680.826**	742.725	888.289	759.905
	Std	**0.238**	37.368	177.392	46.599
	SR	**100%**	**100%**	**100%**	**100%**
	W-t		(+)	(+)	(+)

TABLE 9.6 Statistical Results on CEC2006 Cases (NFE = 500)—Part 2

Problem	Criteria	KTLBO	TLBO	CMODE	FROFI
g10	Best	**7,050.335**	13,413.760	12,842.042	—
	Median	**7,054.236**	17,768.095	13,343.482	—
	Worst	**7,068.933**	22,506.365	14,506.941	-
	Mean	**7,056.459**	18,159.266	13,564.155	—
	Std	**4.772**	2,991.253	854.105	—
	SR	**100%**	45%	15%	0%
	W-t		(+)	(+)	(+)
g12	Best	**−1.000**	−0.999	**−1.000**	**−1.000**
	Median	**−1.000**	−0.987	**−1.000**	**−1.000**
	Worst	**−1.000**	−0.908	−0.964	**−1.000**
	Mean	**−1.000**	−0.979	−0.997	**−1.000**
	Std	**0.000**	0.022	0.008	**0.000**
	SR	**100%**	**100%**	**100%**	**100%**

(*Continued*)

TABLE 9.6 (*Continued*) Statistical Results on CEC2006 Cases (NFE = 500)—Part 2

Problem	Criteria	KTLBO	TLBO	CMODE	FROFI
	W-t		(+)	(+)	(+)
g16	Best	**−1.905**	−1.855	−1.879	−1.702
	Median	**−1.905**	−1.542	−1.483	−1.437
	Worst	**−1.723**	−0.978	−1.167	−1.200
	Mean	**−1.896**	−1.508	−1.508	−1.429
	Std	**0.041**	0.256	0.274	0.172
	SR	**100%**	85%	40%	40%
	W-t		(+)	(+)	(+)
g18	Best	**−0.866**	−0.652	—	—
	Median	**−0.866**	−0.458	—	—
	Worst	**−0.866**	−0.271	—	—
	Mean	**−0.866**	−0.461	—	—
	Std	**0.000**	0.110	—	—
	SR	**100%**	**100%**	0%	0%
	W-t		(+)	(+)	(+)
g19	Best	**32.923**	84.283	342.126	264.017
	Median	**33.682**	248.502	783.799	583.579
	Worst	**35.067**	397.335	2,081.047	1,249.052
	Mean	**33.760**	235.149	851.530	601.482
	Std	**0.520**	87.032	464.958	269.384
	SR	**100%**	**100%**	**100%**	55%
	W-t		(+)	(+)	(+)
g24	Best	**−5.508**	−5.507	−5.507	−5.492
	Median	**−5.508**	−5.499	−5.490	−5.448
	Worst	**−5.508**	−5.377	−5.343	−5.292
	Mean	**−5.508**	−5.485	−5.472	−5.437
	Std	**0.000**	0.035	0.045	0.048
	SR	**100%**	**100%**	**100%**	**100%**
	W-t		(+)	(+)	(+)

on g02, g04, g08, g09, g12 and g24 because these cases have larger feasible space. Tables 9.2, 9.3, 9.5 and 9.6 summarize that SBO algorithms require fewer FEs than metaheuristic algorithms in most cases.

As indicated in Table 9.7, 500 FEs are good enough for KTLBO to find their global optimal solutions. Besides, TLBO always achieves higher SRs and outperforms CMODE and FROFI on SRD, SCBD, TSD and GO. Notably, metaheuristic algorithms are applied directly to computationally expensive and black-box optimization problems, whereas they require

TABLE 9.7 Statistical Results on GO, SE and Engineering Cases (NFE = 500)

Problem	Criteria	KTLBO	TLBO	CMODE	FROFI
SRD	Best	**2,994.471**	3,003.273	3,025.246	3,036.176
	Median	**2,994.471**	3,052.253	3,122.707	3,109.041
	Worst	**2,994.471**	5,574.144	3,899.410	3,457.146
	Mean	**2,994.471**	3,333.409	3,178.711	3,122.580
	Std	**0.000**	615.210	229.481	91.311
	SR	**100%**	**100%**	90%	**100%**
	W-t		(+)	(+)	(+)
SCBD	Best	**62,791.515**	65,966.042	67,788.298	67,238.418
	Median	**62,791.584**	73,772.591	70,641.883	73,076.280
	Worst	**62,791.738**	83,818.265	74,480.140	77,374.617
	Mean	**62,791.598**	73,285.578	71,485.513	72,318.792
	Std	**0.061**	4,698.158	2,055.834	3,124.462
	SR	**100%**	90%	65%	55%
	W-t		(+)	(+)	(+)
TSD	Best	**0.012665**	0.012735	0.012742	0.012965
	Median	**0.012667**	0.013006	0.013769	0.014361
	Worst	**0.012671**	0.015140	0.230292	0.017988
	Mean	**0.012667**	0.013410	0.026247	0.014682
	Std	**0.000002**	0.000830	0.049545	0.001374
	SR	**100%**	**100%**	95%	**100%**
	W-t		(+)	(+)	(+)
GO	Best	**−0.971**	**−0.971**	**−0.971**	**−0.971**
	Median	**−0.971**	−0.968	−0.970	−0.968
	Worst	**−0.971**	−0.867	−0.811	−0.858
	Mean	**−0.971**	−0.947	−0.928	−0.941
	Std	**0.000**	0.036	0.060	0.046
	SR	**100%**	**100%**	**100%**	**100%**
	W-t		(+)	(+)	(+)
SE	Best	**−1.174**	**−1.174**	**−1.174**	−1.172
	Median	**−1.174**	−1.171	−1.173	−1.165
	Worst	**−1.174**	−0.102	−0.580	−1.149
	Mean	**−1.174**	−1.090	−1.128	−1.164
	Std	**0.000**	0.245	0.140	0.006
	SR	**100%**	**100%**	**100%**	**100%**
	W-t		(+)	(+)	(+)

more FEs to achieve convergence. SBO algorithms exploit the predicted information of surrogate models for search guidance, thereby decreasing the NFE. However, SBO methods may exhibit higher sensitivity to the prediction accuracy of surrogate models. Once the surrogate models exhibit worse predicting performance in some cases, SBO may get inefficient immediately. Accordingly, KTLBO combining a metaheuristic searching mechanism and Kriging's predicted information can ensure a robust sampling process, and its results from Table 9.2–9.7 indicate its powerful functionality and significant advantages for EBCPs.

9.5 ENGINEERING APPLICATIONS

Blended-wing-body underwater gliders (BWBUGs) that play an important role in scientific and commercial fields have aroused huge attention over the past few years. In a BWBUG, the pressure shell is an extremely important part that protects the expensive measuring instruments and equipment in a deep-sea environment. In this chapter, to decrease the design cost and meanwhile increase the inner space volume of the BWBUG's pressure shell, this study attempts to improve its buoyancy–weight ratio (BWR) and concurrently satisfy the stress and stability constraints. Figure 9.10 presents the geometric description and defines ten design variables including three thickness parameters (t_1, t_2 and t_3), three radius parameters (R_1, R_2 and R_3), and four size parameters (l_1, l_2, l_3 and l_4).

Furthermore, the specific optimization formula is summarized below:

$$\begin{cases} \max & \dfrac{B}{G}=\dfrac{\rho v(l_1,l_2,l_3,l_4,R_1,R_2,R_3,t_1,t_2,t_3)}{m(l_1,l_2,l_3,l_4,R_1,R_2,R_3,t_1,t_2,t_3)} \\ s.t. & \sigma_{\max}(l_1,l_2,l_3,l_4,R_1,R_2,R_3,t_1,t_2,t_3)\le \gamma\sigma_s \\ & \lambda P_j \le P_{cr}(l_1,l_2,l_3,l_4,R_1,R_2,R_3,t_1,t_2,t_3) \\ & 375\le l_1\le 390 \quad 225\le l_2\le 235 \quad 200\le l_3\le 210 \quad 150\le l_4\le 160 \\ & 65\le R_1\le 85 \quad 80\le R_2\le 100 \quad 20\le R_3\le 30 \\ & 5\le t_1\le 12 \quad 5\le t_2\le 12 \quad 5\le t_3\le 12 \end{cases} \tag{9.17}$$

where B denotes buoyancy, G refers to gravity, is the density of sea water, v is the volume of the whole pressure shell, and m represents the total weight. In the first stress constraint, ρ is the maximal equivalent stress, $\sigma_{\max}$ refers to the yield strength, and σ_s is a safety factor. In the second stability constraint, γ is the computation pressure, $P_j \approx 10$ MPa is the buckling critical load, and P_{cr} is the first-order buckling factor. In this case, the depth of

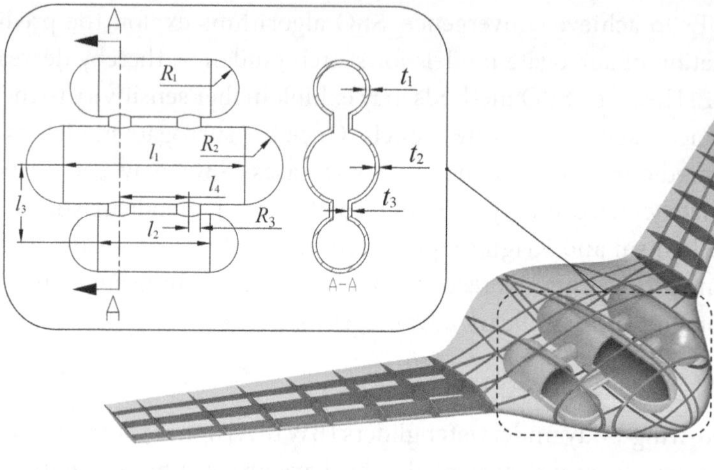

FIGURE 9.10 Illustration of BWBUG's pressure shell.

water is defined as 1,000 m, λ, and $\gamma = 0.8$. Since the aluminum alloy is applied for the pressure shell, $\lambda = 1.5$ is set to 280 MPa. In Eq. (9.17), there are three response values B/G, σ_s and $\sigma_{\max}$ come from the time-consuming simulation model. One simulation analysis takes more than 5 minutes. As revealed from the comparison analyses, SCGOSR and KTLBO exhibit the best performance, so they are employed for this engineering application.

For a fair comparison, KTLBO and SCGOSR adopt the same DoE samples to drive the optimization loop. After 200 simulation analyses, KTLBO identifies a better solution than SCGOSR. Figures 9.11 and 9.12 show the iterative results where the stars represent feasible samples, dots refer to the infeasible ones. In Figure 9.11, the best feasible sample is obtained at the 189th NFE, while in Figure 9.12, the best feasible sample is obtained at the 97th NFE. Intuitively, KTLBO converges after 100 simulation analyses, while SCGOSR seems to get stuck in a local optimal area after 90 calls to the simulation model. Tables 9.8 and 9.9 provide the detailed results. Compared with the best DoE sample, SCGOSR achieves a 21.89% improvement, while KTLBO achieves a 67.40% improvement. Moreover, Figures 9.13–9.15 illustrate the optimal simulation results of DoE, KTLBO and SCGOSR. Obviously, KTLBO is suggested to be more suitable for this engineering case. Table 9.9 and Figures 9.13–9.15 indicate that KTLBO converge to the second constraint bound while SCGOSR remains far away from this constraint bound.

To sum up, KTLBO cannot only deal with benchmark cases, but also efficiently solve simulation-based constrained optimization problems. It

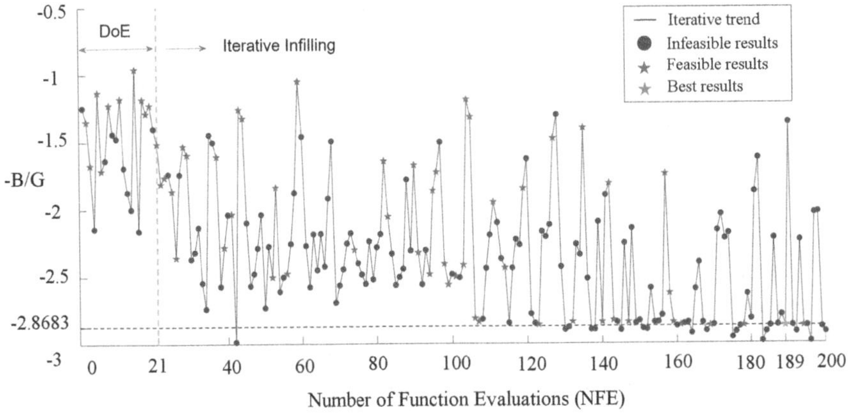

FIGURE 9.11 Iterative results of KTLBO.

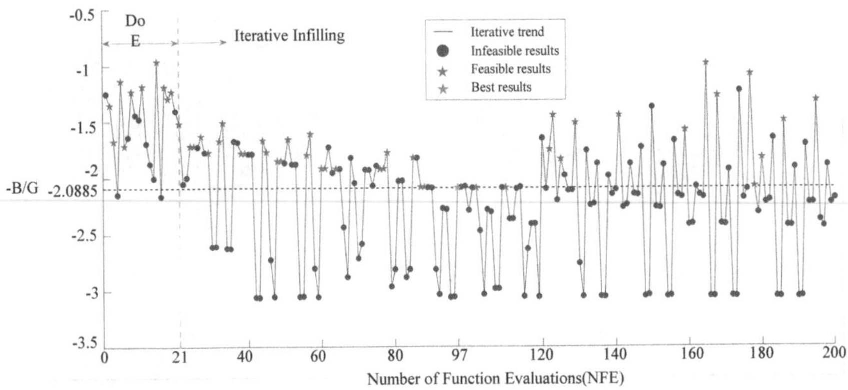

FIGURE 9.12 Iterative results of SCGOSR.

TABLE 9.8 Obtained Best Solutions of BWBUG's Structure Design

	l_1	l_2	l_3	l_4	R_1	R_2	R_3	t_1	t_2	t_3
DoE-opt	388.50	225.50	207.00	158.00	82.00	83.00	28.00	5.70	10.25	8.15
KTLBO-opt	376.55	227.82	200.00	153.27	85.00	93.41	30.00	5.00	5.00	12.00
SCGOSR-opt	385.66	229.14	202.00	159.78	84.63	84.72	26.27	5.55	8.19	6.01

TABLE 9.9 Optimal Response Values for BWBUG's Structure Design

	v/m^3	m/kg	B/G	σ_{max}/MPa	P_{cr}/MPa
DoE-opt	0.0203	13.9904	1.7134	198.2197	69.6338
KTLBO-opt	0.0257	10.4483	**2.8683**	213.0277	15.6479
SCGOSR-opt	0.0223	12.6554	2.0885	215.4628	63.1093

is noteworthy that when each analysis of the simulation model requires several hours or days, fewer calls to the simulation model are significantly critical. KTLBO requires fewer NFE to achieve convergence, which noticeably shortens the design cycle and yields a satisfactory solution for engineers at the simulation phase.

9.6 CHAPTER SUMMARY

In this chapter, an efficient Kriging-assisted TLBO method is proposed to solve computationally expensive constrained optimization problems. By complying with TLBO's two-phase search pattern, two Kriging-assisted sampling strategies are formulated, retaining TLBO's search mechanism while reasonably balancing the exploitation of surrogates and exploration of

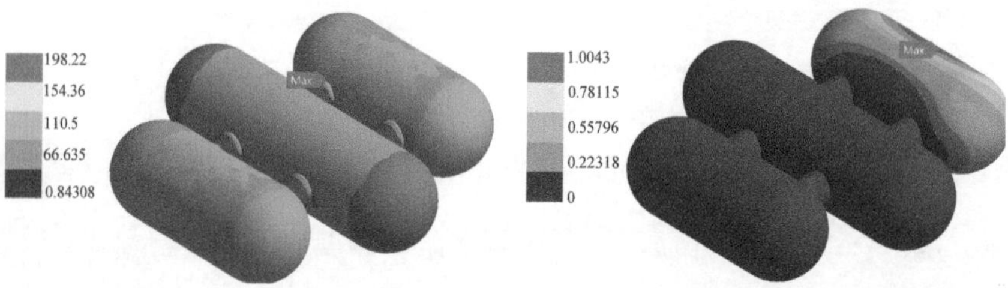

FIGURE 9.13 Equivalent stress and buckling results of DoE's best sample.

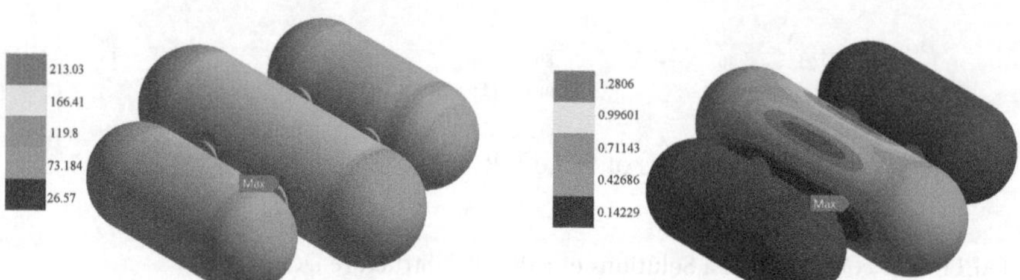

FIGURE 9.14 Equivalent stress and first mode of KTLBO's best sample.

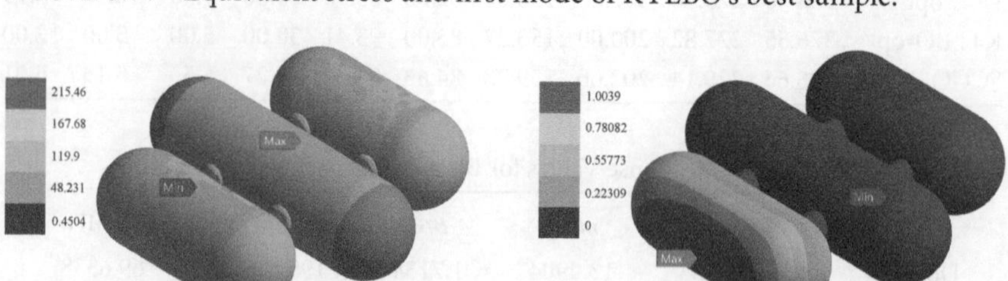

FIGURE 9.15 Equivalent stress and first mode of SCGOSR's best sample.

unknown areas. In KATP, the neighborhoods around the present best solution are sufficiently exploited, and a constrained EI function considering the probability of feasibility is defined as a filter to pick up the promising individuals from the learners. In KALP, a constrained MSE function focusing on Kriging's prediction uncertainty is proposed to choose the learners located at the sparsely sampled feasible region for global exploration. Initial DoE samples and newly generated expensive samples are iteratively sorted based on their penalty function values, and new teachers and brilliant learners are continuously updated until the algorithm identifies the true global optima.

NOTE

1 Based on "Kriging-assisted Teaching-Learning-based Optimization (KTLBO) to Solve Computationally Expensive Constrained Problems," published in [Information Sciences], [2021]. Permission obtained from [Elsevier].

REFERENCES

Akbari, H., & Kazerooni, A. (2020). KASRA: A Kriging-Based Adaptive Space Reduction Algorithm for Global Optimization of Computationally Expensive Black-Box Constrained Problems. *Applied Soft Computing, 90*, Article 106154. https://doi.org/10.1016/j.asoc.2020.106154

Bagheri, S., Konen, W., Emmerich, M., & Baeck, T. (2017). Self-Adjusting Parameter Control for Surrogate-Assisted Constrained Optimization under Limited Budgets. *Applied Soft Computing, 61*, 377–393. https://doi.org/10.1016/j.asoc.2017.07.060

Chen, X., Mei, C., Xu, B., Yu, K., & Huang, X. (2018). Quadratic Interpolation Based Teaching-Learning-Based Optimization for Chemical Dynamic System Optimization. *Knowledge-Based Systems, 145*, 250–263. https://doi.org/10.1016/j.knosys.2018.01.021

Daneshyari, M., & Yen, G. G. (2012). Constrained Multiple-Swarm Particle Swarm Optimization within a Cultural Framework. *IEEE Transactions on Systems Man and Cybernetics Part A-Systems and Humans, 42*(2), 475–490. https://doi.org/10.1109/tsmca.2011.2162498

Dong, H., Li, C., Song, B., & Wang, P. (2018). Multi-Surrogate-Based Differential Evolution with Multi-Start Exploration (MDEME) for Computationally Expensive Optimization. *Advances in Engineering Software, 123*, 62–76. https://doi.org/10.1016/j.advengsoft.2018.06.001

Dong, H., Song, B., Dong, Z., & Wang, P. (2016). Multi-Start Space Reduction (MSSR) Surrogate-Based Global Optimization Method. *Structural and Multidisciplinary Optimization, 54*(4), 907–926. https://doi.org/10.1007/s00158-016-1450-1

Dong, H., Song, B., Dong, Z., & Wang, P. (2018). SCGOSR: Surrogate-Based Constrained Global Optimization Using Space Reduction. *Applied Soft Computing, 65*, 462–477. https://doi.org/10.1016/j.asoc.2018.01.041

Dong, H., Song, B., Wang, P., & Dong, Z. (2018). Surrogate-Based Optimization with Clustering-Based Space Exploration for Expensive Multimodal Problems. *Structural and Multidisciplinary Optimization*, *57*(4), 1553–1577. https://doi.org/10.1007/s00158-017-1826-x

Dong, H., Sun, S., Song, B., & Wang, P. (2019). Multi-Surrogate-Based Global Optimization Using a Score-Based Infill Criterion. *Structural and Multidisciplinary Optimization*, *59*(2), 485–506. https://doi.org/10.1007/s00158-018-2079-z

Farmani, R., & Wright, J. A. (2003). Self-Adaptive Fitness Formulation for Constrained Optimization. *IEEE Transactions on Evolutionary Computation*, *7*(5), 445–455. https://doi.org/10.1109/tevc.2003.817236

Haftka, R. T., Villanueva, D., & Chaudhuri, A. (2016). Parallel Surrogate-Assisted Global Optimization with Expensive Functions - A Survey. *Structural and Multidisciplinary Optimization*, *54*(1), 3–13. https://doi.org/10.1007/s00158-016-1432-3

Jones, D. R., Perttunen, C. D., & Stuckman, B. E. (1993). Lipschitzian Optimization without the Lipschitz Constant. *Journal of Optimization Theory and Applications*, *79*(1), 157–181. https://doi.org/10.1007/bf00941892

Kar, A. K. (2016). Bio Inspired Computing - A Review of Algorithms and Scope of Applications. *Expert Systems with Applications*, *59*, 20–32. https://doi.org/10.1016/j.eswa.2016.04.018

Li, E. (2019). An Adaptive Surrogate Assisted Differential Evolutionary Algorithm for High Dimensional Constrained Problems. *Applied Soft Computing*, *85*, Article 105752. https://doi.org/10.1016/j.asoc.2019.105752

Li, F., Shen, W., Cai, X., Gao, L., & Wang, G. G. (2020). A Fast Surrogate-Assisted Particle Swarm Optimization Algorithm for Computationally Expensive Problems. *Applied Soft Computing*, *92*, Article 106303. https://doi.org/10.1016/j.asoc.2020.106303

Liu, B., Zhang, Q., & Gielen, G. G. E. (2014). A Gaussian Process Surrogate Model Assisted Evolutionary Algorithm for Medium Scale Expensive Optimization Problems. *IEEE Transactions on Evolutionary Computation*, *18*(2), 180–192. https://doi.org/10.1109/tevc.2013.2248012

Liu, H., Xu, S., Chen, X., Wang, X., & Ma, Q. (2017). Constrained Global Optimization via a DIRECT-Type Constraint-Handling Technique and an Adaptive Metamodeling Strategy. *Structural and Multidisciplinary Optimization*, *55*(1), 155–177. https://doi.org/10.1007/s00158-016-1482-6

Liu, H., Xu, S., Wang, X., Wu, J., & Song, Y. (2015). A Global Optimization Algorithm for Simulation-Based Problems Via the Extended DIRECT Scheme. *Engineering Optimization*, *47*(11), 1441–1458. https://doi.org/10.1080/0305215x.2014.971777

Mavrovouniotis, M., Li, C., & Yang, S. (2017). A Survey of Swarm Intelligence for Dynamic Optimization: Algorithms and Applications. *Swarm and Evolutionary Computation*, *33*, 1–17. https://doi.org/10.1016/j.swevo.2016.12.005

Mezura-Montes, E., & Coello Coello, C. A. (2011). Constraint-Handling in Nature-Inspired Numerical Optimization: Past, Present and Future. *Swarm and Evolutionary Computation*, *1*(4), 173–194. https://doi.org/10.1016/j.swevo.2011.10.001

Miranda-Varela, M.-E., & Mezura-Montes, E. (2018). Constraint-Handling Techniques in Surrogate-Assisted Evolutionary Optimization. An Empirical Study. *Applied Soft Computing, 73*, 215–229. https://doi.org/10.1016/j.asoc.2018.08.016

Muller, J., & Woodbury, J. D. (2017). GOSAC: Global Optimization with Surrogate Approximation of Constraints. *Journal of Global Optimization, 69*(1), 117-136. https://doi.org/10.1007/s10898-017-0496-y

Ororbia, M. E., Chhabra, J. P. S., Warn, G. P., Miller, S. W., Yukish, M. A., & Qiu, T. (2020). Increasing the Discriminatory Power of Bounding Models Using Problem-Specific Knowledge When Viewing Design as a Sequential Decision Process. *Structural and Multidisciplinary Optimization, 62*(2), 709–728. https://doi.org/10.1007/s00158-020-02528-0

Rao, R. V., Savsani, V. J., & Vakharia, D. P. (2011). Teaching-Learning-Based Optimization: A Novel Method for Constrained Mechanical Design Optimization Problems. *Computer-Aided Design, 43*(3), 303–315. https://doi.org/10.1016/j.cad.2010.12.015

Regis, R. G. (2011). Stochastic Radial Basis Function Algorithms for Large-Scale Optimization Involving Expensive Black-Box Objective and Constraint Functions. *Computers & Operations Research, 38*(5), 837–853. https://doi.org/10.1016/j.cor.2010.09.013

Wang, Y., & Cai, Z. (2012). Combining Multiobjective Optimization with Differential Evolution to Solve Constrained Optimization Problems. *IEEE Transactions on Evolutionary Computation, 16*(1), 117–134. https://doi.org/10.1109/tevc.2010.2093582

Wang, Y., Wang, B.-C., Li, H.-X., & Yen, G. G. (2016). Incorporating Objective Function Information into the Feasibility Rule for Constrained Evolutionary Optimization. *IEEE Transactions on Cybernetics, 46*(12), 2938–2952. https://doi.org/10.1109/tcyb.2015.2493239

Wang, Y., Yin, D.-Q., Yang, S., & Sun, G. (2019). Global and Local Surrogate-Assisted Differential Evolution for Expensive Constrained Optimization Problems with Inequality Constraints. *IEEE Transactions on Cybernetics, 49*(5), 1642–1656. https://doi.org/10.1109/tcyb.2018.2809430

Wright, J. A., & Farmani, R. (2001). Genetic algorithms: a fitness formulation for constrained minimization. *Proceedings of the Genetic and Evolutionary Computation Conference (GECCO-2001), Sixth Annual Genetic Programming Conference (GP-2001), Tenth International Conference on Genetic Algorithms (ICGA-2001)*, San Francisco, California, USA, July 7–11, 2001.

Wu, Y., Yin, Q., Jie, H., Wang, B., & Zhao, J. (2018). A RBF-Based Constrained Global Optimization Algorithm for Problems with Computationally Expensive Objective and Constraints. *Structural and Multidisciplinary Optimization, 58*(4), 1633–1655. https://doi.org/10.1007/s00158-018-1987-2

Yang, Z., Qiu, H., Gao, L., Cai, X., Jiang, C., & Chen, L. (2020). Surrogate-Assisted Classification-Collaboration Differential Evolution for Expensive Constrained Optimization Problems. *Information Sciences, 508*, 50–63.

Yu, H., Tan, Y., Sun, C., & Zeng, J. (2019). A Generation-Based Optimal Restart Strategy for Surrogate-Assisted Social Learning Particle Swarm Optimization. *Knowledge-Based Systems, 163*, 14–25. https://doi.org/10.1016/j.knosys.2018.08.010

CHAPTER 10

KDGO

Kriging-Assisted Discrete Global Optimization for Black-Box Problems with Costly Objective and Constraints[1]

10.1 INTRODUCTION

With the rapid progress of computer technology, high-fidelity simulation has become an indispensable tool in modern industry applications, which can effectively reduce design budgets and bring higher economic benefits (Dong et al., 2017; Jiang et al., 2019; Zhou et al., 2018). Simultaneously, when the accuracy requirement continuously increases, the computation cost of simulation analysis may get huge, causing difficulty in optimization design (Dong, Li, et al., 2018; H. Liu et al., 2018; Stander et al., 2016). Besides, many real-world applications, such as management, scheduling, logistics, structure design and pattern recognition, involve discrete domains (Ekel & Neto, 2006; Lawler, 1972; Sayadi et al., 2013) and time-demanding simulation analysis (Dede, 2014). Therefore, discrete and computationally intensive global optimization problems are challenging, and have begun to gain more attention in recent years.

For discrete optimization problems, the branch and bound (BB) algorithm (Land & Doig, 2010) that recursively divides the solution set and evaluates the bound values can find the optimal combination of these discrete

 DOI: 10.1201/9781003636267-10

values. For example, Demeulemeester and Herroelen (1992) employed a BB procedure for multiple resource-constrained project scheduling. Nakariyakul and Casasent (2007) proposed an adaptive BB algorithm to select the optimal subsets of features in pattern recognition applications. However, BB appears inappropriate for computationally expensive global optimization problems, because it has to construct a relaxed problem whose global optimum must be found to identify the lower bound, which will cause many calls of the costly functions especially for multimodal problems. Variable neighborhood search (VNS) presented by Mladenović and Hansen is an effective tool for global combinatorial optimization problems (Mladenović & Hansen, 1997). VNS can systematically explore the possible neighborhood structures to identify the local optima, and further find the global optimum with the help of perturbation. VNS has been extensively applied in various fields like artificial intelligence, clustering analysis, scheduling and so on (Adibi et al., 2010; Kytöjoki et al., 2007; Polacek et al., 2004). VNS was primarily developed for box-constrained integer optimization problems, but it cannot be directly used for nonlinear-constrained problems. Nonsmooth optimization by mesh adaptive direct search (NOMAD) (Abramson et al., 2009) was developed for computationally expensive and black-box optimization problems. NOMAD is a derivative-free optimization method and is applicable for continuous, integer and mixed design domains. Moreover, NOMAD is also good at handling nonlinear-constrained optimization problems, making it suitable for most real-world applications. However, there are no extensive numerical studies on NOMAD's capability that deal with computationally expensive optimization problems. It is worth mentioning that in the existing literature there is another type of algorithm to deal with discrete and black-box global optimization problems, that is, swarm intelligence and evolutionary computation (Anghinolfi & Paolucci, 2009; Guendouz et al., 2017; Zhang et al., 2015). Generally, swarm/evolution-based algorithms are inspired by some natural phenomenon and can generate a population in each cycle to randomly search the design space. With the population updated and the objective function evaluated many times, promising solutions can be gradually acquired. Most of these discrete metaheuristic algorithms have been applied to real-world applications. For example, Li et al. (2019) proposed a discrete particle swarm optimization algorithm (DPSO-PDM) for community detection in complex networks. DPSO-PDM redefines the particle velocity and position, and adds the evolutionary operation in discretization to avoid getting trapped in local optima.

Surrogate-assisted optimization (SAO) (Dong, Song, et al., 2018; Shi et al., 2020; Zhou et al., 2021) plays an important role in simulation-based engineering applications, because it is rather efficient for computationally expensive problems. Surrogate modeling techniques like Kriging, radial basis functions (RBF) or polynomial response surface can effectively organize the obtained data to predict the potential solutions, considerably decreasing the number of costly function evaluations. However, most of the existing literature in this field emphasizes the methods for the continuous design domain and seldom focuses on discrete cases. Müller et al. (2013) presented a surrogate-based global optimization algorithm for mixed-integer black-box problems (SO-MI). SO-MI utilizes RBF to select candidate samples from discrete and continuous domains. In each cycle, four groups of cheap points are generated, where three of them are generated around the present best solution and one is randomly distributed in the design space. Thereafter, the most promising points are separately selected from the four sample sets to update the RBF model. It is worth noting that SO-MI needs at least one feasible point to drive the algorithm for constrained problems. Therefore, it is difficult for SO-MI to solve the constrained problems with a smaller feasible space. Furthermore, Müller et al. (2014) introduced a surrogate-based algorithm SO-I for expensive nonlinear integer programming problems, in which the RBF value and the distance to the known samples are synthetically considered to evaluate a potential point. SO-I shows excellent ability when dealing problems with costly objective and constraints, and also has impressive performance on practical engineering applications like hydropower generation and throughput maximization. J. Liu et al. (2018) extended the multi-start space reduction (MSSR) (Dong et al., 2016) algorithm for a hybrid energy storage system with integer and continuous design variables. In the extended MSSR, the discrete variables of those promising samples were rounded to integers for simulation analysis in each cycle and showed absolute advantages over the genetic algorithm. Similarly, some other SAO algorithms have been improved or extended to solve computationally expensive and discrete/mixed-variable engineering applications (Holmström et al., 2008; Rashid et al., 2013). However, most of the above methods are developed for a certain type of actual problems (e.g. binary, integer, unimodal, multimodal, box-constrained), and less literature has introduced widely applicable algorithms.

Inspired by SO-I that combines the RBF's prediction values and the distance between samples, we expect to develop a Kriging-based global optimization method for computationally expensive problems with generalized discrete space that allows binary, integer, noninteger, uni/multimodal and box/inequality-constrained types. To make KDGO widely applicable for most discrete cases, a data matrix with a discrete structure is proposed to reflect the original design domain. Besides, a multi-start knowledge mining process is carried out to acquire the promising samples in each cycle, specifically including four steps: optimization, projection, sampling and selection. First, a multi-start optimization is used to capture the promising solutions in the continuous design range. All these potential solutions are projected to the discrete matrix and a grid sampling method applicable for low- and high-dimensional space is proposed to get the promising discrete samples. Thereafter, the *k*-nearest neighbors (KNN) search strategy and expected improvement (EI) criterion are jointly used to select the supplementary samples. KDGO keeps running to update Kriging and find the most potential samples until a satisfactory solution is obtained. KDGO is mainly used to solve various discrete problems including binary, integer, noninteger, unimodal, multimodal, equality and inequality-constrained problems.

10.2 DISCRETE OPTIMIZATION CONSTRUCTION

More precisely, the problem this chapter concentrates on is described below:

$$\begin{aligned} &\min f(\boldsymbol{x}) \\ &s.t. \quad g_i(\boldsymbol{x}) \le 0, \quad \forall i=1,\ldots,m \\ &\qquad -\infty < x_k^l \le x_k \le x_k^u < \infty, \quad \forall k=1,\ldots,d \\ &\qquad x_k \in \Gamma_k \subset \mathbb{R} \end{aligned} \tag{10.1}$$

where $f(\boldsymbol{x})$ denotes the computationally intensive black-box objective; $g_i(\boldsymbol{x})$ is the ith costly black-box constraint; m and d represent the number of constraints and design variables, respectively. Besides, x_k is the kth discrete variable and $\Gamma_k$is its corresponding discrete set. It is a remarkable fact that $|\Gamma_{k1}|$ can be different from $|\Gamma_{k2}|$ if $k1 \ne k2$. It is also assumed that $\infty > |\Gamma_k| \ge 2, \forall k=1,\ldots,d$. What is more, the values of each discrete set $\Gamma_k, \forall k=1,\ldots,d$ are allowed to have uneven distributions as well. Correspondingly, a 2d illustration is provided in Figure 10.1, and three representative cases with different characteristics are introduced to make it more intuitive.

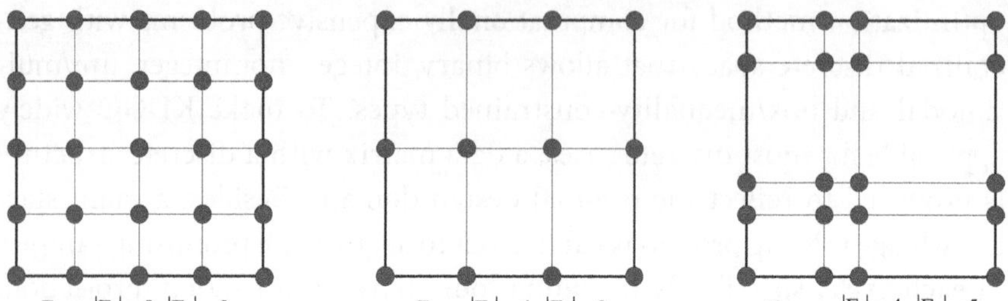

FIGURE 10.1 Different discrete design spaces.

To mathematically express a generalized discrete space, a matrix $\boldsymbol{D}$ is created to save these discrete sets as the preprocessing step for the subsequent surrogate-based optimization. This proposed matrix considers all the possible situations including even or uneven distributions and the same or different sizes of discrete sets at each dimension.

$$D_{M\times d}^{Init} = \begin{bmatrix} \infty & \infty & \infty & \cdots & \infty \\ \infty & \infty & \infty & \cdots & \infty \\ \vdots & \ddots & \vdots & \cdots & \vdots \\ \infty & \infty & \infty & \cdots & \infty \end{bmatrix} \Rightarrow$$

$$D_{M\times d} = \begin{bmatrix} r_1^1 & r_2^1 & \cdots & r_k^1 & \cdots & r_d^1 \\ r_1^2 & r_2^2 & \cdots & r_k^2 & \cdots & r_d^2 \\ \vdots & \ddots & \cdots & \vdots & \cdots & \vdots \\ \infty & \infty & \cdots & r_k^M & \cdots & \infty \end{bmatrix} \tag{10.2}$$

$$M=\max\left(|\Gamma_1|,\ldots,|\Gamma_d|\right),\quad r_k^i\in\Gamma_k,\quad \forall i=1,\ldots,\ |\Gamma_k|,\quad \forall k=1,\ldots,d$$

where $D_{M\times d}^{Init}$ denotes the initial $\boldsymbol{D}$ matrix that is assigned with $M\times d$ infinity values. Thereafter, the discrete sets $\Gamma_k, \forall k=1,\ldots,d$are saved into this initial $\boldsymbol{D}$ matrix. r_k^i is the ith element of Γ_k, and M refers to the maximal size of these discrete sets at different dimensions. Thereafter, the points generated using the design of experiments (DoE) method are correspondingly approximated to their closest discrete values in $\boldsymbol{D}$, and their objective and constraints values are calculated, respectively. Furthermore, the initial Kriging models of objective and constraints are separately built using these DoE samples. In the following Kriging-assisted optimization process, new samples selected from the continuous space will be projected into the defined matrix $\boldsymbol{D}$ to get the promising discrete samples.

10.2.1 Multi-Start Knowledge Mining on Kriging

As mentioned above, Kriging can build a continuous mathematical model to predict the landscape of the original discrete problem. Hence, efficient search or sampling strategies for continuous optimization problems can still be utilized to mine the useful discrete information from surrogate models. Generally, the conventional surrogate-based sampling strategies consider the most promising positions in the continuous space as candidate points, like the maximal "expectation of improvement (EI)" point or the minimal prediction (MP) point. For discrete optimization problems, these new samples from the continuous space can be approximated to be the discrete individuals of set Γ, to drive the subsequent optimization. However, the search may pay too much attention to the gap between two discrete values of set Γ, decreasing the optimization efficiency, and sometimes no new discrete samples will be supplemented to update the surrogate models, making the program get stuck. Therefore, a multi-start knowledge mining approach is presented to capture the promising discrete samples, which involves four main steps: multi-start optimization, projection, grid sampling and selection. Correspondingly, Figures 10.2–10.5 give a 2d illustration to describe this process clearly.

As we all know, Kriging can approximate nonlinear problems and always generate multiple predicted local optima. Multi-start optimization can identify these potential local positions, realizing the global search. Mathematically, the predicted local optimal solutions can be expressed as below

$$\hat{f}\left(\boldsymbol{x}_{lo}^{i}\right) \leq \hat{f}(\boldsymbol{x})$$
$$\forall \boldsymbol{x} \in V_i\left(\boldsymbol{x}_{lo}^{i}\right) \subset \Omega, \quad \forall i \in 1,\ldots,q \tag{10.3}$$

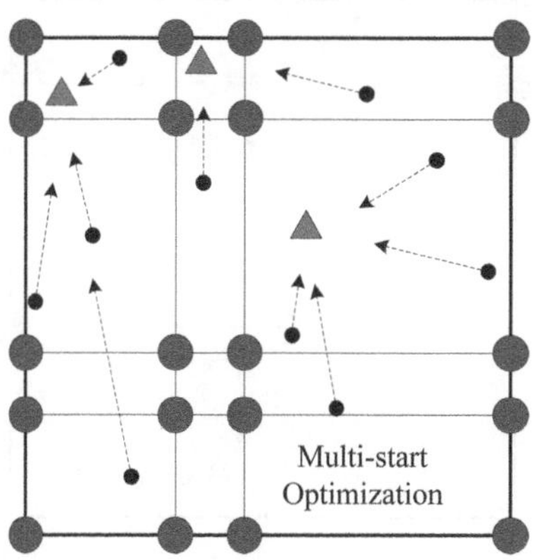

FIGURE 10.2 Step1: Multi-start optimization.

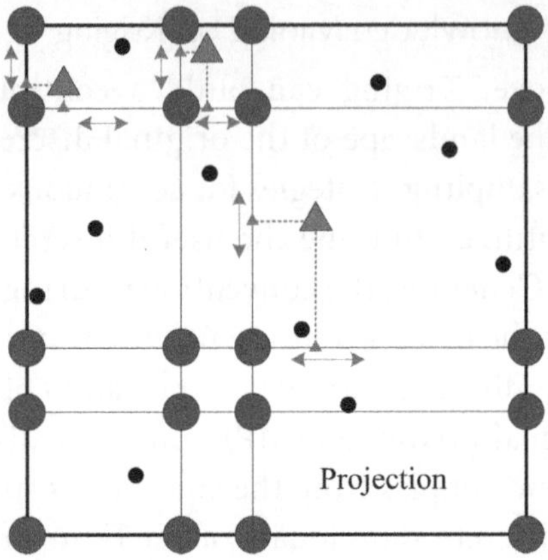

FIGURE 10.3 Step2: Projection to matrix *D*.

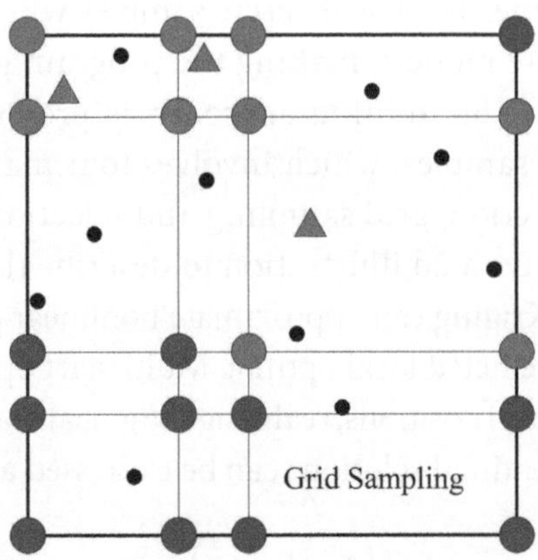

FIGURE 10.4 Step3: Grid sampling.

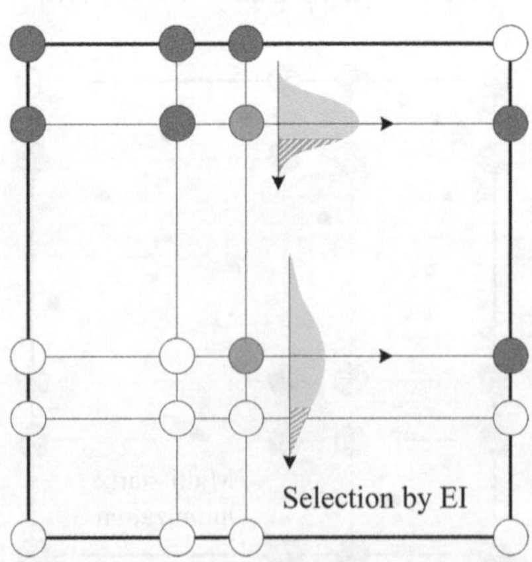

FIGURE 10.5 Step4: Selection by *EI*.

where $\hat{f}\left(\boldsymbol{x}_{lo}^{i}\right)$ refers to the Kriging value at the ith predicted local optimal location $\boldsymbol{x}_{lo}^{i}$, V_i denotes the ith vicinity region around $\boldsymbol{x}_{lo}^{i}$, Ω is the variable range, and q is the number of local optimal locations. An efficient way to get these $\boldsymbol{x}_{lo}$ is to assign a group of starting points that evenly cover the continuous design space and then run local optimization sequentially. The search from the starting points that are located in the same region probably converges to the same optimal solution. In other words, the number of optima is generally smaller than the number of starting points. Figure 10.2 shows that ten starting points (small black dots) will converge to three local optimal solutions $\boldsymbol{x}_{lo}$ (triangles). Additionally, a reduced space enclosing the present best solution is used to improve the computational efficiency of multi-start optimization.

$$RS = \left[\boldsymbol{x}_{pbest} - \frac{dis}{2}, \boldsymbol{x}_{pbest} + \frac{dis}{2}\right] \cap [a,b]$$
$$dis = \xi \times (b-a) \tag{10.4}$$

where RS refers to a neighborhood of the present best solution $\boldsymbol{x}_{pbest}$, $[a, b]$ represents the original design space, and ξ is a coefficient defined as 0.1. When the number of iterations reaches an even number, the reduced space RS is used; otherwise, the original design range is used for the multi-start optimization. The specific pseudo-code of the multi-start optimization is shown in Algorithm 10.1(a).

Algorithm 10.1(a): Multi-start Knowledge Mining: Multi-start Optimization

Input: Kriging model, Original design space [$\boldsymbol{a}$, $\boldsymbol{b}$], Number of iteration *iter*

Output: Predicted local optimal solutions$\boldsymbol{X}_{lo}$

(01) **If** ***iter***/**2∈Z**
(02) ***Range***←Build the reduced space *RS*
(03) ***h*** ← Define the number of starting points as 3.
(04) **Else**
(05) ***Range***←[$\boldsymbol{a}$, $\boldsymbol{b}$].
(06) ***h*** ← Define the number of starting points as 10.
(07) **End if**
(08) ***SP*** ←Employ LHS to get ***h*** starting points in ***Range***.

For i from 1 to $\boldsymbol{h}$
  $\boldsymbol{x}_{lo}^{i} \leftarrow$ Run local optimizer on Kriging to get the local optima in ***Range***.
End for
  $\boldsymbol{X}_{lo} \leftarrow$ Delete the repeated solutions and save $\boldsymbol{q}$ local optimal solutions.
Return $\boldsymbol{X}_{lo}$

Intuitively, these $\boldsymbol{X}_{lo}$ locate in the continuous space, which cannot be chosen as the candidate discrete samples directly. Therefore, projection is suggested to obtain the promising discrete samples. As Eq. (10.2) describes, the $\boldsymbol{D}$ matrix has saved the discrete sets. Project $\boldsymbol{x}_{lo}^{i}$ to each column of D and then find its closest lower and upper discrete values $\boldsymbol{lb}_{lo}^{i}$ and $\boldsymbol{ub}_{lo}^{i}$ in $\Gamma_k, \forall k=1,\ldots,d$. Thereafter, the discrete boundary values $\left[\boldsymbol{lb}_{lo}^{i}, \boldsymbol{ub}_{lo}^{i}\right]$ of each $\boldsymbol{x}_{lo}^{i}$ in $\boldsymbol{D}$ are identified as input for grid sampling. Algorithm 10.1(b) describes the projection process clearly.

Algorithm 10.1(b): Multi-start Knowledge Mining: Projection

Input: Predicted local optimal solutions $\boldsymbol{X}_{lo} = \left\{\boldsymbol{x}_{lo}^{1}, \boldsymbol{x}_{lo}^{2}, \cdots, \boldsymbol{x}_{lo}^{q}\right\}$, $\boldsymbol{D}$ matrix

Output: Discrete boundary values $\left[\boldsymbol{lb}_{lo}^{1}, \boldsymbol{lb}_{lo}^{2}, \cdots, \boldsymbol{lb}_{lo}^{q}\right]$, $\left[\boldsymbol{ub}_{lo}^{1}, \boldsymbol{ub}_{lo}^{2}, \cdots, \boldsymbol{ub}_{lo}^{q}\right]$

For $\boldsymbol{i}$ from 1 to $\boldsymbol{q}$
  For $\boldsymbol{k}$ from 1 to $\boldsymbol{d}$
    Index $\leftarrow$ Employ KNN search to find the index of the nearest individual to $\boldsymbol{x}_{lo}^{i}(k)$ in the kth column Γ_k of $\boldsymbol{D}$.
    If $\Gamma_k^{index} > \boldsymbol{x}_{lo}^{i}(k)$ /* the nearest discrete value is larger than $\boldsymbol{x}_{lo}^{i}(k)$*/
    $\boldsymbol{lb}_{lo}^{i}(k) \leftarrow \Gamma_k^{(index\text{-}1)}$; $\boldsymbol{ub}_{lo}^{i}(k) \leftarrow \Gamma_k^{(index)}$.
    Else if $\Gamma_k^{index} < \boldsymbol{x}_{lo}^{i}(k)$ /* the nearest discrete value is smaller than $\boldsymbol{x}_{lo}^{i}(k)$*/
      $\boldsymbol{lb}_{lo}^{i}(k) \leftarrow \Gamma_k^{(index)}$; $\boldsymbol{ub}_{lo}^{i}(k) \leftarrow \Gamma_k^{(index+1)}$.
    Else /* the nearest discrete value equals to $\boldsymbol{x}_{lo}^{i}(k)$*/
      If ***Index*** =1
        $\boldsymbol{lb}_{lo}^{i}(k) \leftarrow \Gamma_k^{(index)}$; $\boldsymbol{ub}_{lo}^{i}(k) \leftarrow \Gamma_k^{(index+1)}$.
    Else
      $\boldsymbol{lb}_{lo}^{i}(k) \leftarrow \Gamma_k^{(index\text{-}1)}$; $\boldsymbol{ub}_{lo}^{i}(k) \leftarrow \Gamma_k^{(index)}$.
    End if

(14) **End if**
(15) **End for**
(16) **End for**
(17) **Return** $\left[\boldsymbol{lb}_{lo}^{1},\boldsymbol{lb}_{lo}^{2},\cdots,\boldsymbol{lb}_{lo}^{q}\right],\left[\boldsymbol{ub}_{lo}^{1},\boldsymbol{ub}_{lo}^{2},\cdots,\boldsymbol{ub}_{lo}^{q}\right]$

Figures 10.2 and 10.3 show the projection and grid sampling process in a two-dimensional space. When all the promising discrete grid samples are collected together, the repeated points are deleted and nine big darker dots are flagged as the candidate point set (see Figure 10.4). It is noteworthy that the number of grid sampling is 2^d, which will dramatically increase when the dimension d gets larger. For example, if there is a 20-dimensional problem, it will generate 1,048,576 grid sampling points. It is time-consuming to call the 20-dimensional Kriging model 1,048,576 times in each cycle, which greatly decreases KDGO's search efficiency. Therefore, a probability-based grid sampling approach is proposed to get the high-dimensional ($d > 8$) candidate samples. Specifically, the mathematical expression is described below as:

$$P^{i}(k)=\frac{x_{lo}^{i}(k)-lb_{lo}^{i}(k)}{ub_{lo}^{i}(k)-lb_{lo}^{i}(k)}$$

$$\begin{cases} c_j(k)=ub_{lo}^{i}(k) & \textit{if } R<P^{i}(k) \\ c_j(k)=lb_{lo}^{i}(k) & \textit{if } R\geq P^{i}(k) \end{cases} \tag{10.5}$$

$$\forall i=1,2,\ldots,q. \quad \forall k=1,2,\ldots,d. \quad \forall j=1,2,\ldots,m.$$

where $P^{i}(k)$ is the probability threshold value, $x_{lo}^{i}(k)$ is the kth dimension of the ith local optimal solution, and its corresponding discrete boundary values are $lb_{lo}^{i}(k)$ and $ub_{lo}^{i}(k)$. Besides, a random variable R between [0, 1] is defined to be compared with $P^{i}(k)$ for selection. It will have a higher probability to select one of the boundary values $lb_{lo}^{i}(k)$ or $ub_{lo}^{i}(k)$ that is closer to the continuous point $x_{lo}^{i}(k)$, and the selected discrete points are saved in a candidate sample set $\mathbf{C}=\{c_1,c_2,\ldots,c_m\}$. Equation (10.5) guarantees that KDGO can extract the most potential points from the complete grid sampling sets and control the number of points to avoid generating a large computational cost in each cycle. The pseudo-code of grid sampling is summarized in Algorithm 10.1(c)

Algorithm 10.1(c): Multi-start Knowledge Mining: Grid Sampling

Input: Discrete boundary values $\left[\boldsymbol{lb}_{lo}^{1}, \boldsymbol{lb}_{lo}^{2}, \cdots, \boldsymbol{lb}_{lo}^{q}\right]$, $\left[\boldsymbol{ub}_{lo}^{1}, \boldsymbol{ub}_{lo}^{2}, \cdots, \boldsymbol{ub}_{lo}^{q}\right]$, Predicted local optimal solutions $\boldsymbol{X}_{lo} = \left\{\boldsymbol{x}_{lo}^{1}, \boldsymbol{x}_{lo}^{2}, \cdots, \boldsymbol{x}_{lo}^{q}\right\}$, Search region ***Range***

Output: Discrete candidate samples $\boldsymbol{C} = \{c_1, c_2, \cdots, c_m\}$

$\boldsymbol{C} \leftarrow \varnothing$ /* Initialize the candidate sample set*/
If $\boldsymbol{d} < 8$ /* if this is a lower dimensional problem*/
  Delete the repeated samples in $\left[\boldsymbol{lb}_{lo}^{1}, \boldsymbol{lb}_{lo}^{2}, \cdots, \boldsymbol{lb}_{lo}^{q}\right]$, $\left[\boldsymbol{ub}_{lo}^{1}, \boldsymbol{ub}_{lo}^{2}, \cdots, \boldsymbol{ub}_{lo}^{q}\right]$.
  For $\boldsymbol{i}$ from 1 to $\boldsymbol{q}$
    Temp ← Generate the grid samples using $\left[\boldsymbol{lb}_{lo}^{i}, \boldsymbol{ub}_{lo}^{i}\right]$.
    $\boldsymbol{C} \leftarrow \boldsymbol{C} \cup \boldsymbol{Temp}$ /* Update $\boldsymbol{C}$ */
  End for
Else /* if this is a higher dimensional problem*/
  For $\boldsymbol{i}$ from 1to $\boldsymbol{q}$
    For $\boldsymbol{j}$ from 1 to $\boldsymbol{m}$ /* $\boldsymbol{m}$ equals to $100\boldsymbol{d}$ */
      For $\boldsymbol{k}$ from 1 to $\boldsymbol{d}$
        Temp ← Generate the grid samples based on Eq. (10.5).
        $\boldsymbol{C} \leftarrow \boldsymbol{C} \cup \boldsymbol{Temp}$ /* Update $\boldsymbol{C}$ */
      End for
    End for
  End for
End if
C ← Delete the repeated grid samples and update **C**.
If C is $\varnothing$
  C ← Generate $10\boldsymbol{d}$ rounded samples in $[\boldsymbol{a}, \boldsymbol{b}]$ by LHS
End if
Return $\boldsymbol{C} = \{c_1, c_2, \cdots, c_m\}$

As Figure 10.5 shows, the further mining is necessary to get the most valuable points (two light colored dots with a normal distribution) from **C**. In KDGO, the KNN search is used to check the conflict of the known sample pool **S** and the candidate points **C**. Specifically, the judgment conditions are summarized below as:

$$\begin{cases} \textit{feasible} & \textit{if } knn(S, c_i) \neq 0 \\ \textit{infeasible} & \textit{if } knn(S, c_i) = 0 \end{cases} \tag{10.6}$$

$$\forall i = 1, \ldots, m$$

where m refers to the number of candidate points, and c_i is the ith candidate point. If a candidate is infeasible, this implies it has appeared in the expensive sample pool **S**, and it will not be considered here. Furthermore, the *EI* criterion is employed to sort the remaining feasible points in $\mathbf{C}=\{c_1,c_2,\ldots,c_p\}$, $p\leq m$, and the top n samples with larger EI values will be selected to update Kriging. According to the basic theory of Kriging, a candidate sample c_i can be regarded as a random variable $Y_i(\boldsymbol{x})$ with mean value $\hat{y}_i(\boldsymbol{x})$ and variance $\hat{s}_i^2(\boldsymbol{x})$. Naturally, the improvement of the new candidate sample beyond the present best sample y_{best} from the sample pool **S** can be expressed below as:

$$I_i(\boldsymbol{x})=\max\left(y_{best}-Y_i(\boldsymbol{x}),0\right) \tag{10.7}$$

Obviously, $I_i(\boldsymbol{x})$ is a random variable, and its mathematical expectation is formulated as follows:

$$EI_i(\boldsymbol{x})=\begin{cases}\left(y_{best}-\hat{y}_i(\boldsymbol{x})\right)\Phi\left(\dfrac{y_{best}-\hat{y}_i(\boldsymbol{x})}{\hat{s}_i(\boldsymbol{x})}\right)+\hat{s}_i(\boldsymbol{x})\phi\left(\dfrac{y_{best}-\hat{y}_i(\boldsymbol{x})}{\hat{s}_i(\boldsymbol{x})}\right), & \hat{s}_i(\boldsymbol{x})\neq 0\\ 0 & \hat{s}_i(\boldsymbol{x})=0\end{cases}$$

$$\forall i=1,2,\ldots,p$$

$$\mathbf{EI}=\left\{EI_1\geq EI_2\geq\cdots\geq EI_n\geq\cdots\geq EI_p\right\} \tag{10.8}$$

where ϕ and Φ represent the probability density and cumulative density functions, respectively. More precisely, the detailed pseudo-code is given.

Algorithm10.1(d): Multi-start Knowledge Mining: Selection

Input: Discrete candidate samples $\mathbf{C}=\{c_1,c_2,\cdots,c_m\}$, Expensive sample pool **S**, Kriging model, Number of sampling per cycle ***n***

Output: Promising samples $\boldsymbol{PS}=\{ps_1,ps_2,\cdots,ps_n\}$

(01) ***PS***$\leftarrow\varnothing$ /* Initialize the promising sample set*/
(02) $\mathbf{C}=\{c_1,c_2,\cdots,c_p\}\leftarrow$ Utilize Eq. (10.8) to update the candidate sample set.
(03) **For *i* from 1 to *p***
(04) $EI_i\leftarrow$ Utilize Eq. (10.8) to get the corresponding EI value.

(05) **End for**
(06) **If** $n<p$
(07) $PS \leftarrow$ Sort $\boldsymbol{C}$ and Select the top $\boldsymbol{n}$ promising samples based on *EI* value from $\boldsymbol{C}$
(08) **else**
(09) $PS \leftarrow C$
(10) **End if**
(11) Return $\boldsymbol{PS}=\{ps_1, ps_2, \cdots, ps_n\}$

Additionally, when the reduced space is used to speed up the multi-start search, there is some possibility that the promising sample set ***PS*** may be empty. Once it happens, 100*d* cheap points are generated by LHS in the original design space, and their corresponding *EI* values are calculated. The point with the maximal *EI* value will be selected and approximated to the discrete values in matrix ***D***, making the loop continue working.

10.2.2 Constraint Handling

Computationally intensive inequality constraints are also considered in KDGO. Each constraint function $g_i(\boldsymbol{x})$ is approximated by Kriging and will be updated with iteration continuing. In the multi-start optimization, the local search needs to meet the constraint conditions as follows:

$$\begin{gathered}\hat{f}\left(\boldsymbol{x}_{lo}^{i}\right) \leq \hat{f}(\boldsymbol{x}) \\ \hat{g}_{j}\left(\boldsymbol{x}_{lo}^{i}\right) \leq 0, \quad \forall j \in 1,\ldots,m \\ \forall \boldsymbol{x} \in V_{i}\left(\boldsymbol{x}_{lo}^{i}\right) \subset \Omega, \quad \forall i \in 1,\ldots,q\end{gathered} \tag{10.9}$$

where m refers to the number of constraints and q is the number of local optima. Besides, the corresponding constraint information of each sample is supplemented to the expensive sample pool ***S*** and a penalty function is used to fuse the objective and constraints.

$$F(\boldsymbol{x}) = f(\boldsymbol{x}) + P \times \sum_{i=1}^{m} \max\left(g_i(\boldsymbol{x}), 0\right) \tag{10.10}$$

where P is a penalty coefficient with a large value 1e10. $F(\boldsymbol{x})$ will replace $f(\boldsymbol{x})$ to identify the current best location and value in Eqs. (10.4) and (10.8). Moreover, the *EI* criterion has been modified for constrained problems.

$$EI(\boldsymbol{x})=\begin{cases}\left(F_{best}-\hat{f}_P(\boldsymbol{x})\right)\Phi\left(\dfrac{F_{best}-\hat{f}_P(\boldsymbol{x})}{\hat{s}(\boldsymbol{x})}\right)+\hat{s}(\boldsymbol{x})\phi\left(\dfrac{F_{best}-\hat{f}_P(\boldsymbol{x})}{\hat{s}(\boldsymbol{x})}\right), & \hat{s}(\boldsymbol{x})\neq 0\\ 0 & \hat{s}(\boldsymbol{x})=0\end{cases}$$

$$\hat{f}_P(\boldsymbol{x})=\hat{f}(\boldsymbol{x})+P\times\sum_{i=1}^{m}\max\left(\hat{g}_i(\boldsymbol{x}),0\right),\quad \boldsymbol{x}\in\Omega$$

$$F_{best}=\min F(\boldsymbol{s}),\quad \boldsymbol{s}\in\boldsymbol{S} \tag{10.11}$$

where $\hat{f}_P(\boldsymbol{x})$ is the penalty function of the predicted objective and constraints and F_{best} is the current best value in sample pool **S**.

10.3 OVERALL OPTIMIZATION FRAMEWORK

In this chapter, the whole optimization flow and the detailed steps of KDGO are provided. Figure 10.6 shows the detailed optimization flow. It is clear that KDGO mainly includes two parts: one is the initialization, and the other one is the proposed multi-start knowledge mining. Specifically, the steps are summarized below as:

- Step 1: Initialize the Matrix $\boldsymbol{D}$ and some basic parameters including design range $[\boldsymbol{a}, \boldsymbol{b}]$, the dimension d, the number of DoE points N_{DoE}, the maximal number of sampling in each cycle n, the number of starting points in local and global ranges h, the coefficient of local range ξ, and the number of iterations $iter = 0$.
- Step 2: Carry out OLHS sampling to get N_{DoE} initial sample points and project these points to the Matrix $\boldsymbol{D}$ for the discrete samples.
- Step 3: Calculate their objective and constraint function values, sort them according to Eq. (10.10), and build the initial Kriging models for the objective and constraints.
- Step 4: As Algorithm 10.1(a) shows, carry out "**Multi-start Optimization**" and get the predicted local optimal solutions $\boldsymbol{X}_{lo}=\left\{\boldsymbol{x}_{lo}^{1},\boldsymbol{x}_{lo}^{2},\ldots,\boldsymbol{x}_{lo}^{q}\right\}$.
- Step 5: As Algorithm 10.1(b) shows, carry out "**Projection**" and get the closest bounds $\left[\boldsymbol{lb}_{lo}^{1},\boldsymbol{lb}_{lo}^{2},\ldots,\boldsymbol{lb}_{lo}^{q}\right],\left[\boldsymbol{ub}_{lo}^{1},\boldsymbol{ub}_{lo}^{2},\ldots,\boldsymbol{ub}_{lo}^{q}\right]$ of these predicted local optimal solutions from $\boldsymbol{D}$.

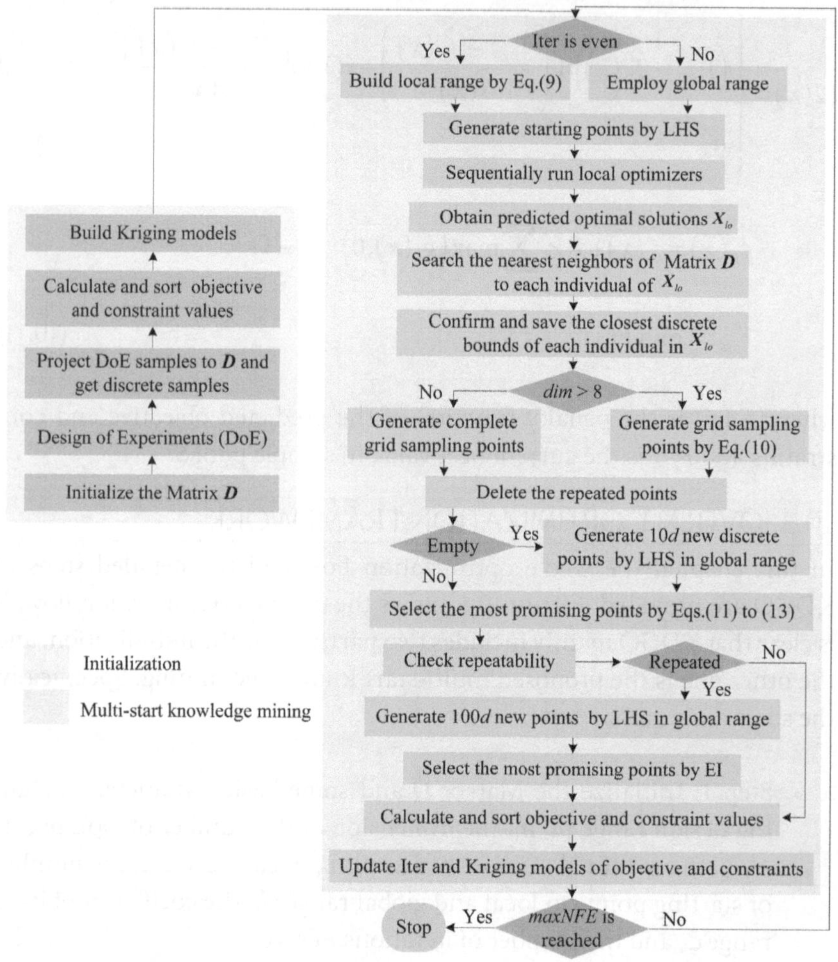

FIGURE 10.6 Overall optimization flow of KDGO.

- Step 6: As Algorithm 10.1(c) shows, carry out "**Grid Sampling**" and use these bounds to create grid points $\boldsymbol{C}=\{c_1,c_2,\ldots,c_m\}$ as the candidate sample points.

- Step 7: As Algorithm 10.1(d) shows, carry out "**Selection**" and select top n promising individuals $\boldsymbol{PS}=\{ps_1,ps_2,\ldots,ps_n\}$ from the candidate sample points.

- Step 8: If ***PS*** is empty, generate 100*d* LHS points in the design range and select the best one with maximal *EI* value. Moreover, find its approximate solution in ***D***.

- Step 9: Calculate the objective and constraint values of these promising points, and sort them.
- Step 10: Update the number of iterations $iter = iter + 1$, and update the Kriging models.
- Step 11: If the number of function evaluations (NFE) reaches *maxNFE*, KDGO stops. Otherwise, the algorithm goes back to Step 4 and continues this loop.

10.4 ALGORITHMIC TEST

10.4.1 Mathematical Example Tests

To sufficiently verify the KDGO's ability, 20 representative benchmark cases with different characteristics are used for test runs, including five box-constrained problems, eight inequality-constrained problems and seven black-box engineering applications. All these mathematical functions are regarded as black-box models, meaning that only input and output data are extracted to complete the optimization search. Table 10.1 shows the specific information of these test cases, where dim refers to the number of design variables. Moreover, LO, UMO, MMO and BBO individually represent linear, unimodal, multimodal and black-box objectives, and LC and NLC are linear and nonlinear constraints, respectively. In engineering applications, H1p1, H1p2, H1p3, H2p1, H2p2 and H2p3 are six subproblems about optimization design of hydropower generation. Since large hydropower facilities are designed using different generators, the goal of these applications is to maximize the power output during 1 day for hydropower plants with five types of generating units. More details can be found in the reference (Li et al., 2013). Besides, The three-stage buffer allocation problem (TP) is an application problem of throughput maximization (Pichitlamken et al., 2006), where the total buffer size and service rate are restricted. The goal of TP is to maximize the average output rate in a flow line with 12 stations that will generate 11 variables about buffer storage and 12 variables about service rate.

Furthermore, six algorithms including genetic algorithm (GA), NOMAD, SO-I, local-SO-I, SO-MI, and VNS are used as comparisons to demonstrate KDGO's powerful ability. Tables 10.2–10.6 directly come from Müller's work (Müller et al., 2014). For a fair comparison, KDGO uses the same termination criterion that the algorithm will stop after 400 function

TABLE 10.1 Specific Information about the Test Cases

Types	ID	Cases	*dim*	Design Space	Description
Box-constrained problems	1	Cf	8	$[-10, 10]^8$	UMO
	2	Nvs	10	$[3, 9]^{10}$	UMO
	3	Anvs	10	$[3, 99]^{10}$	MMO
	4	Rast01	12	$[-1, 3]^{12}$	MMO
	5	Rast02	12	$[-10, 30]^{12}$	MMO
Inequality-constrained problems	6	G6	2	$[13, 100] \times [0, 100]$	MMO, 2NLC
	7	Ex	5	$[0, 10]^{3} \times [0, 1]^2$	LO, 2NLC, 3LC
	8	G4	5	$[78, 102] \times [33, 45] \times [27, 45]^3$	UMO, 6NLC
	9	Aex	5	$[0, 10]^{3} \times [0, 1]^2$	LO, 3LC
	10	G9	7	$[-10, 10]^7$	UMO, 4NLC
	11	G1	13	$[0, 1]^{10} \times [0, 100]^3$	MMO, 9LC
	12	G1m	13	$[0, 100]^{13}$	MMO, 9LC
	13	Hmi	16	$[0, 1]^{16}$	MMO, 7NLC
Engineering applications	14	H1p1	5	$[0, 10]^5$	BBO, 1NLC
	15	H1p2	5	$[0, 10]^5$	BBO, 1NLC
	16	H1p3	5	$[0, 10]^5$	BBO, 1NLC
	17	H2p1	5	$[0, 10]^5$	BBO, 2NLC
	18	H2p2	5	$[0, 10]^5$	BBO, 2NLC
	19	H2p3	5	$[0, 10]^5$	BBO, 2NLC
	20	TP	23	$[1, 20]^{23}$	BBO, 2LC

evaluations, and the mean of the best values and the standard errors of the means (SEM) after 30 runs are recorded in Tables 10.2–10.6. It is worth noting that there are two versions of VNS for inequality-constrained problems. VNS-i uses a similar strategy as SO-I that minimizes the constraint violation function to find the first feasible point, while VNS-ii directly uses SO-I to find the feasible solutions as the starting points. Moreover, the rank in Tables 10.2–10.6 lists the ranks of all the algorithms and NF refers to the number of test runs that cannot get the feasible solutions. NF is the first priority to determine the performance of one algorithm, mean and SEM are the second and third, respectively. If the algorithm A has a smaller NF value than the algorithm B, it indicates A is better than B. If NF values of A and B are the same and the mean of A is smaller than B, it suggests that A is better than B. Similarly, if the NF and mean values of A and B are the same, the better algorithm should have a smaller SEM.

Table 10.2 provides the comparison results on box-constrained cases. For the two unimodal problems Cf and Nvs, all seven algorithms can find

TABLE 10.2 Comparison Results on Box-Constrained Cases

Cases	Algorithms	#NF	Mean(SEM)	Rank
Cf	KDGO	0	**0.00** (0.00)	1
	GA	0	2.72 (0.95)	7
	SO-I	0	**0.00** (0.00)	1
	local-SO-I	0	**0.00** (0.00)	1
	SO-MI	0	**0.00** (0.00)	1
	NOMAD	0	**0.00** (0.00)	1
	VNS-i/ VNS-ii	0	**0.00** (0.00)	1
Nvs	KDGO	0	**−43.13** (0.00)	1
	GA	0	**−43.13** (0.00)	1
	SO-I	0	**−43.13** (0.00)	1
	local-SO-I	0	**−43.13** (0.00)	1
	SO-MI	0	**−43.13** (0.00)	1
	NOMAD	0	**−43.13** (0.00)	1
	VNS-i/VNS-ii	0	**−43.13** (0.00)	1
Anvs	KDGO	0	−9,591.72 (0.00)	1
	GA	0	−9,289.87 (81.24)	6
	SO-I	0	−9,591.72 (0.00)	1
	local-SO-I	0	−9,591.72 (0.00)	1
	SO-MI	0	−9,591.72 (0.00)	1
	NOMAD	0	−9,591.72 (0.00)	1
	VNS-i/VNS-ii	0	−5,448.97 (358.19)	7
Rast01	KDGO	0	**−12.00** (0.00)	1
	GA	0	−10.87 (0.19)	6
	SO-I	0	**−12.00** (0.00)	1
	local-SO-I	0	−10.03 (0.77)	7
	SO-MI	0	**−12.00** (0.00)	1
	NOMAD	0	**−12.00** (0.00)	1
	VNS-i/VNS-ii	0	**−12.00** (0.00)	1
Rast02	KDGO	0	**−12.00** (0.00)	1
	GA	0	33.83 (4.52)	7
	SO-I	0	**−12.00** (0.00)	1
	local-SO-I	0	**−12.00** (0.00)	1
	SO-MI	0	**−12.00** (0.00)	1
	NOMAD	0	−10.67 (1.33)	5
	VNS-i/VNS-ii	0	16.47 (22.67)	6

the global optimal solutions within 400 function evaluations except GA on Cf. Besides, GA and VNS seem to have difficulty in dealing with Anvs, while others can successfully find its global optimum. Rast02, an extended version of Rast01, has a larger design space, increasing the search difficulty

TABLE 10.3 Comparison Results on Inequality-Constrained Cases-Part1

Cases	Algorithms	#NF	Mean(SEM)	Rank
G6	KDGO	0	**−3,971.00** (0.00)	1
	GA	2	−3,971.00 (0.00)	5
	SO-I	0	**−3,971.00** (0.00)	1
	local-SO-I	19	−3,971.00 (0.00)	7
	SO-MI	0	**−3,971.00** (0.00)	1
	NOMAD	30	NA	8
	VNS-i	14	−3,971.00 (0.00)	6
	VNS-ii	0	**−3,971.00** (0.00)	1
Ex	KDGO	0	**0.00** (0.00)	1
	GA	0	0.72 (0.14)	7
	SO-I	0	**0.00** (0.00)	1
	local-SO-I	0	**0.00** (0.00)	1
	SO-MI	0	**0.00** (0.00)	1
	NOMAD	0	**0.00** (0.00)	1
	VNS-i	10	0.00 (0.00)	8
	VNS-ii	0	0.03 (0.03)	6
G4	KDGO	0	**−30,456.91** (2.76)	1
	GA	0	−30,073.77 (43.30)	5
	SO-I	0	−30,303.66 (31.17)	2
	local-SO-I	0	−29,069.70 (106.61)	8
	SO-MI	0	−30,075.73 (53.15)	4
	NOMAD	0	−30,192.67 (35.29)	3
	VNS-i	0	−29,574.12 (92.80)	6
	VNS-ii	0	−29,486.62 (68.83)	7
Aex	KDGO	0	**−8.00** (0.00)	1
	GA	0	−7.10 (0.16)	5
	SO-I	0	**−8.00** (0.00)	1
	local-SO-I	0	−6.88 (0.21)	6
	SO-MI	0	**−8.00** (0.00)	1
	NOMAD	8	−8.00 (0.00)	7
	VNS-i	16	−7.75 (0.11)	8
	VNS-ii	0	−7.93 (0.04)	4

of GA, NOMAD and VNS. Intuitively, the three surrogate-based algorithms KDGO, SO-I and SO-MI almost have the same performance on these box-constrained cases, and outperform the other four algorithms.

For inequality-constrained problems, NF is an important indicator to evaluate the performance of these algorithms. For example, local-SO-I, NOMAD and VNS-i have worse Rank on G6, because they frequently get

TABLE 10.4 Comparison Results on Inequality-Constrained Cases-Part2

Cases	Algorithms	#NF	Mean (SEM)	Rank
G9	KDGO	0	**744.80** (8.93)	1
	GA	0	896.53 (29.21)	4
	SO-I	0	771.40 (14.97)	2
	local-SO-I	0	997.10 (246.89)	5
	SO-MI	0	812.17 (12.46)	3
	NOMAD	0	1,770.50 (462.58)	6
	VNS-i	4	8,906.35 (6,161.61)	8
	VNS-ii	0	2,097.17 (1,367.02)	7
G1	KDGO	0	−14.57 (0.15)	2
	GA	0	−6.07 (0.59)	6
	SO-I	0	**−14.83** (0.10)	1
	local-SO-I	0	−12.00 (0.00)	4
	SO-MI	0	−12.00 (0.32)	5
	NOMAD	0	−5.97 (0.03)	7
	VNS-i	30	NA	8
	VNS-ii	0	−14.37 (0.24)	3
G1m	KDGO	0	**−50,197.70** (1.34)	1
	GA	1	−40,105.07 (3,175.53)	7
	SO-I	0	−40,687.10 (3,145.90)	5
	local-SO-I	0	−42,185.67 (1,966.68)	4
	SO-MI	0	−50,024.17 (36.40)	2
	NOMAD	0	−48,363.03 (1,197.02)	3
	VNS-i	30	NA	8
	VNS-ii	0	−35,687.03 (3,252.89)	6
Hmi	KDGO	0	**13.20** (0.14)	1
	GA	8	17.73 (0.86)	3
	SO-I	14	14.00 (0.68)	7
	local-SO-I	8	20.96 (3.11)	4
	SO-MI	14	13.50 (0.50)	6
	NOMAD	22	13.00 (0.00)	8
	VNS-i	4	13.73 (0.44)	2
	VNS-ii	14	13.00 (0.00)	5

stuck in infeasible regions. In particular, NOMAD cannot succeed once on G6, thus gets Rank 8. Compared with Aex, Ex has two additional nonlinear constraints. However, most algorithms can find satisfactory solutions on Ex, while having difficulty on Aex. VNS-i and VNS-ii use different manners to find feasible solutions, thus they have different search efficiency. Intuitively, VNS-ii seems better than VNS-i, because VNS-i always has

TABLE 10.5 Comparison Results on Engineering Applications-Part1

Cases	Algorithms	#NF	Mean(SEM)	Rank
H1p1	KDGO	0	**758.25** (0.00)	1
	GA	0	735.34 (6.75)	7
	SO-I	0	**758.25** (0.00)	1
	local-SO-I	0	681.38 (10.25)	8
	SO-MI	0	754.38 (0.88)	4
	NOMAD	0	744.00 (0.96)	6
	VNS-i	0	753.83 (1.54)	5
	VNS-ii	0	755.04 (1.12)	3
H1p2	KDGO	0	**2,021.67** (0.22)	1
	GA	0	2,008.83 (4.68)	5
	SO-I	0	2,020.67 (1.33)	2
	local-SO-I	0	1,835.14 (24.94)	8
	SO-MI	0	2,015.46 (1.51)	3
	NOMAD	0	2,003.83 (5.01)	7
	VNS-i	0	2,010.83 (3.81)	4
	VNS-ii	0	2,006.46 (6.91)	6
H1p3	KDGO	0	4,116.39 (4.31)	3
	GA	0	4,108.84 (4.49)	5
	SO-I	0	4,114.63 (2.50)	4
	local-SO-I	0	3,890.61 (20.74)	8
	SO-MI	0	4,117.98 (2.58)	2
	NOMAD	0	**4,125.75** (12.06)	1
	VNS-i	0	4,075.42 (10.44)	7
	VNS-ii	0	4,099.17 (6.69)	6

larger NF values on Ex and Aex. G4 has a larger feasible space ratio, thus all eight algorithms can successfully identify the feasible area and go close to the global optimum. It is obvious that KDGO has the best mean value –30,456.91 on G4 and is much better than others. Similarly, KDGO still gets rank 1 on G9 and SO-I has satisfactory performance as well. However, VNS-i, VNS-ii and NOMAD perform so badly on G9 that their mean values are much larger than 1,000. Although GA sometimes may fail to get feasible solutions, it has better global exploration ability and can always obtain acceptable results. For example, GA gets the mean value 896.53 on G9, much better than local-SO-I.

G1m is an extended version of G1 and has a larger search space. There is no doubt that SO-I outperforms others on G1 and gets the best mean value –14.83. Relatively, KDGO finds the true global optimum –15 for

TABLE 10.6 Comparison Results on Engineering Applications-Part2

Cases	Algorithms	#NF	Mean(SEM)	Rank
H2p1	KDGO	0	**1,679.05** (1.91)	1
	GA	0	1,560.36 (25.96)	6
	SO-I	0	1,677.17 (2.10)	2
	local-SO-I	0	1,443.12 (46.42)	7
	SO-MI	0	1,657.08 (3.75)	4
	NOMAD	0	1,626.18 (19.64)	5
	VNS-i	4	1,653.65 (13.92)	8
	VNS-ii	0	1,671.50 (2.92)	3
H2p2	KDGO	0	**4,124.70** (7.08)	1
	GA	0	4,016.17 (23.14)	5
	SO-I	0	4,097.40 (11.86)	2
	local-SO-I	0	3,668.60 (50.60)	7
	SO-MI	0	4,095.50 (6.96)	3
	NOMAD	0	3,899.40 (25.57)	6
	VNS-i	2	4,000.93 (28.66)	8
	VNS-ii	0	4,070.83 (16.29)	4
H2p3	KDGO	0	**8,302.33** (7.99)	1
	GA	0	8,220.67 (22.22)	4
	SO-I	0	8,299.00 (12.84)	2
	local-SO-I	0	7,550.17 (73.20)	8
	SO-MI	0	8,253.17 (9.62)	3
	NOMAD	0	8,122.33 (32.05)	5
	VNS-i	0	8,055.83 (49.83)	6
	VNS-ii	0	7,996.33 (64.34)	7
TP	KDGO	0	**4.18** (0.19)	1
	GA	0	3.10 (0.17)	4
	SO-I	0	3.15 (0.20)	3
	local-SO-I	0	2.07 (0.22)	6
	SO-MI	0	3.82 (0.13)	2
	NOMAD	0	0.89 (0.06)	7
	VNS-i	26	1.74 (0.39)	8
	VNS-ii	0	2.52 (0.20)	5

22 times and also gets a satisfactory mean −14.57. On the other hand, KDGO becomes the only method that can obtain a mean result smaller than -−50,000 on G1m. Furthermore, according to the statistical results, KDGO successfully reaches the global optimum −50,200 for 21 times on G1m, showing its superior robustness. When it comes to Hmi, all the algorithms except KDGO seem to encounter some troubles. This is because

Hmi is a binary problem with high dimension and seven nonlinear constraints. However, KDGO can accurately find the global optimum 13 for 28 times and get the impressive mean 13.20, once again demonstrating its high efficiency.

For the constrained engineering cases, it seems that all the algorithms except VNS-i can easily find feasible solutions but they have different convergence abilities. KDGO and SO-I can always find the global optimum 758.25 on H1p1. Besides, KDGO maintains the first on H1p2, H2p1, H2p2 and H2p3. And SO-I follows KDGO closely in most cases. According to the results in Tables 10.5 and 10.6, it can be found that KDGO not only has a better mean but also has a smaller SEM, demonstrating its excellent stability. For H1p3, NOMAD wins the competition and SO-MI has impressive results as well.

It is clear from Table 10.1 that TP is a high-dimensional case that generally needs more function evaluations to explore the design space. After 30 test runs, KDGO undoubtedly acquires the biggest mean 4.18 that has a 9% improvement over SO-MI. In the 20 test cases, KDGO gets 18 Rank 1, 1 Rank 2 and 1 Rank 3; SO-I gets 10 Rank 1, 6 Rank 2 and 1 Rank 3; SO-MI acquires 8 Rank 1, 3 Rank 2 and 4 Rank 3. To sum it up, KDGO is not only good at dealing with mathematical benchmark cases, but also has extraordinary ability to solve actual engineering applications. The results in Tables 10.2–10.6 verify KDGO's functionality and demonstrate its superior performance.

10.4.2 Practical Engineering Application

In this chapter, the presented KDGO is used for structure optimization of a blended-wing-body underwater glider (BWBUG). When the BWBUG is lifted from the water, stress concentration may arise in the skeleton structure because of the vertical downward force on the two wings, which involves the gravities of skeleton, equipment and buoyancy material. For the equipment, the total gravity is defined as 1,500N that depends on the specific tasks and functions of this BWBUG. On the other hand, for the buoyancy material, the density $\rho_{buoyancy}$ is 500 kg/m^3, the occupied volume is $V_{airfoil} \approx 0.11\,\text{m}^3$, and the weight of the material is 55 kg. Therefore, the total gravity of the buoyancy material is $G_{buoyancy} = W_{buoyancy}g \approx 550N$. To get the lightest weight and meanwhile satisfy the stress and deformation constraints, the specific design parameters and optimization formula are summarized below.

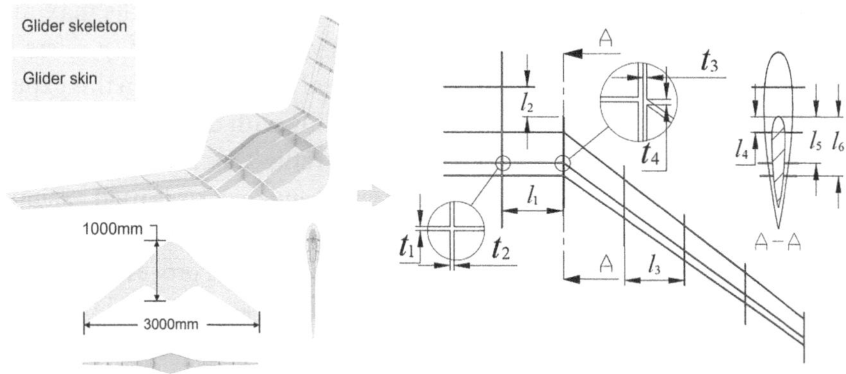

FIGURE 10.7 Structure parameters and illustration.

Figure 10.7 shows the ten design variables including: four thickness parameters t_1, t_2, t_3 and t_4; six relative position parameters l_1, l_2, l_3, l_4, l_5 and l_6. The total length and width of this BWBUG are 1,000 and 3,000 mm, respectively. Besides, the numbers of transverse and longitudinal beams in the body are constants 4 and 2, and those in the wings are constants 3 and 5, respectively. The design objective is to minimize the skeleton weight and meanwhile need to be subject to the equivalent stress and total deformation constraints.

$$
\begin{cases}
\min W_{skeleton} \\
s.t. \quad \sigma_{\max} \le \sigma_s / \gamma \\
\qquad\;\; d_{\max} \le 50\,\text{mm} \\
\qquad\;\; \left.\begin{array}{l} 4 \le t_1 \le 10 \\ 4 \le t_2 \le 10 \\ 3 \le t_3 \le 7 \\ 3 \le t_4 \le 7 \end{array}\right\} \Rightarrow \textit{discrete interval}\ 0.05 \\
\qquad\;\; \left.\begin{array}{l} 255 \le l_1 \le 345 \\ 50 \le l_2 \le 120 \\ 250 \le l_3 \le 320 \end{array}\right\} \Rightarrow \textit{discrete interval}\ 0.5 \\
\qquad\;\; \left.\begin{array}{l} 0.10 \le l_4 \le 0.35 \\ 0.45 \le l_5 \le 0.55 \\ 0.65 \le l_6 \le 0.90 \end{array}\right\} \Rightarrow \textit{discrete interval}\ 0.01
\end{cases}
\tag{10.12}
$$

where $W_{skeleton}$ is the weight of the skeleton structure, $\sigma_{\max}$ is the maximal equivalent stress, σ_s is the tensile/compressive yield strength, γ refers to safety factor and $d_{\max}$ denotes the maximal total deformation. In this experiment, the structure material is aluminum alloy with density 2,770 kg/m^3, Young's modulus 71,000 MPa, and Poisson's ratio 0.33. Besides, the safety factor is 1.6 and σ_s is 280 MPa. The finite element analysis is used to simulate this actual case, and Figure 10.8 shows the specific structure mesh.

Furthermore, Figure 10.9 shows the detailed iterative process, where the stars refer to the feasible solutions, the dots represent infeasible designs, and the best feasible solution is located at the 89th NFE. From Figure 10.9, it is clear that the initial samples from the DoE phase have a wide distribution in the design space, while the efficient infilling strategy makes KDGO find the feasible and optimal regions rapidly. After several iterations, the search focuses on the boundary of the deformation constraint. Intuitively,

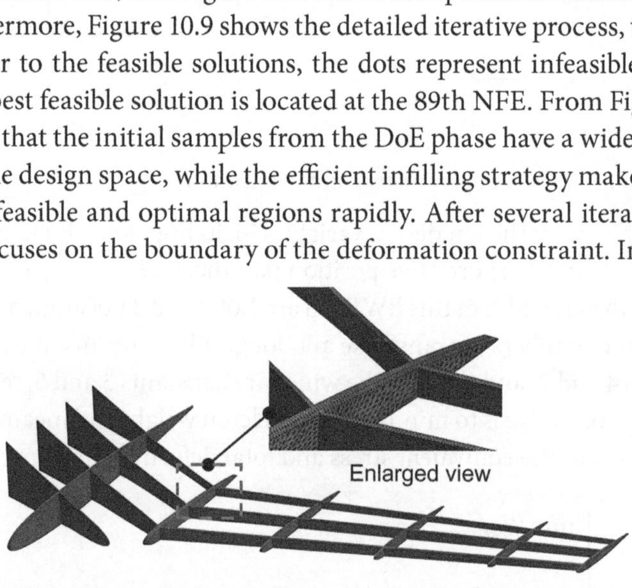

FIGURE 10.8 Illustration of mesh generation.

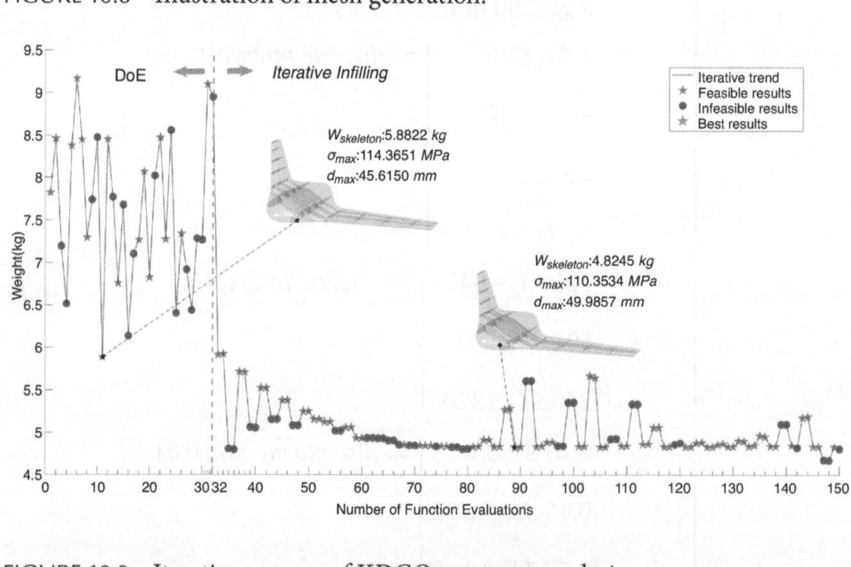

FIGURE 10.9 Iterative process of KDGO on structure design.

TABLE 10.7 Best Solutions in Different Phases

Solutions	t_1	t_2	t_3	t_4	l_1	l_2	l_3	l_4	l_5	l_6
DoE-opt	5.15	4	5.2	6.75	278	72.5	263.5	0.15	0.47	0.77
Final-opt	4	4	3	4.7	255	120	250	0.17	0.51	0.65

TABLE 10.8 Best Response Values in Different Phases

Response	$W_{skeleton}$ (kg)	d_{max} (mm)	σ_{max} (MPa)
DoE-opt	5.8822	45.6150	114.3651
Final-opt	4.8245	49.9857	110.3534

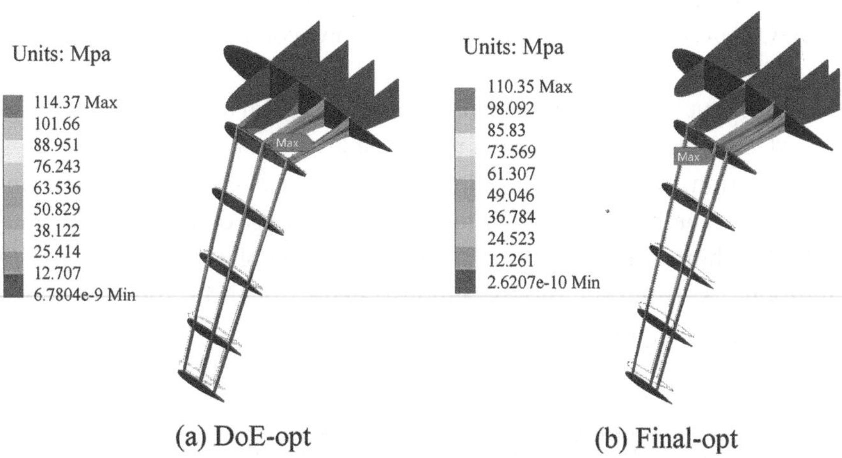

FIGURE 10.10 Equivalent stress diagram. (a) DoE-opt. (b) Final-opt.

KDGO begins to converge after 80 simulations and finally identifies the best objective value after 88 simulations.

Additionally, the best results obtained in different phases (the DoE phase and the final phase) are summarized in Tables 10.7 and 10.8. Here, DoE-opt refers to the best result obtained after DoE and Final-opt denotes the final best result. The final weight has an 18% improvement after the iterative infilling process. Correspondingly, the equivalent stress and total deformation diagrams are provided in Figures 10.10 and 10.11. In summary, KDGO cannot only deal with complex mathematical cases but can also efficiently tackle the actual engineering application.

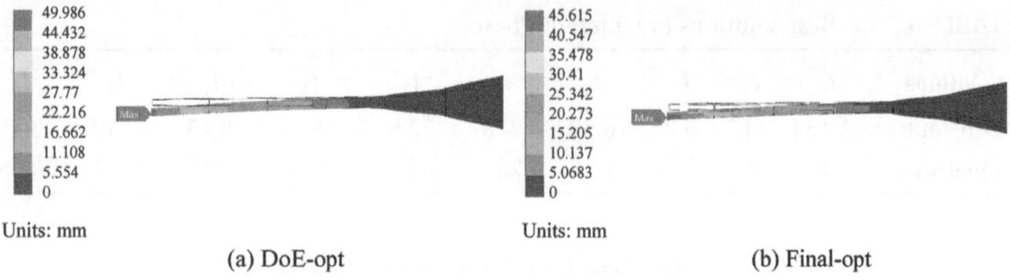

FIGURE 10.11 Total deformation diagram. (a) DoE-opt. (b) Final-opt.

10.5 CHAPTER SUMMARY

In this chapter, a novel discrete global optimization method named KDGO is presented, which can effectively solve computationally expensive black-box problems. In KDGO, an efficient infilling criterion is proposed to iteratively supplement new expensive samples, which involves a multi-start knowledge mining process. The new samples are generated by four steps: optimization, projection, sampling and selection. And the greatest advantage of this method is its ability to solve a wide range of discrete problems, including binary, integer and discrete number set black-box problems. In addition, its strong stability is demonstrated through its application in a wide range of engineering problems.

NOTE

1 Based on "Kriging-assisted Discrete Global Optimization (KDGO) for black-box problems with costly objective and constraints," published in [Applied Soft Computing], [2020]. Permission obtained from [Elsevier].

REFERENCES

Abramson, M. A., Audet, C., Chrissis, J. W., & Walston, J. G. (2009). Mesh Adaptive Direct Search Algorithms for Mixed Variable Optimization. *Optimization Letters*, *3*, 35–47.

Adibi, M. A., Zandieh, M., & Amiri, M. (2010). Multi-Objective Scheduling of Dynamic Job Shop Using Variable Neighborhood Search. *Expert Systems with Applications*, *37*(1), 282–287.

Anghinolfi, D., & Paolucci, M. (2009). A New Discrete Particle Swarm Optimization Approach for the Single-Machine Total Weighted Tardiness Scheduling Problem with Sequence-Dependent Setup Times. *European Journal of Operational Research*, *193*(1), 73–85.

Dede, T. (2014). Application of Teaching-Learning-Based-Optimization Algorithm for the Discrete Optimization of Truss Structures. *KSCE Journal of Civil Engineering, 18*, 1759–1767.

Demeulemeester, E., & Herroelen, W. (1992). A Branch-and-Bound Procedure for the Multiple Resource-Constrained Project Scheduling Problem. *Management Science, 38*(12), 1803–1818.

Dong, H., Li, C., Song, B., & Wang, P. (2018). Multi-Surrogate-Based Differential Evolution with Multi-Start Exploration (MDEME) for Computationally Expensive Optimization. *Advances in Engineering Software, 123*, 62–76.

Dong, H., Song, B., Dong, Z., & Wang, P. (2016). Multi-Start Space Reduction (MSSR) Surrogate-Based Global Optimization Method. *Structural and Multidisciplinary Optimization, 54*, 907–926.

Dong, H., Song, B., Dong, Z., & Wang, P. (2018). SCGOSR: Surrogate-Based Constrained Global Optimization Using Space Reduction. *Applied Soft Computing, 65*, 462–477.

Dong, H., Song, B., & Wang, P. (2017). Kriging-Based Optimization Design for a New Style Shell with Black Box Constraints. *Journal of Algorithms & Computational Technology, 11*(3), 234–245.

Ekel, P. Y., & Neto, F. H. S. (2006). Algorithms of Discrete Optimization and Their Application to Problems with Fuzzy Coefficients. *Information Sciences, 176*(19), 2846–2868.

Guendouz, M., Amine, A., & Hamou, R. M. (2017). A Discrete Modified Fireworks Algorithm for Community Detection in Complex Networks. *Applied Intelligence, 46*, 373–385.

Holmström, K., Quttineh, N.-H., & Edvall, M. M. (2008). An Adaptive Radial Basis Algorithm (ARBF) for Expensive Black-Box Mixed-Integer Constrained Global Optimization. *Optimization and Engineering, 9*, 311–339.

Jiang, C., Wang, D., Qiu, H., Gao, L., Chen, L., & Yang, Z. (2019). An Active Failure-Pursuing Kriging Modeling Method for Time-Dependent Reliability Analysis. *Mechanical Systems and Signal Processing, 129*, 112–129. https://doi.org/10.1016/j.ymssp.2019.04.034

Kytöjoki, J., Nuortio, T., Bräysy, O., & Gendreau, M. (2007). An Efficient Variable Neighborhood Search Heuristic for Very Large Scale Vehicle Routing Problems. *Computers & Operations Research, 34*(9), 2743–2757.

Land, A. H., & Doig, A. G. (2010). *An automatic method for solving discrete programming problems*. Springer.

Lawler, E. L. (1972). A Procedure for Computing the k Best Solutions to Discrete Optimization Problems and Its Application to the Shortest Path Problem. *Management Science, 18*(7), 401–405.

Li, F.-F., Shoemaker, C. A., Wei, J.-H., & Fu, X.-D. (2013). Estimating Maximal Annual Energy Given Heterogeneous Hydropower Generating Units with Application to the Three Gorges System. *Journal of Water Resources Planning and Management, 139*(3), 265–276.

Li, X., Wu, X., Xu, S., Qing, S., & Chang, P.-C. (2019). A Novel Complex Network Community Detection Approach Using Discrete Particle Swarm Optimization with Particle Diversity and Mutation. *Applied Soft Computing, 81*, 105476.

Liu, H., Ong, Y.-S., & Cai, J. (2018). A Survey of Adaptive Sampling for Global Metamodeling in Support of Simulation-Based Complex Engineering Design. *Structural and Multidisciplinary Optimization, 57*, 393–416.

Liu, J., Dong, H., Jin, T., Liu, L., Manouchehrinia, B., & Dong, Z. (2018). Optimization of Hybrid Energy Storage Systems for Vehicles with Dynamic On-Off Power Loads Using a Nested Formulation. *Energies, 11*(10), 2699.

Mladenović, N., & Hansen, P. (1997). Variable Neighborhood Search. *Computers & Operations Research, 24*(11), 1097–1100.

Müller, J., Shoemaker, C. A., & Piché, R. (2013). SO-MI: A Surrogate Model Algorithm for Computationally Expensive Nonlinear Mixed-Integer Black-Box Global Optimization Problems. *Computers & Operations Research, 40*(5), 1383–1400.

Müller, J., Shoemaker, C. A., & Piché, R. (2014). SO-I: A Surrogate Model Algorithm for Expensive Nonlinear Integer Programming Problems Including Global Optimization Applications. *Journal of Global Optimization, 59*, 865–889.

Nakariyakul, S., & Casasent, D. P. (2007). Adaptive Branch and Bound Algorithm for Selecting Optimal Features. *Pattern Recognition Letters, 28*(12), 1415–1427.

Pichitlamken, J., Nelson, B. L., & Hong, L. J. (2006). A Sequential Procedure for Neighborhood Selection-of-the-Best in Optimization via Simulation. *European Journal of Operational Research, 173*(1), 283–298.

Polacek, M., Hartl, R. F., Doerner, K., & Reimann, M. (2004). A Variable Neighborhood Search for the Multi Depot Vehicle Routing Problem with Time Windows. *Journal of Heuristics, 10*, 613–627.

Rashid, K., Ambani, S., & Cetinkaya, E. (2013). An Adaptive Multiquadric Radial Basis Function Method for Expensive Black-Box Mixed-Integer Nonlinear Constrained Optimization. *Engineering Optimization, 45*(2), 185–206.

Sayadi, M. K., Hafezalkotob, A., & Naini, S. G. J. (2013). Firefly-Inspired Algorithm for Discrete Optimization Problems: An Application to Manufacturing Cell Formation. *Journal of Manufacturing Systems, 32*(1), 78–84.

Shi, R., Liu, L., Long, T., Wu, Y., & Gary Wang, G. (2020). Multi-Fidelity Modeling and Adaptive Co-Kriging-Based Optimization for All-Electric Geostationary Orbit Satellite Systems. *Journal of Mechanical Design, 142*(2), 021404.

Stander, J. N., Venter, G., & Kamper, M. J. (2016). High Fidelity Multidisciplinary Design Optimisation of an Electromagnetic Device. *Structural and Multidisciplinary Optimization, 53*, 1113–1127.

Zhang, S., Lee, C. K. M., Chan, H. K., Choy, K. L., & Wu, Z. (2015). Swarm Intelligence Applied in Green Logistics: A Literature Review. *Engineering Applications of Artificial Intelligence, 37*, 154–169.

Zhou, Q., Rong, Y., Shao, X., Jiang, P., Gao, Z., & Cao, L. (2018). Optimization of Laser Brazing onto Galvanized Steel Based on Ensemble of Metamodels. *Journal of Intelligent Manufacturing, 29*(7), 1417–1431. https://doi.org/10.1007/s10845-015-1187-5

Zhou, Q., Wu, J., Xue, T., & Jin, P. (2021). A Two-Stage Adaptive Multi-Fidelity Surrogate Model-Assisted Multi-Objective Genetic Algorithm for Computationally Expensive Problems. *Engineering with Computers, 37*, 623–639.

CHAPTER 11

SAGWO

Surrogate-Assisted Gray Wolf Optimization for High-Dimensional, Computationally Expensive Black-Box Problems[1]

11.1 INTRODUCTION

In searching for the optimal solution, surrogate models are commonly employed to approximate the objective function, thereby replacing expensive simulations and significantly reducing the number of evaluations of costly functions (Dong et al., 2016; Forrester & Keane, 2009). Two types of surrogate-based optimization strategies are typically utilized during the optimization search process. The first is the direct offline optimization approach (Goel et al., 2007; Guo et al., 2018; Hajikolaei & Gary Wang, 2014), which focuses on constructing an accurate surrogate model using a set of well-distributed expensive sample points. Subsequent evolutionary computation (EC) or swarm intelligence (SI) searches are then performed on the surrogate model without further evaluations of the expensive objective function. However, it is challenging to construct a globally accurate surrogate model with a limited number of samples, especially for multimodal or high-dimensional optimization problems. The second is the dynamic or online optimization approach (Dong et al., 2018a; Liu et al., 2017; Long et al., 2015; Müller et al., 2014; Regis & Shoemaker, 2013), which begins with a coarse surrogate model and adaptively refines

DOI: 10.1201/9781003636267-11

it by adding new expensive samples according to certain infill strategies during each iteration of the search process. The key challenges in online optimization lie in designing effective infill strategies and balancing the exploration of unknown regions with the exploitation of the current model (Haftka et al., 2016).

These stated search methods could sufficiently utilize the predictive information of surrogate models and perform well on lower dimensional ($D < 10$) problems, but the algorithms also encountered challenges in the higher dimensional ($D \geq 10$), computationally expensive optimizations. One reason is that the high-dimensional problems have much larger exploration space and more local optima, leading to the difficulty in global optimization search. The other contributing factor is that the current approximation techniques generate huge errors in search of a high-dimensional problem, mistakenly guiding the search and wasting a large amount of computational effort. Too much dependence on surrogate models makes the search method inefficient, ineffective and even infeasible for solving high-dimensional optimization problems (Dong et al., 2018b; Shan & Wang, 2010).

Surrogate-assisted EC or SI algorithms (SAEC/SIAs) are different from the methods discussed above. Although the SAEC/SIAs still need intelligent infill sampling to update new individuals and the generation of points, they do not overly rely on the prediction information coming from the surrogates. SAEC/SIAs retain the metaheuristic characteristics that stochastically capture new samples around the present best solution or go to the unknown area for global exploration. Generally, SAEC/SIAs utilize the surrogate models as the prescreening tools to select promising individuals, which makes SAEC/SIAs more suitable for high-dimensional, computationally expensive global optimization. The strategies for managing surrogates in SAEC/SIAs can be classified into generation-based, individual-based and population-based methods. In the generation-based methods, the points of some generations are created using surrogates, while the others are still produced by evaluating the expensive fitness/objective function. In the individual-based strategies, surrogates are used to evaluate the fitness of some individual points in each generation. In the population-based methods, each subpopulation has its surrogate, and some of the subpopulations can use surrogates for fitness evaluations to reduce computation costs. Recently, considerable progress has been made in improving the SAEC/SIAs search schemes. Lim et al. (2009) used an ensemble model composed of several different surrogates to mitigate prediction error and applied polynomial response surface (PRS) to acquire a smooth function with fewer local minima. Training data for building the surrogates is chosen in the vicinity of each individual, and initial individuals are gradually replaced by higher quality solutions from the proposed surrogates. Liu et al. (2013) developed a surrogate-based evolutionary algorithm (GPEME) for expensive optimization problems with 20–50 design variables. The Gaussian process surrogate model assisted evolutionary algorithm for medium-scale computationally expensive optimization problems (GPEME) utilizes GP to build surrogates and adaptively coordinates the exploitation of surrogates and evolutionary search. Besides, Sammon mapping is used to reduce the design dimension so that GP can generate more accurate surrogates in a low-dimensional space. Regis (2014) introduced an RBF (radial basis function)-assisted particle swarm optimization (PSO) algorithm for 30–36 dimensional problems, where RBF is used to identify the best trial in each swarm, and the present best trial needs to be redefined by a possible trial in its vicinity. Sun et al. (2017) presented the surrogate-assisted cooperative swarm optimization (SA-COSO) method for 50–100 dimensional, expensive optimization problems, in which the surrogate-assisted PSO and social learning-based PSO (SL-PSO) schemes are cooperatively used to search for the global optimum. In SA-COSO, a fitness estimation strategy was also presented to assist the PSO search to generate more promising individuals. Furthermore, Yu et al. (2018) developed a surrogate-assisted hierarchical PSO algorithm (SHPSO) that also combines PSO and SL-PSO to enhance the global and local search, and SHPSO had an impressive performance on 30-, 50- and 100-dimensional cases. Recently, Wang et al. (2019) introduced the novel evolutionary sampling assisted optimization (ESAO) algorithm that builds two surrogate models for global and local searches, respectively. Expensive samples were used to build the global model, while several better individuals were collected to construct the local model. The ESAO has shown excellent performance in the tests using 20–200 dimensional benchmark cases.

This chapter introduces a new search method, called surrogate-assisted gray wolf optimization (SAGWO), which uses RBF to assist the gray wolf optimization (GWO) (Mirjalili et al., 2014) algorithm in solving high-dimensional computationally expensive black-box problems. SAGWO operates in three phases: initial exploration, RBF-assisted metaheuristic exploration and knowledge mining on the RBF model. In the initial exploration phase, a group of well-distributed samples is generated using the design of experiments (DoE) to roughly approximate the high-dimensional design space, and the original wolf pack and leaders are sequentially identified. Furthermore, knowledge mining on the RBF model combines global search using GWO and multi-start local search around promising regions. During the RBF-assisted metaheuristic exploration phase, the predictive information from the RBF model is utilized to guide the generation of wolf leaders in each iteration, and the positions of the wolf pack dynamically change following the wolf leaders, thus achieving a balance between global exploration and local exploitation.

11.2 GRAY WOLF OPTIMIZATION

Since GWO was presented by Mirjalili et al. (2014), the method has received considerable attention and has been successfully applied in various engineering applications. For example, Sánchez et al. (2017) proposed a gray wolf optimizer for modular granular neural network (MGNN) that was applied to human recognition. Compared with other algorithms, GWO could find the optimal architecture parameters of MGNN more efficiently. Rodríguez et al. (2017) proposed a new hierarchical transformation operator with five variants in the hunting process of GWO. Through a large amount of tests, they proved that the fuzzy hierarchical operator can maximize the improvement of GWO's performance. Moreover, Majumder and Eldho (2020) utilized an artificial neural network (ANN) to build the surrogate model for the groundwater flow and solute transport processes. The comparative study demonstrated that GWO could successfully identify the optimal solution of the ANN model and had better stability and convergence behavior. In recent years, how to improve GWO and how to apply GWO to solve certain problems have become research hotspots. Due to GWO's high efficiency and strong stability, this chapter expects to draw support from GWO's search mechanism to solve high-dimensional expensive black-box optimization problems.

GWO is a nature-inspired GO algorithm, mathematically describing the gray wolves' social hierarchy and hunting mechanism. In GWO, the wolf pack mainly includes four hierarchies: the fittest solution alpha (α), the second and third best solutions beta (β) and delta (δ), and the others omega (ω). Alpha, beta and delta will guide omega to hunt the prey that is the global optimal solution. Generally, gray wolves will track and encircle the prey before the attack, and the general formulation of the approach is summarized below:

$$D = \left| C \cdot X_p(t) - X(t) \right| \tag{11.1}$$

$$X(t+1) = X_p(t) - A \cdot D \tag{11.2}$$

$$A = 2a \cdot r_1 - a, \quad C = 2 \cdot r_2 \tag{11.3}$$

where $X_p(t)$ refers to the prey's position in the present iteration, r_1 and r_2 are two random vectors, and a is a parameter that linearly decrease from 2 to 0. It is worth noting that A and C are two random factors for exploitation and exploration, respectively. To simulate the hunting behavior mathematically, all the wolves update their positions with the guidance of alpha, beta and delta. The formulas are summarized as follows:

$$\begin{aligned} D_\alpha &= \left| C_1 \cdot X_\alpha(t) - X(t) \right| \\ D_\beta &= \left| C_2 \cdot X_\beta(t) - X(t) \right| \\ D_\delta &= \left| C_3 \cdot X_\delta(t) - X(t) \right| \end{aligned} \tag{11.4}$$

$$\begin{aligned} X_1 &= X_\alpha - A_1 \cdot (D_\alpha) \\ X_2 &= X_\beta - A_2 \cdot (D_\beta) \\ X_3 &= X_\delta - A_3 \cdot (D_\delta) \end{aligned} \tag{11.5}$$

$$X(t+1) = \frac{X_1 + X_2 + X_3}{3} \tag{11.6}$$

where $X(t)$ is the position of a wolf in the current iteration, $X(t+1)$ is the corresponding new position in the next iteration. X_1, X_2 and X_3 are three updated positions based on the wolf leaders alpha, beta and delta. The random factors C_i and A_i in Eqs. (11.4)–(11.6) are independent.

11.3 SURROGATE-ASSISTED GWO

Surrogate-assisted EC and SI algorithms have shown a superior capability in dealing with higher dimensional, computation-expensive optimization problems, and GWO is a widely used, efficient swarm intelligent GO algorithm. In this chapter, the RBF with a simple structure and very efficient model-building mechanism for high-dimensional problems is used as the surrogate model to assist the search in the GWO algorithm. The specific introduction and expression of RBF can be seen in Chapter 6. The new

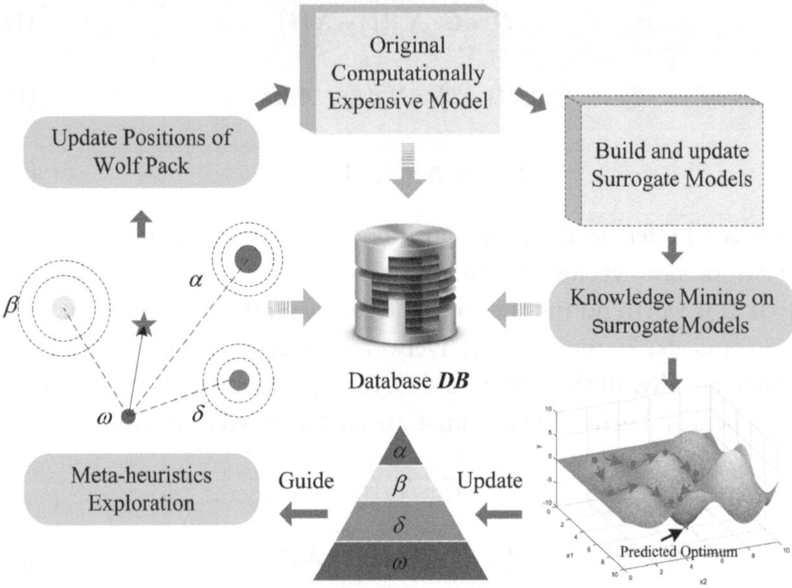

FIGURE 11.1 Surrogate-assisted gray wolf optimization.

SAGWO algorithm integrates RBF and GWO in the optimization loop to explore the high-dimensional design space, as shown in Figure 11.1.

Details of the new search SAGWO method are further illustrated using the following algorithm descriptions.

Algorithm 11.1 Surrogate-Assisted Gray Wolf Optimization

(01) Design of Experiments: Employing Latin Hypercube Sampling (LHS) to choose m wolves $\boldsymbol{S}$ in the design space, evaluating their function values $\boldsymbol{Y}$, and saving them in an archive $\boldsymbol{DB}$.
(02) Initialize Wolf Pack: Sorting $\boldsymbol{S}$ based on $\boldsymbol{Y}$, and choose the top n wolves as the initial Wolf Pack WP_{init}.
(03) Initial knowledge mining on RBF: Training an RBF model by $\boldsymbol{DB}$, getting the predicted best solution f_{pbest} from RBF, and saving it into $\boldsymbol{DB}$.
(04) Generate three best wolves Alpha, Beta, Delta from WP_{init}.
(05) iteration ← 1;
(06) **Repeat**
(07) Run **Surrogate-assisted** meta**heuristic exploration**
(08) Run **Knowledge mining on surrogate models**
(09) Update Alpha, Beta, Delta.

(10) Sort ***DB*** based on function values and find the best sample ***Best*** in ***DB***.
(11) *iteration* ← *iteration* +1;
(12) **Until** the termination criterion is satisfied.
(13) **Return *Best***.

In Algorithm 11.1, a database ***DB*** is created to store expensive samples. In the beginning, $2(d+1)$ sample points are generated using LHS, where top n samples are selected as the initial positions of the wolf pack based on their function values. Here, d refers to the dimension, and n is the size of the wolf pack. The initial knowledge mining is carried out on the linear RBF model to get the predicted best solution and the corresponding function value is calculated to update ***DB***. After the initial alpha, beta and delta are identified from the wolf pack, the entire optimization loop begins. Figure 11.2 shows how the first wolf pack is generated in the initial process. More details on the RBF-assisted metaheuristic exploration and knowledge mining on RBF are provided in the following sections.

11.3.1 Surrogate-Assisted Metaheuristic Exploration

Assume that an experienced wolf coming from other wolf packs or getting special training by more intelligent creatures. Naturally, this experienced wolf may better guide other wolves to hunt prey. As per the previous discussion, RBF can collect the hunting data of the wolf pack in each cycle, and provide an approximate prediction, to generate an "experienced wolf." From Algorithm 11.1 and Figure 11.1, it is clear that the database ***DB*** includes two types of information: the iterative positions of the wolf pack and the predicted samples of RBF. Intuitively, one way to find experienced

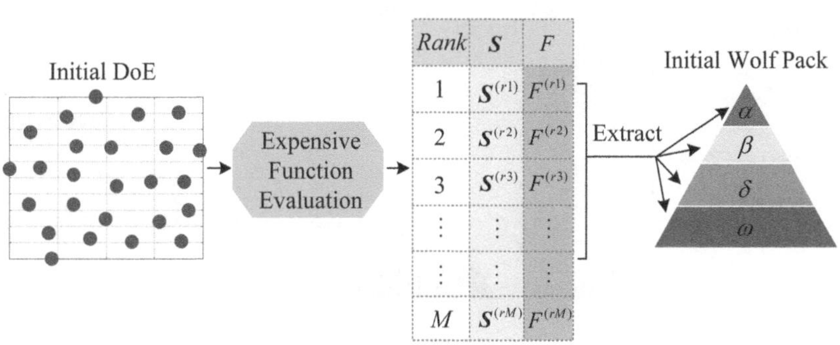

FIGURE 11.2 Generation of initial wolf pack.

leaders for the wolf pack is to choose promising solutions from ***DB*** to update alpha, beta and delta. Equation (11.7) provides the formulation.

$$\begin{aligned}
&\boldsymbol{X}_{\alpha}(t+1)=\arg\min_{x} f(\boldsymbol{x}), \quad \boldsymbol{x}\in\left\{\boldsymbol{WP}\cup \boldsymbol{S}_{rbf}\right\}\\
&\boldsymbol{X}_{\beta}(t+1)=\arg\min_{x} f(\boldsymbol{x}), \quad \boldsymbol{x}\in\left\{\boldsymbol{WP}\cup \boldsymbol{S}_{rbf}-\boldsymbol{X}_{\alpha}(t+1)\right\}\\
&\boldsymbol{X}_{\delta}(t+1)=\arg\min_{x} f(\boldsymbol{x}), \quad \boldsymbol{x}\in\left\{\boldsymbol{WP}\cup \boldsymbol{S}_{rbf}-\boldsymbol{X}_{\alpha}(t+1)-\boldsymbol{X}_{\beta}(t+1)\right\}\\
&\qquad\qquad \boldsymbol{WP}=\bigcup_{i=1}^{t}\boldsymbol{WP}(i), \quad \boldsymbol{S}_{rbf}=\bigcup_{j=1}^{t}\boldsymbol{S}_{rbf}(j)
\end{aligned} \tag{11.7}$$

where $\boldsymbol{X}_{\alpha}(t+1)$, $\boldsymbol{X}_{\beta}(t+1)$ and $\boldsymbol{X}_{\delta}(t+1)$ are the updated alpha, beta and delta, respectively. $\boldsymbol{WP}(i)$ are the positions of the wolf pack in the ith iteration, $\boldsymbol{S}_{rbf}(j)$ are the predicted samples from RBF in the jth iteration, and $f(x)$ is the objective function. From Eq. (11.7), it is easy to find that the new leaders of the wolf pack possess more knowledge that not only comes from the experience of the wolf pack, but also comes from the prediction by the RBF. After the wolf leaders are obtained by Eq. (11.7), Eqs. (11.4)–(11.6) are continuously used to update the whole wolf pack. To make it clearer, an illustration about the data flow of the proposed metaheuristic exploration is shown in Figure 11.3.

Moreover, another way to fuse the wolves' experience and the prediction of RBF is also presented for comparison. The method used in the subsequent experiments is named SAGWO_M. Here, alpha, beta and delta are

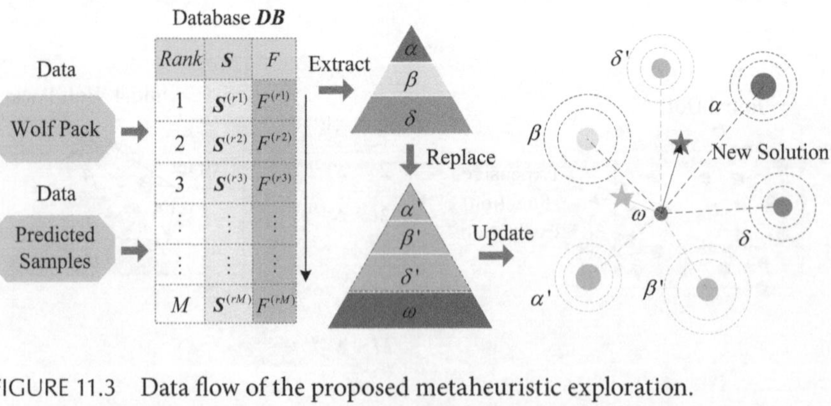

FIGURE 11.3 Data flow of the proposed metaheuristic exploration.

updated using the original way, and the present best solution in ***DB*** is used to guide others, leading to the following formulations:

$$X_{Best}(t) = \arg\min_{x} f(\boldsymbol{x}), \quad \boldsymbol{x} \in \{\boldsymbol{WP} \cup \mathbf{S}_{rbf}\} \tag{11.8}$$

$$\boldsymbol{D}_{Best} = |\boldsymbol{C}_4 \cdot \boldsymbol{X}_{Best}(t) - \boldsymbol{X}(t)| \tag{11.9}$$

$$\boldsymbol{X}_4 = \boldsymbol{X}_{Best} - \boldsymbol{A}_4 \cdot (\boldsymbol{D}_{Best}) \tag{11.10}$$

$$\boldsymbol{X}(t+1) = \frac{\boldsymbol{X}_1 + \boldsymbol{X}_2 + \boldsymbol{X}_3 + \boldsymbol{X}_4}{4} \tag{11.11}$$

where $\boldsymbol{X}_{\text{Best}}(t)$ refers to the present best solution, and $\boldsymbol{C}_4$ and $\boldsymbol{A}_4$ are two independent random factors. To explain the search process of SAGWO, the search steps are listed in Algorithm 11.2. It is worth noting that SAGWO and SAGWO_M have the same optimization flow except that different equations to get $\boldsymbol{X}(t+1)$ and wolf leaders are used.

Algorithm 11.2 Surrogate-Assisted MetaHeuristic Exploration

(01) Update Alpha, Beta and Delta based on (11).
(02) **for** $i \leftarrow 1$ to n (Here, n refers to the wolf pack size)
(03)   **for** $j \leftarrow 1$ to *dim* (Here, *dim* refers to dimension)
(04)   Use (8) to (10);
(05)   On the *j*th dimension, Generate $\boldsymbol{X}_1$ based on the *i*th wolf and Alpha;
(06)   On the *j*th dimension, Generate $\boldsymbol{X}_2$ based on the *i*th wolf and Beta;
(07)   On the *j*th dimension, Generate $\boldsymbol{X}_3$ based on the *i*th wolf and Delta;
(08)   Update the *j*th dimension of the *i*th wolf's position by $\boldsymbol{X}_1$, $\boldsymbol{X}_2$, $\boldsymbol{X}_3$. (Using (11.6))
(09)   **endfor**
(10) Make sure the *i*th wolf's position inside the original range.
(11) **endfor**
(12) Evaluate the function values at the new positions of the wolf pack.
(13) Save all the positions and function values of wolf pack into ***DB***.
(14) Update the RBF model using the samples in ***DB***.
(15) **Return** ***DB*** and an updated RBF model.

11.3.2 Knowledge Mining on Surrogate Models

In general, it is difficult to build a globally accurate surrogate model and it is easier to make accurate predictions in a local trust region. Therefore, this work focuses on a small region around the present best solution using the following formulations:

$$
\begin{aligned}
& \boldsymbol{L}_{b\_local} = \max\left[\left(\boldsymbol{Best}_{pos} - w \cdot (\boldsymbol{U}_b - \boldsymbol{L}_b)\right), \boldsymbol{L}_b\right] \\
& \boldsymbol{U}_{b\_local} = \min\left[\left(\boldsymbol{Best}_{pos} + w \cdot (\boldsymbol{U}_b - \boldsymbol{L}_b)\right), \boldsymbol{U}_b\right] \\
& Local\_region = \begin{bmatrix} \boldsymbol{L}_{b\_local} \\ \boldsymbol{U}_{b\_local} \end{bmatrix}
\end{aligned}
\tag{11.12}
$$

where $\boldsymbol{Best}_{pos}$ is the present best solution, $\boldsymbol{L}_b$ and $\boldsymbol{U}_b$ are the lower and upper bounds of the original design region, and w is a scaling factor. To acquire useful knowledge from the RBF model, a combination search of global optimization and multi-start local optimizations is conducted. The global optimizer is used to get the predicted best solutions $Gbest_{global}$ and $Gbest_{local}$ in the original space and in the local region, respectively. The multi-start optimization process is carried out in the local region to capture the predicted local optimal solutions $Lbest_{local}$. In this algorithm, the gray wolf optimizer is employed as the global optimizer, and the sequential quadratic programming (SQP) is used as the local optimizer.

In the multi-start optimization, several starting points are generated using LHS over the defined region, and local optimization is then conducted using these starting points. After the predictive local optimal solutions are obtained from RBF, a separation distance is used to avoid the obtained points getting too close to the known samples.

In Figure 11.4, the method of multi-start optimization is illustrated using a 1-D example graphically. In the diagram, the darkest black dots represent selected promising solutions, the lightest gray dots correspond to known samples in the database, and the medium gray dots indicate inappropriate local optima (including repeated points and those positioned too close to the known sample points). A multi-start optimization process can find several local optima of a surrogate model, but the method cannot determine which ones are appropriate to be retained. To extract the representative local optimal solutions, eliminate redundant points, and avoid increasing the number of function evaluations, the defined distance given in Eq. (11.13) is used in an iterative process to select promising

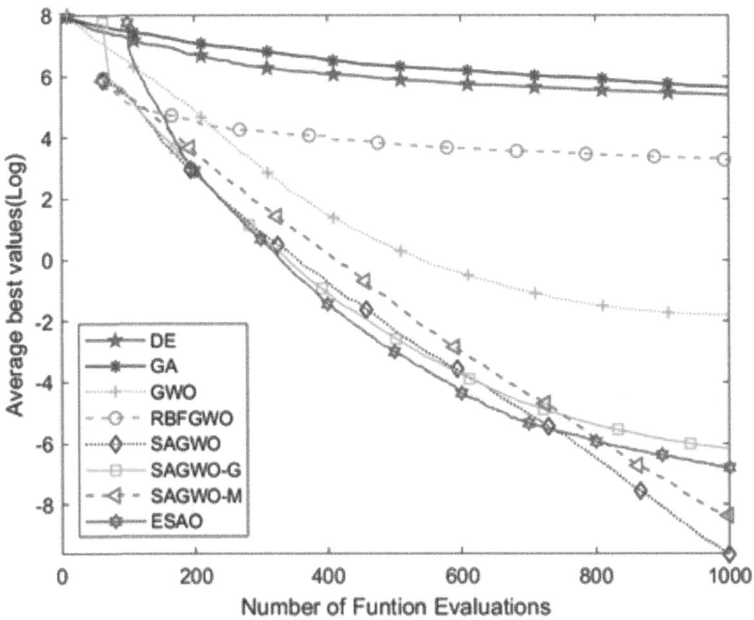

FIGURE 11.4 Demonstration of multi-start optimization search.

optimal solutions. The pseudo-codes for this *knowledge mining* search process are provided in Algorithm 11.3.

$$Dist = \varepsilon \cdot \sqrt{\sum_{i=1}^{dim} \left(U_b(i) - L_b(i) \right)^2} \tag{11.13}$$

where *dim* refers to dimension, and ε is a scaling coefficient.

Algorithm 11.3 Knowledge Mining on Surrogate Models

(01) $\boldsymbol{Best_{pos}}$ ← Acquire the best solution from the database ***DB***;
(02) $\boldsymbol{Gbest_{global}}$ ← Search the original space to get the predicted best solution from RBF by a global optimizer;
(03) Evaluate the function value of $\boldsymbol{Gbest_{global}}$ and update ***DB*** and RBF;
(04) ***Local_region*** ← Create the local search region based on (16);
(05) $\boldsymbol{Gbest_{local}}$ ← Search ***Local_region*** to get the predicted best solution from RBF by a global optimizer;
(06) Evaluate the function value of $\boldsymbol{Gbest_{local}}$ and update ***DB*** and RBF;
(07) ***Dist*** ← Define the separation distance based on (17);

(08) ***Start_point*** ← Generate ***M*** sample points in ***Local_region*** by LHS;
(09) **for** $i \leftarrow 1$ to ***M***
(10) ***Predict_best***$_{local}$ **(*i*)** ← At the starting point ***Start_point(i)***, call a local optimizer to get the *i*th local optimal solution from RBF in ***Local_region***;
(11) **endfor**
(12) ***Temp*** ← Define a temporary variable that initially equals to the sample points in ***DB***;
(13) ***Local***$_{Predict} \leftarrow \varnothing$;
(14) **for** $i \leftarrow 1$ to ***M***
(15) ***Min_Dist*** ← Find the closest point to ***Predict_best***$_{local}$**(*i*)** from ***Temp*** and calculate the minimum distance between them;
(16) **if *Min_Dist*>*Dist***
(17) ***Local***$_{Predict}$ ← ***Local***$_{Predict} \cup$ ***Predict_best***$_{local}$ **(*i*)**;
(18) ***Temp*** ← ***Temp*** $\cup$ ***Predict_best***$_{local}$**(*i*)**;
(19) **endif**
(20) **endfor**
(21) Sort the samples in ***Local***$_{Predict}$ according their RBF values;
(22) **if |*Local*$_{Predict}$|>*Local_sample_num***
(23) ***Lbest***$_{local}$ ← Choose the top ***Local_sample_num*** samples from ***Local***$_{Predict}$;
(24) **else**
(25) ***Lbest***$_{local}$ ← ***Local***$_{Predict}$;
(26) **endif**
(27) Evaluate the function values of ***Lbest***$_{local}$ and update ***DB*** and RBF;
(28) **Return *DB*** and an updated RBF.

Intuitively, the *knowledge mining* process includes global search and local search. In the global search (Algorithm 11.3, Lines 2–3) $Gbest_{global}$ is obtained, and in the local search (Algorithm 11.3, Lines 4–24) $Gbest_{local}$ and $Lbest_{local}$ are identified to refine the RBF model. Specifically, Lines 7–18 of the algorithm describe how the multi-start optimization works, and Lines 19–23 explain how to select the promising samples. The scaling factor w is defined as 0.05, and the scaling coefficient ε is set as 1e-5 in Eqs. (11.12) and (11.13). The algorithm returns the updated database ***DB*** and RBF model that have collected all valuable information.

11.3.3 Optimization Flow

The previous sections discussed the three contributing elements of the new SAGWO algorithm, *initial exploration*, *RBF-assisted metaheuristic*

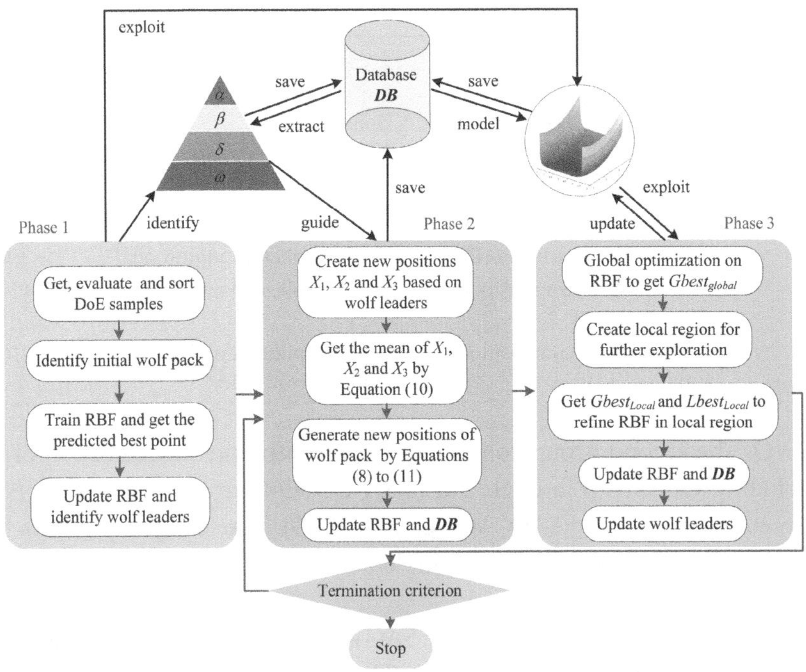

FIGURE 11.5 Flow chart of SAGWO.

exploration and *knowledge mining on RBF*. The database, ***DB***, which stores all expensive sample points, plays as the link to each of these three parts. Initial exploration identifies the initial parameters and carries out an initial global search on the original RBF model. RBF-assisted metaheuristic exploration gives SAGWO effective exploration capability. Knowledge mining on RBF sufficiently exploits the RBF to guide the metaheuristic exploration and accomplishes the balance between exploration and exploitation. The flowchart of the SAGWO algorithm is shown in Figure 11.5.

11.4 EXPERIMENTS AND DISCUSSION

The new SAGWO algorithm is tested using 21 benchmark test cases with 30, 50 and 100 design variables, which have been frequently used for evaluating computationally expensive high-dimensional optimization search algorithms. These include seven representative functions with different characteristics, as listed in Table 11.1. Besides, comparisons of search efficiency and robustness between the new SAGWO algorithm and three groups of other well-known advanced GO search algorithms have been made. The first group includes the well-known EC and SI algorithms, including the genetic algorithm (GA), differential evolution (DE) and

TABLE 11.1 Benchmark Test Functions

Cases	Description	Characteristics	Global Optimum
F1	Ellipsoid	Unimodal	0
F2	Rosenbrock	Multimodal with narrow valley	0
F3	Ackley	Multimodal	0
F4	Griewank	Multimodal	0
F5	Shifted rotated rastrigin (F10)	Very complicated multimodal	−330
F6	Rotated hybrid composition function (F16)	Very complicated multimodal	120
F7	Rotated hybrid composition function (F19)	Very complicated multimodal with narrow valley	10

GWO. The second group consists of the recently introduced SAEC/SIAs, including GPEME (Liu et al., 2013), SA-COSO (Sun et al., 2017), SHPSO (Yu et al., 2018) and ESAO (Wang et al., 2019). The last group covers the SAGWO method with different implementations, including SAGWO_M, SAGWO_G and RBFGWO.

As previously discussed, the SAGWO and SAGWO_M have the same search strategies, except that SAGWO uses Eqs. (11.4)–(11.7) and SAGWO_M uses Eqs. (11.4)–(11.5) and (11.8)–(11.11) to update the wolf leaders and the new positions. The SAGWO_G is nearly the same as SAGWO, but SAGWO_G does not conduct knowledge mining on RBF in its *initial exploration* and does not use a local search strategy in Algorithm 11.3. The RBFGWO does not include metaheuristic exploration, and it just uses the GWO method to produce the best solution from RBF in each cycle.

During the test runs, the number of function evaluations (NFE) that represents the computational cost for a computation-expensive optimization problem is monitored and set to be less than its maximum of 1,000. The population sizes for GA, DE, GWO, SAGWO, SAGWO_M and SAGWO_G are set as 10. In the SAGWO, the number of starting points (*M* in Algorithm 11.3) and the number of sampling (*Local_sample_num* in Algorithm 11.3) in the multi-start optimization are defined as 5 and 2, respectively. Besides, the gray wolf optimizer uses the default parameters as its original paper, while the size of the population is 20 and the number of the generation is set as 500.

The statistical results come from 20 independent runs, and the Wilcoxon rank-sum tests (W-test) were calculated at a significance level of 5%. In the statistical tables, "≈" means no significant difference between the two groups of results, "+" indicates that SAGWO is relatively better, and "−" denotes that SAGWO is worse. Since the statistical results of GPEME,

SA-COSO and SHPSO directly came from the original research references, "*" was used to indicate that their Wilcoxon test results cannot be provided.

Table 11.2 presents the statistical optimization results of the ten algorithms using the 30-variable test examples, and Figure 11.6 shows the

TABLE 11.2 Statistical Results on 30-Dimensional Test Functions

Case	Approach	Best	Worst	Mean	Std	Rank	W-t
F1	ESAO	8.6562e-05	2.7820e-01	2.7470e-02	6.9640e-02	4	*
	SHPSO	4.4782e-02	7.2024e-01	2.1199e-01	1.5229e-01	7	*
	GPEME	1.5500e-02	1.6470e-01	7.6200e-02	4.0100e-02	5	*
	GA	1.7420e+02	5.0388e+02	2.8109e+02	8.9302e+01	10	+
	DE	1.2174e+02	3.3717e+02	2.1891e+02	5.8507e+01	9	+
	GWO	4.1835e-02	3.6898e-01	1.6522e-01	9.0830e-02	6	+
	RBFGWO	6.2568e+00	1.0768e+02	2.6374e+01	2.4171e+01	8	+
	SAGWO	9.8207e-06	3.3038e-04	**6.5846e-05**	7.5113e-05	1	+
	SAGWO_M	2.0920e-05	6.0433e-04	2.3151e-04	1.7282e-04	2	+
	SAGWO_G	2.8334e-04	5.4808e-03	2.1042e-03	1.5143e-03	3	
F2	**ESAO**	2.2158e+01	2.9404e+01	**2.5036e+01**	1.5701e+00	1	*
	SHPSO	2.7726e+01	2.9290e+01	2.8566e+01	4.0441e-01	5	*
	GPEME	2.6262e+01	8.8233e+01	4.6177e+01	2.5520e+01	7	*
	GA	3.7213e+02	1.1200e+03	6.5968e+02	2.0312e+02	10	+
	DE	2.0792e+02	5.5223e+02	3.7956e+02	1.1401e+02	9	+
	GWO	2.8257e+01	3.0637e+01	2.9461e+01	6.9142e-01	6	+
	RBFGWO	8.9374e+01	1.7144e+02	1.2920e+02	2.5974e+01	8	+
	SAGWO	2.6790e+01	2.8826e+01	2.8297e+01	5.1705e-01	2	≈
	SAGWO_M	2.7340e+01	2.8889e+01	2.8454e+01	4.4128e-01	3	≈
	SAGWO_G	2.7493e+01	3.0209e+01	2.8510e+01	6.4083e-01	4	
F3	ESAO	7.8000e-02	3.9096e+00	2.5213e+00	8.3960e-01	6	*
	SHPSO	5.6091e-01	2.9574e+00	1.4418e+00	7.7404e-01	4	*
	GPEME	1.9491e+00	4.9640e+00	3.0105e+00	9.2500e-01	7	*
	GA	1.2686e+01	1.6785e+01	1.4571e+01	1.1448e+00	10	+
	DE	1.1868e+01	1.6831e+01	1.4546e+01	1.3243e+00	9	+
	GWO	9.4736e-01	3.3947e+00	1.8725e+00	6.8009e-01	5	+
	RBFGWO	5.1997e-01	7.9820e+00	4.2738e+00	2.6978e+00	8	+
	SAGWO	7.9048e-14	2.4603e-13	**1.4371e-13**	4.1280e-14	1	+
	SAGWO_M	7.1114e-08	1.2881e-05	3.1803e-06	3.6527e-06	2	+
	SAGWO_G	2.1652e-07	9.0396e-05	1.6106e-05	2.1110e-05	3	
F4	ESAO	7.8600e-01	1.0221e+00	9.5340e-01	5.0370e-02	5	*
	SHPSO	7.0609e-01	1.0275e+00	9.2053e-01	8.8062e-02	4	*
	GPEME	7.3680e-01	1.0761e+00	9.9690e-01	1.0800e-01	6	*
	GA	3.2320e+01	9.6362e+01	6.3395e+01	1.9597e+01	9	+
	DE	4.3282e+01	1.3185e+02	7.1151e+01	2.3785e+01	10	+
	GWO	7.4976e-01	1.2102e+00	1.0177e+00	9.5911e-02	7	+
	RBFGWO	1.9313e+00	9.9980e+00	3.8270e+00	1.8501e+00	8	+
	SAGWO	1.3153e-06	1.3466e-01	**1.5756e-02**	3.1977e-02	1	+
	SAGWO_M	5.9291e-05	1.7021e-01	2.7857e-02	4.4472e-02	3	+
	SAGWO_G	2.5690e-04	5.8268e-02	1.6397e-02	1.7899e-02	2	

(Continued)

TABLE 11.2 (*Continued*) Statistical Results on 30-Dimensional Test Functions

Case	Approach	Best	Worst	Mean	Std	Rank	W-t
F5	ESAO	−3.5780e+01	9.0332e+01	6.3250e+00	2.6477e+01	7	*
	SHPSO	−1.3297e+02	−5.9993e+01	−9.2830e+01	2.2544e+01	4	*
	GPEME	−5.7068e+01	1.8033e+01	−2.1861e+01	3.6449e+01	6	*
	GA	−7.1404e+01	1.4600e+02	1.7739e+01	5.9584e+01	8	+
	DE	8.5727e+01	3.3614e+02	1.7987e+02	6.3066e+01	10	+
	GWO	−3.7374e+01	1.4639e+02	5.3641e+01	5.6215e+01	9	+
	RBFGWO	−1.5782e+02	−5.2813e+01	−9.6542e+01	2.5925e+01	3	+
	SAGWO	−1.7600e+02	−5.8706e+01	**−1.2881e+02**	3.0823e+01	1	
	SAGWO_M	−1.5603e+02	−4.6490e+01	−1.1389e+02	2.5695e+01	2	≈
	SAGWO_G	−1.2844e+02	−2.7194e+01	−7.1915e+01	2.5471e+01	5	+
F6	SHPSO	3.2715e+02	6.4948e+02	4.6433e+02	8.5125e+01	2	*
	GPEME	—	—	—	—	—	*
	GA	4.4815e+02	1.1268e+03	5.9053e+02	1.6047e+02	5	+
	DE	5.7205e+02	9.7630e+02	7.0275e+02	9.9422e+01	8	+
	GWO	3.9666e+02	7.8791e+02	6.2881e+02	1.2028e+02	6	+
	RBFGWO	4.1579e+02	8.0178e+02	6.3440e+02	1.2117e+02	7	+
	SAGWO	3.4843e+02	6.7579e+02	4.8985e+02	1.2882e+02	3	
	SAGWO_M	3.5066e+02	6.6762e+02	**4.3004e+02**	7.4478e+01	1	≈
	SAGWO_G	3.7243e+02	7.1133e+02	5.1102e+02	1.1015e+02	4	≈
F7	**ESAO**	9.2335e+02	9.5389e+02	**9.3167e+02**	8.9417e+00	1	*
	SHPSO	9.2248e+02	9.6363e+02	9.3961e+02	9.0177e+00	2	*
	GPEME	9.3316e+02	9.9286e+02	9.5859e+02	2.5695e+01	3	*
	GA	9.8180e+02	1.2008e+03	1.0565e+03	5.5053e+01	7	+
	DE	1.0485e+03	1.2358e+03	1.1345e+03	4.9333e+01	9	+
	GWO	1.0118e+03	1.1926e+03	1.1048e+03	4.6367e+01	8	+
	RBFGWO	1.1028e+03	1.2123e+03	1.1541e+03	3.1691e+01	10	+
	SAGWO	9.4251e+02	1.0158e+03	9.7323e+02	1.8469e+01	4	
	SAGWO_M	8.8750e+02	1.0190e+03	9.8662e+02	2.9923e+01	5	+
	SAGWO_G	9.6278e+02	1.1059e+03	1.0407e+03	3.9036e+01	6	+

convergence of the programs at different iterations. In Table 11.2, "Rank" is made according to the "Mean" values of the results. The SAGWO outperformed others on F1, F3, F4 and F5; and showed superior performance on F1 and F3, getting close to the global optima after 1,000 NFEs. SAGWO achieved satisfactory results, although the algorithm performed less well than SHPSO on F6 and F7. ESAO performed best on F2 and F7, while SAGWO showed similar capability in these two cases. The W-test showed that SAGWO and SAGWO_M performed similarly on F6, and SAGWO is superior to others on F7. For the pure metaheuristic algorithms, GWO outperforms GA and DE on F1 to F4; and GA showed better performance on F5 to F7. For the SAEC/SIAs, ESAO and SHPSO demonstrated excellent performance on most test cases, but SAGWO outperformed others.

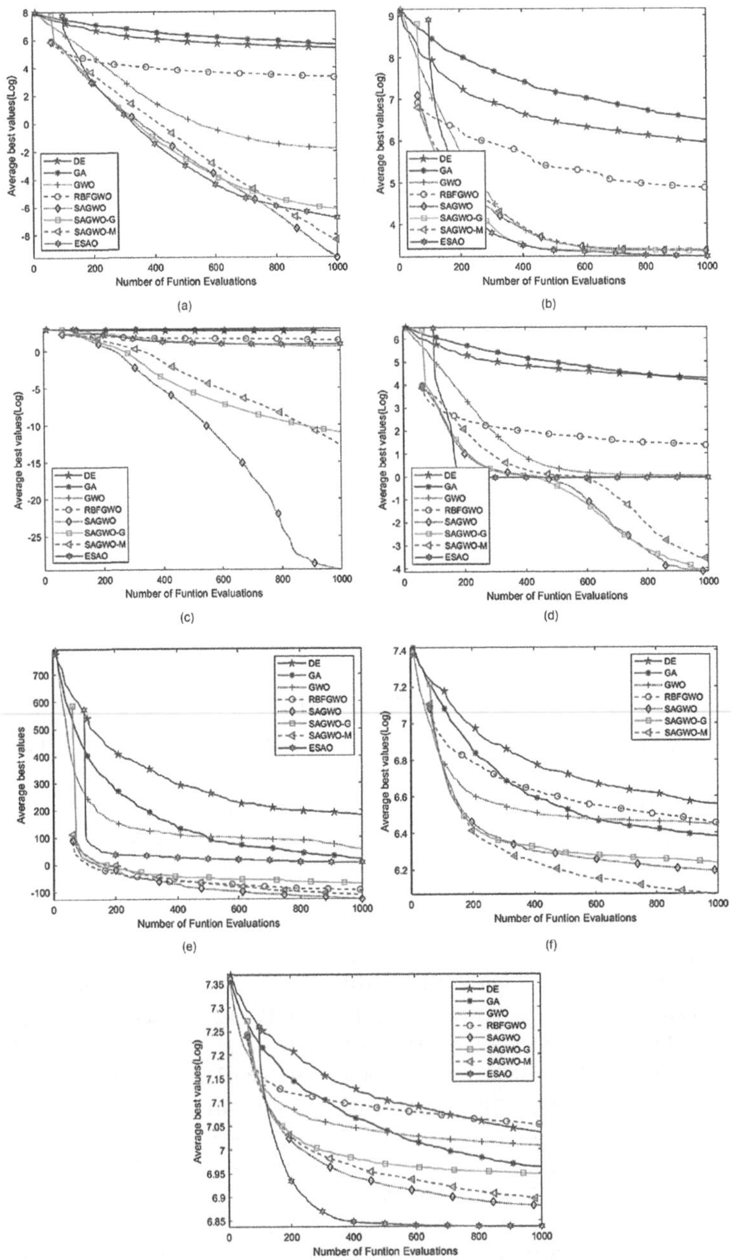

FIGURE 11.6 Iteration graph on 30-dimensional cases. (a) F1 Ellipsoid function. (b) F2 Rosenbrock function. (c) F3 Ackley function. (d) F4 Griewank function. (e) F5 shifted rotated Rastrigin function. (f) F6 rotated hybrid composition function. (g) F7 rotated hybrid composition function.

These tests and results of RBFGWO also showed that pure exploitation of surrogates could not produce satisfactory performance. The combination of metaheuristic exploration and knowledge mining on RBF is more effective in producing an efficient and robust global optimization method. The comparisons of SAGWO and SAGWO_G showed that the introduced local search could improve search efficiency.

In short, based on the tests using the 30-dimensional benchmark examples, SAEC/SIAs could get better results within 1,000 function evaluations, and SAGWO demonstrated top performance among all tested algorithms. Figure 11.6 supports the same conclusion that SAGWO, SAGWO_M and SAGWO_G converge faster.

Table 11.3 presents the statistical optimization results of the 11 algorithms using the 50-variable test examples, and Figure 11.7 shows the convergence of the programs at different iterations. The first group of algorithms, GA, DE, GWO, showed poor performance and appeared to need more function evaluations to get close to the global optima. Among the recently published SAEC/SIAs, ESAO and SHPSO are more efficient and

TABLE 11.3 Statistical Results on 50-Dimensional Test Functions

Case	Approach	Best	Worst	Mean	Std	Rank	W-t
F1	ESAO	1.6460e-01	2.2644e+00	7.3950e-01	5.5490e-01	4	*
	SA-COSO	—	—	5.1475e+01	1.6246e+01	8	*
	SHPSO	—	—	4.0281e+00	2.0599e+00	6	*
	GPEME	1.3407e+02	3.7256e+02	2.2108e+02	8.1612e+01	9	*
	GA	9.3344e+02	2.2346e+03	1.5104e+03	2.8574e+02	11	+
	DE	6.0249e+02	1.4331e+03	1.0032e+03	2.2722e+02	10	+
	GWO	1.5149e+00	6.0444e+00	3.4329e+00	1.1829e+00	5	+
	RBFGWO	5.8383e+00	3.1042e+01	1.3503e+01	5.9945e+00	7	+
	SAGWO	6.8653e-04	1.5296e-02	**4.0117e-03**	3.5801e-03	1	
	SAGWO_M	9.1396e-04	3.9234e-02	1.0930e-02	9.5852e-03	2	+
	SAGWO_G	1.3819e-02	1.5799e-01	5.0418e-02	3.7407e-02	3	+
F2	**ESAO**	4.3122e+01	4.9249e+01	**4.7391e+01**	1.7118e+00	1	*
	SA-COSO	—	—	2.5258e+02	4.0744e+01	8	*
	SHPSO	—	—	5.0800e+01	3.0305e+00	5	*
	GPEME	1.7235e+02	4.0142e+02	2.5828e+02	8.0188e+01	9	*
	GA	1.0121e+03	2.4886e+03	1.7525e+03	3.7181e+02	11	+
	DE	5.8820e+02	1.5955e+03	9.7703e+02	3.0630e+02	10	+
	GWO	5.0603e+01	6.5986e+01	5.5470e+01	4.5469e+00	6	+
	RBFGWO	1.1727e+02	1.6160e+02	1.3764e+02	1.3016e+01	7	+
	SAGWO	4.8349e+01	4.9936e+01	4.9055e+01	4.4925e-01	4	
	SAGWO_M	4.8011e+01	4.9356e+01	4.8813e+01	3.3765e-01	2	≈
	SAGWO_G	4.8368e+01	5.0528e+01	4.8983e+01	4.4391e-01	3	≈

(*Continued*)

TABLE 11.3 (*Continued*) Statistical Results on 50-Dimensional Test Functions

Case	Approach	Best	Worst	Mean	Std	Rank	W-t
F3	ESAO	1.0571e+00	2.4326e+00	1.4311e+00	2.4910e-01	5	*
	SA-COSO	–	–	8.9318e+00	1.0668e+00	8	*
	SHPSO	–	–	1.8389e+00	5.6370e-01	6	*
	GPEME	9.2524e+00	1.4934e+01	1.3233e+01	1.5846e+00	9	*
	GA	1.5595e+01	1.9068e+01	1.7102e+01	7.6469e-01	11	+
	DE	1.4801e+01	1.7466e+01	1.5737e+01	6.7673e-01	10	+
	GWO	2.6962e+00	3.9506e+00	3.5012e+00	3.0424e-01	7	+
	RBFGWO	4.3642e-10	6.8794e+00	1.3882e+00	2.5183e+00	4	+
	SAGWO	2.0735e-11	5.6329e-11	**4.0079e-11**	1.0122e-11	1	
	SAGWO_M	7.3469e-10	2.2275e-05	2.7050e-06	5.7588e-06	3	+
	SAGWO_G	1.3714e-09	3.1482e-06	5.2386e-07	9.7825e-07	2	+
F4	ESAO	8.5180e-01	1.0207e+00	9.4040e-01	4.2090e-02	4	*
	SA-COSO	—	—	6.0062e+00	1.1043e+00	8	*
	SHPSO	—	—	9.4521e-01	6.1404e-02	5	*
	GPEME	2.2546e+01	6.4977e+01	3.6646e+01	1.3176e+01	9	*
	GA	1.5005e+02	2.7782e+02	2.1681e+02	2.8582e+01	11	+
	DE	1.0250e+02	2.3169e+02	1.6610e+02	3.6249e+01	10	+
	GWO	1.2701e+00	3.5371e+00	1.7563e+00	5.3188e-01	6	+
	RBFGWO	1.6733e+00	4.1050e+00	2.4182e+00	7.3815e-01	7	+
	SAGWO	3.4783e-05	2.2988e-01	**2.5573e-02**	5.8155e-02	1	
	SAGWO_M	1.9460e-03	7.6486e-01	9.2928e-02	1.6997e-01	2	+
	SAGWO_G	7.1163e-03	7.6487e-01	2.7410e-01	2.4844e-01	3	+
F5	ESAO	1.1625e+02	2.8909e+02	1.9861e+02	4.5825e+01	5	*
	SA-COSO	—	—	1.9716e+02	3.0599e+01	4	*
	SHPSO	—	—	1.3442e+02	3.2256e+01	3	*
	GPEME	—	—	—	—	—	*
	GA	2.9296e+02	5.6739e+02	4.3421e+02	7.6263e+01	9	+
	DE	5.9319e+02	9.4458e+02	7.7043e+02	1.1676e+02	10	+
	GWO	2.5640e+02	5.6726e+02	4.0821e+02	8.6890e+01	8	+
	RBFGWO	1.8959e+02	3.2630e+02	2.5815e+02	3.2843e+01	7	+
	SAGWO	-1.6634e+01	1.6151e+02	**9.8391e+01**	4.6901e+01	1	
	SAGWO_M	3.4501e+01	1.8412e+02	1.0542e+02	3.8417e+01	2	≈
	SAGWO_G	1.1560e+02	2.5694e+02	2.0888e+02	3.2617e+01	6	+
F6	SA-COSO	—	—	1.0809e+03	3.2859e+01	9	*
	SHPSO	—	—	**4.7438e+02**	4.2029e+01	1	*
	GPEME	—	—	—	—	—	*
	GA	5.5945e+02	7.2480e+02	6.5803e+02	5.0251e+01	5	+
	DE	6.4938e+02	1.0490e+03	8.8082e+02	1.1662e+02	8	+
	GWO	5.7645e+02	1.0145e+03	7.3131e+02	1.1967e+02	7	+
	RBFGWO	5.5029e+02	8.2425e+02	6.6000e+02	6.6359e+01	6	+
	SAGWO	4.3018e+02	5.6424e+02	5.0206e+02	4.5251e+01	2	
	SAGWO_M	3.9391e+02	6.0399e+02	5.1080e+02	6.0870e+01	3	≈
	SAGWO_G	5.0321e+02	7.5871e+02	5.8543e+02	5.7061e+01	4	+

(*Continued*)

TABLE 11.3 (*Continued*) Statistical Results on 50-Dimensional Test Functions

Case	Approach	Best	Worst	Mean	Std	Rank	W-t
F7	**ESAO**	9.4099e+02	1.0499e+03	**9.7532e+02**	3.7110e+01	1	*
	SA-COSO	—	—	—	—	—	*
	SHPSO	—	—	9.9660e+02	2.2145e+01	2	*
	GPEME	—	—	—	—	—	*
	GA	1.0730e+03	1.2872e+03	1.1593e+03	5.2797e+01	7	+
	DE	1.1714e+03	1.3582e+03	1.2741e+03	4.9794e+01	9	+
	GWO	1.1087e+03	1.2296e+03	1.1723e+03	3.4390e+01	8	+
	RBFGWO	9.1022e+02	1.2186e+03	1.1583e+03	8.2127e+01	6	+
	SAGWO	9.1000e+02	1.1320e+03	1.0441e+03	4.0828e+01	3	
	SAGWO_M	1.0251e+03	1.0917e+03	1.0610e+03	1.5866e+01	4	+
	SAGWO_G	1.0940e+03	1.1889e+03	1.1369e+03	2.2134e+01	5	+

performed best on F2, F6 and F7. RBFGWO with just knowledge mining showed slower convergence in most cases. On the other hand, SAGWO showed superior performance on F1, F3 and F4; and SAGWO achieved satisfactory results on all the seven cases. Although SAGWO is ranked second and third on F6 and F7, its results are much closer to the minimum. SAGWO_M also showed good performance, although it was not a match for SAGWO, especially in cases F1 and F3.

The results from SAGWO and SAGWO_G also indicated that the introduced local search strategy in Algorithm 3 played an important role in search efficiency. The iterative curves of SAGWO, SAGWO_M and SAGWO_G descended more quickly, as shown in Figure 11.7. According to the W-test results, SAGWO is good at solving these 50-dimensional cases.

Table 11.4 presents the statistical optimization results of the ten algorithms using the 100-variable test examples, and Figure 11.8 shows the convergence of the programs at different iterations. Compared with DE and GA, GWO had performed better in cases F1–F4 and F7; and for F5 and F6, GA performed better. SAEC/SIAs used fewer function evaluations to get satisfactory results on these 100-dimensional problems. For F6, SHPSO and SAGWO had very close results. However, SAGWO performed much better than SHPSO in all other cases. Similarly, ESAO outperformed SAGWO on F5, but SAGWO was more robust, considering its overall performance.

In these tests, SAGWO_M and SAGWO_G could not be always as efficient as SAGWO, and they showed advantages in some cases. For example, SAGWO_M ranked first on F7, and SAGWO_G ranked first on F2.

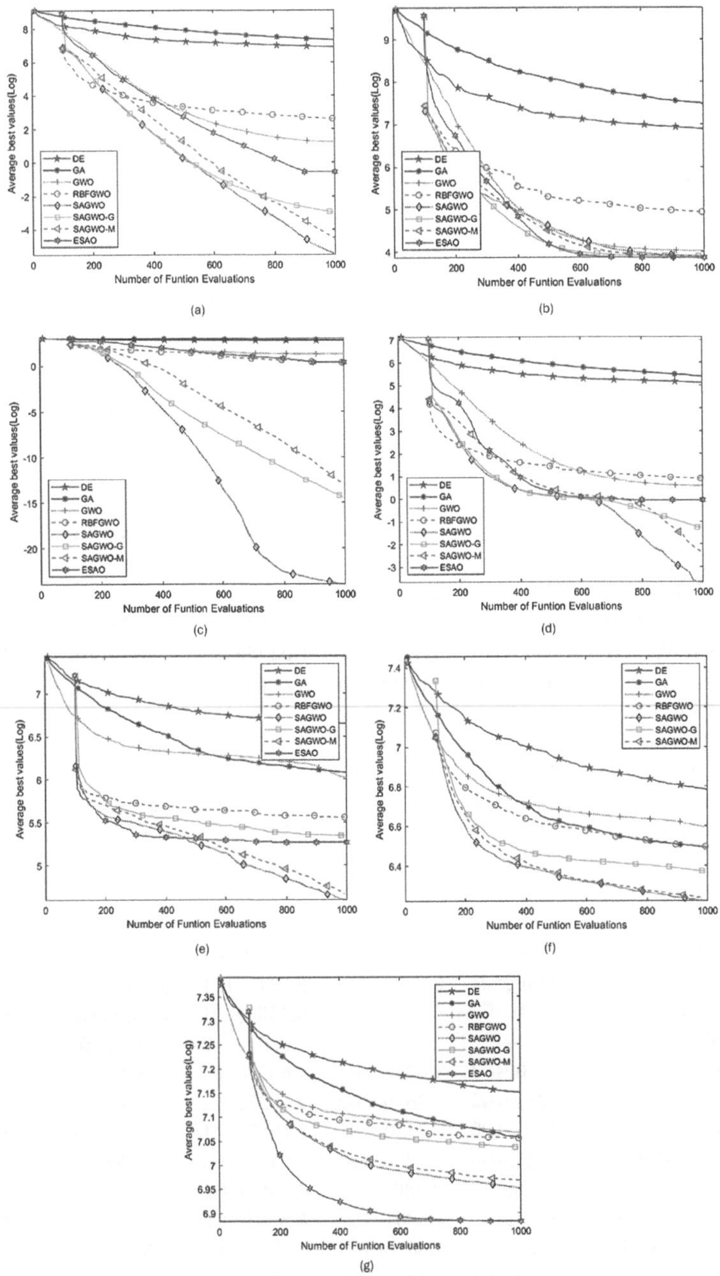

FIGURE 11.7 Iteration graph on 50-dimensional cases. (a) F1 Ellipsoid function. (b) F2 Rosenbrock function. (c) F3 Ackley function. (d) F4 Griewank function. (e) F5 shifted rotated Rastrigin function. (f) F6 rotated hybrid composition function. (g) F7 rotated hybrid composition function.

TABLE 11.4 Statistical Results on 100-Dimensional Test Functions

Case	Approach	Best	Worst	Mean	Std	Rank	W-t
F1	ESAO	1.1023e+03	1.5388e+03	1.2829e+03	1.3439e+02	8	*
	SA-COSO	—	—	1.0332e+03	3.1718e+02	7	*
	SHPSO	—	—	7.6106e+01	2.1447e+01	5	*
	GA	9.6266e+03	1.3324e+04	1.1443e+04	1.1186e+03	10	+
	DE	4.3560e+03	7.7354e+03	5.9378e+03	9.7446e+02	9	+
	GWO	7.8078e+01	2.4459e+02	1.4172e+02	4.7117e+01	6	+
	RBFGWO	9.9654e+00	2.8113e+01	1.4063e+01	4.0008e+00	4	+
	SAGWO	1.6621e-02	3.7119e-01	**1.3996e-01**	9.6807e-02	1	
	SAGWO_M	2.5521e-01	1.4308e+00	6.4491e-01	2.7070e-01	2	+
	SAGWO_G	5.7410e-01	2.1091e+00	1.3740e+00	5.4002e-01	3	+
F2	ESAO	5.2120e+02	6.7324e+02	5.7884e+02	4.4767e+01	7	*
	SA-COSO	—	—	2.7142e+03	1.1702e+02	8	*
	SHPSO	—	—	1.6559e+02	2.6366e+01	4	*
	GA	6.1550e+03	9.6522e+03	8.1846e+03	1.0429e+03	10	+
	DE	1.7335e+03	4.1449e+03	2.9532e+03	5.8400e+02	9	+
	GWO	1.3736e+02	3.5146e+02	2.0982e+02	5.7589e+01	6	+
	RBFGWO	1.5642e+02	2.0095e+02	1.7642e+02	1.2410e+01	5	+
	SAGWO	1.0490e+02	1.4481e+02	1.2338e+02	1.1021e+01	3	
	SAGWO_M	1.0097e+02	1.3276e+02	1.0981e+02	7.6818e+00	2	–
	SAGWO_G	1.0000e+02	1.0658e+02	**1.0228e+02**	1.8874e+00	1	–
F3	ESAO	9.9664e+00	1.0732e+01	1.0364e+01	2.1130e-01	7	*
	SA-COSO	—	—	1.5756e+01	5.0245e-01	8	*
	SHPSO	—	—	4.1134e+00	5.9247e-01	5	*
	GA	1.8575e+01	1.9567e+01	1.9114e+01	2.5621e-01	10	+
	DE	1.5880e+01	1.7640e+01	1.6727e+01	5.0897e-01	9	+
	GWO	4.8145e+00	7.5527e+00	5.7254e+00	6.6842e-01	6	+
	RBFGWO	3.7299e-07	7.1981e-07	5.6679e-07	8.1550e-08	2	+
	SAGWO	3.0570e-08	7.4842e-08	**5.4035e-08**	1.2163e-08	1	
	SAGWO_M	1.6243e-07	1.8590e-06	6.1486e-07	3.8472e-07	3	+
	SAGWO_G	5.2082e-07	1.1662e-06	7.7398e-07	1.9042e-07	4	+
F4	ESAO	4.7346e+01	6.9225e+01	5.7342e+01	5.8387e+00	7	*
	SA-COSO	—	—	6.3353e+01	1.9021e+01	8	*
	SHPSO	—	—	1.0704e+00	2.0485e-02	4	*
	GA	6.8970e+02	1.0325e+03	8.6827e+02	1.0941e+02	10	+
	DE	3.3230e+02	5.2619e+02	4.1035e+02	5.3397e+01	9	+
	GWO	6.0071e+00	1.7320e+01	1.1922e+01	2.7013e+00	6	+
	RBFGWO	1.3520e+00	1.9886e+00	1.5518e+00	1.5690e-01	5	+
	SAGWO	2.0766e-04	2.2883e-01	**2.3993e-02**	5.1906e-02	1	
	SAGWO_M	4.1044e-01	1.0941e+00	8.8984e-01	1.9976e-01	2	+
	SAGWO_G	9.7929e-01	1.0898e+00	1.0394e+00	3.6336e-02	3	+
F5	**ESAO**	6.6263e+02	7.5881e+02	**7.1347e+02**	2.6454e+01	1	*
	SA-COSO	—	—	1.2731e+03	1.1719e+02	7	*
	SHPSO	—	—	8.0173e+02	7.2252e+01	3	*
	GA	1.3010e+03	2.0001e+03	1.6525e+03	1.7493e+02	8	+
	DE	1.7739e+03	2.3571e+03	2.0889e+03	1.3163e+02	10	+
	GWO	1.5030e+03	2.0142e+03	1.7658e+03	1.2086e+02	9	+

(*Continued*)

TABLE 11.4 (*Continued*) Statistical Results on 100-Dimensional Test Functions

Case	Approach	Best	Worst	Mean	Std	Rank	W-t
	RBFGWO	1.0018e+03	1.2626e+03	1.1238e+03	6.4233e+01	6	+
	SAGWO	6.7665e+02	9.1895e+02	8.0016e+02	7.9265e+01	2	
	SAGWO_M	7.0889e+02	1.2225e+03	8.9599e+02	1.1499e+02	4	+
	SAGWO_G	9.8444e+02	1.2294e+03	1.0976e+03	6.0589e+01	5	+
F6	SA-COSO	—	—	1.3657e+03	3.0867e+01	9	*
	SHPSO	—	—	**5.1619e+02**	3.2060e+01	1	*
	GA	6.4216e+02	8.5115e+02	7.0946e+02	5.2281e+01	6	+
	DE	8.7437e+02	1.2478e+03	1.0626e+03	9.1659e+01	8	+
	GWO	6.9914e+02	1.0099e+03	8.3791e+02	7.8673e+01	7	+
	RBFGWO	6.5325e+02	7.6724e+02	6.9796e+02	3.3667e+01	5	+
	SAGWO	4.8201e+02	5.5528e+02	5.1866e+02	2.0540e+01	2	
	SAGWO_M	4.7606e+02	6.3633e+02	5.4038e+02	3.6162e+01	3	+
	SAGWO_G	5.5642e+02	6.6950e+02	6.1328e+02	2.7442e+01	4	+
F7	ESAO	1.3218e+03	1.4271e+03	1.3724e+03	2.7539e+01	4	*
	SA-COSO	—	—	—	—	—	*
	SHPSO	—	—	1.4198e+03	3.8238e+01	6	*
	GA	1.3964e+03	1.5606e+03	1.4760e+03	4.1399e+01	9	+
	DE	1.4037e+03	1.4734e+03	1.4400e+03	2.1206e+01	8	+
	GWO	1.3729e+03	1.4896e+03	1.4306e+03	2.9696e+01	7	+
	RBFGWO	1.3339e+03	1.4079e+03	1.3761e+03	2.2113e+01	5	≈
	SAGWO	9.1015e+02	1.4372e+03	1.3500e+03	1.0747e+02	2	
	SAGWO_M	9.4134e+02	1.4302e+03	**1.3326e+03**	1.1856e+02	1	≈
	SAGWO_G	1.3236e+03	1.4273e+03	1.3634e+03	2.2508e+01	3	≈

Furthermore, the W-test results showed that SAGWO was more capable of solving these 100-dimensional problems.

To better illustrate the results from this comparative study, the performances of all these GO algorithms on the 21 test cases are summarized in Tables 11.5 and 11.6. The SAGWO algorithm won the first place rank (rank 1) most frequently and had the best average rank value of 1.8095. The SAGWO_M algorithm obtained an average rank of 2.5238, a little bit behind the SAGWO, SAGWO_G, SHPSO and ESAO received much closer average ranks. The search methods that only used knowledge mining, like RBFGWO, or only employed metaheuristic exploration, like GA, DE and GWO, had worse average rank values. SAGWO considerably outperformed the GPEME and SA_COSO algorithms in these test cases. Table 11.6 shows the average rank values of all the 11 algorithms on the three groups of test cases. The performance of GPEME declines when the dimension of the GO problem increases. Conversely, RBFGWO performed better regardless of the increase in the problem dimension. SHPSO, GWO,

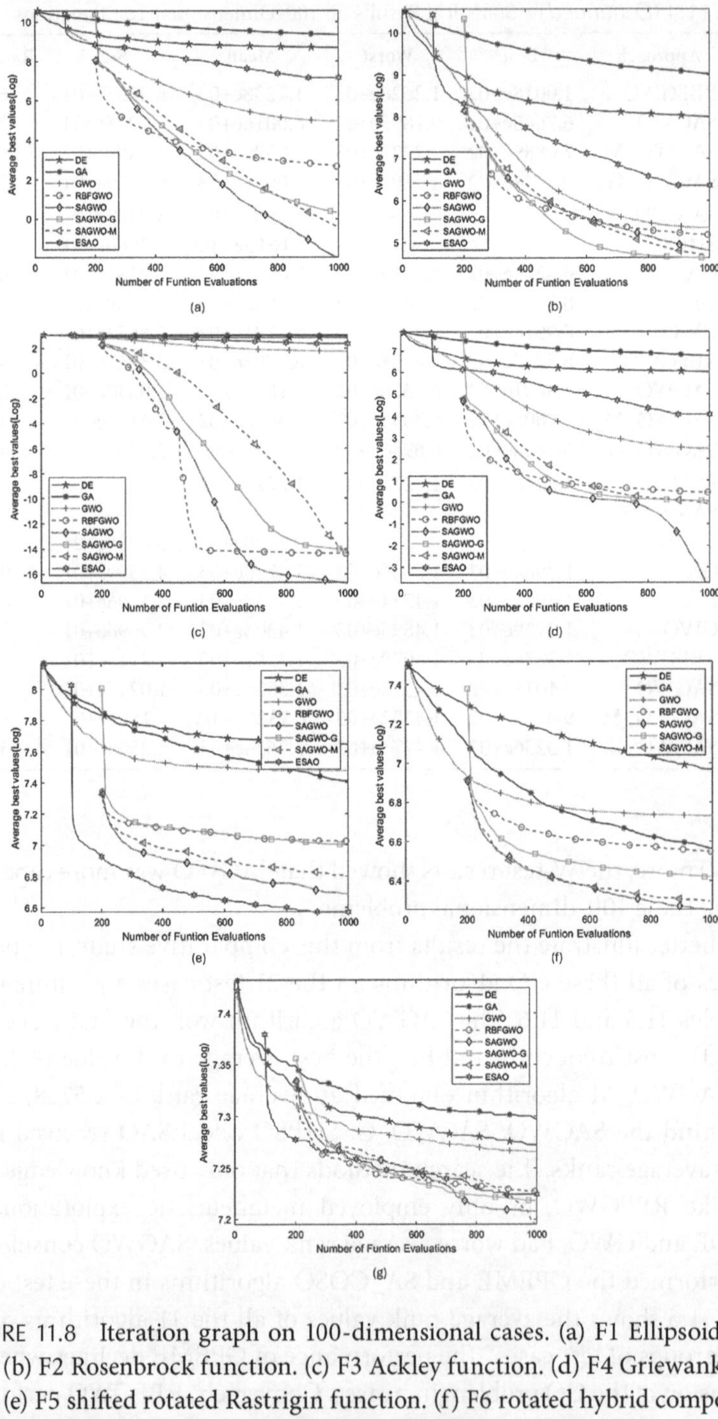

FIGURE 11.8 Iteration graph on 100-dimensional cases. (a) F1 Ellipsoid function. (b) F2 Rosenbrock function. (c) F3 Ackley function. (d) F4 Griewank function. (e) F5 shifted rotated Rastrigin function. (f) F6 rotated hybrid composition function. (g) F7 rotated hybrid composition function.

TABLE 11.5 Summary of Ranks

Algorithms	Cases No.	Sum of Rank	Rank 1 No.	Ave. Rank
ESAO	18	78	5	4.3333
SA-COSO	12	92	0	7.6667
SHPSO	21	84	2	4.0000
GPEME	10	70	0	7.0000
GA	21	187	0	8.9048
DE	21	193	0	9.1905
GWO	21	141	0	6.7143
RBFGWO	21	128	0	6.0952
SAGWO	21	38	11	1.8095
SAGWO_M	21	53	2	2.5238
SAGWO_G	21	76	1	3.6190

TABLE 11.6 Summary of Ranks on Different Cases

Algorithms	Ave. Rank on 30 dim	Ave. Rank on 50 dim	Ave. Rank on 100 dim
ESAO	4.0000	3.3333	5.6667
SA-COSO	NA	7.5000	7.8333
SHPSO	4.0000	4.0000	4.0000
GPEME	5.6667	9.0000	NA
GA	8.4286	9.2857	9.0000
DE	9.1429	9.5714	8.8571
GWO	6.7143	6.7143	6.7143
RBFGWO	7.4286	6.2857	4.5714
SAGWO	1.8571	1.8571	1.7143
SAGWO_M	2.5714	2.5714	2.4286
SAGWO_G	3.8571	3.7143	3.2857

SAGWO, SAGWO_M and SAGWO_G showed stable performance in all three groups of cases.

In the presented SAGWO, the computation complexity mainly consists of five parts, that is, the computation time for initial search, surrogate modeling, function evaluations, global search and local search. In this chapter, we empirically compare the computation time required by these algorithms on the benchmark case ellipsoid. Different numbers of function evaluations and variables are used to form nine cases for comparative study. All the algorithms were implemented on a computer with two 2.40-GHz processors and 32-GB RAM, and the average computation time of 20 runs was summarized in Table 11.7. There is

TABLE 11.7 Average Computation Time of Different Algorithms

Parameters		CPU Time (s) of Different Algorithms						
Dim.	NFE	GA	DE	GWO	RBFGWO	SAGWO	SAGWO_G	SAGWO_M
30d	300	0.555	0.043	0.013	91.34	27.67	9.02	28.73
	600	0.508	0.038	0.014	311.23	87.78	31.28	89.97
	1,000	0.644	0.063	0.018	821.56	226.72	87.78	233.53
50d	300	0.648	0.020	0.007	115.54	54.83	11.01	54.20
	600	0.804	0.042	0.013	436.74	175.06	43.92	180.05
	1,000	0.975	0.075	0.022	1,166.65	428.46	120.80	443.99
100d	300	1.267	0.023	0.010	112.49	80.15	10.89	74.52
	600	1.515	0.047	0.019	660.65	420.17	66.54	419.12
	1,000	1.876	0.078	0.033	1,912.84	1,099.27	195.49	1,125.28

no doubt that the conventional metaheuristic algorithms GA, DE and GWO require less time than the surrogate-based algorithms RBFGWO, SAGWO, SAGWO_M and SAGWO_G. Moreover, when the dimension and NFE increase, the required computation time for GA, DE and GWO still stays at a lower level. On the contrary, these surrogate-based algorithms are dramatically affected by the two factors NFE and dimension. This is because a higher dimension and larger NFE will greatly increase the computation time for surrogate modeling and optimization search on surrogate models. Among these surrogate-based algorithms, SAGWO has performed similar to SAGWO_M, SAGWO_G requires the least computation time and RBFGWO spends the most CPU time. Compared with SAGWO, SAGWO_G lacks the initial search and local search that increase the computation complexity, thus it can run faster. On the other side, since RBFGWO purely exploits RBF to capture new samples per cycle, it needs more calls to the surrogate models. Thus, RBFGWO runs slower and is more sensitive to dimension and NFE. It is worth noting that the computation time for function evaluations can be ignored in this experiment because one run for the mathematical expression takes less than 1e-2 seconds. However, the required time for an actual expensive problem may be several minutes, hours or even days. For the time-consuming engineering problems, the time for running the algorithm itself can be ignored and the total computation cost will mainly come from the NFEs.

In summary, the newly proposed SAGWO algorithm showed superior search efficiency and outstanding robustness on all 21 benchmark test cases; and the algorithm is able to solve high-dimensional, computation-expensive, black-box global optimization problems.

11.5 CHAPTER SUMMARY

In this chapter, a novel RBF-assisted, metaheuristic algorithm, surrogate-assisted gray wolf optimization (SAGWO), for solving high-dimensional, computation-expensive, black-box global optimization problems is presented. The new algorithm conducts the search in three successive phases, *initial exploration*, *RBF-assisted metaheuristic exploration* and *knowledge mining on RBF*.

In the "initial exploration," a group of DoE samples is generated and stored in a database, ***DB***, to capture the overall feature of the design space. After that, the initial wolf pack with better fitness function values is selected from the ***DB***, and the wolf leaders are identified. In the "knowledge mining on the surrogate," the RBF model is dynamically updated and is sufficiently exploited by a dedicated optimization process consisting of a global optimization search and a multi-start optimization search. A small region around the present best solution is also created for the local search to speed up convergence. In the "RBF-assisted metaheuristic exploration," the precious knowledge from RBF is used to assist the generation of wolf leaders that will guide the whole wolf pack to explore the design space.

Representative test cases and published data from four top-rated surrogate-assisted evolutionary algorithms are used for a comparative study in this work to test the functionality and verify the performance of the new SAGWO algorithm. The comparison experiments on 21 test cases, ranging from 30 to 100 design variables showed that the SAGWO has superior computation efficiency and robustness.

For now, SAGWO can be directly used for computationally expensive constrained problems by the penalty-function method. However, when the number of expensive constraints increases, SAGWO may have difficulty in finding feasible solutions by the penalty function. In future work, it is of interest to extend SAGWO's capability to solve high-dimensional optimization problems with multiple costly inequality constraints that are another huge challenge in the engineering optimization field. Moreover, SAGWO will be used for the large-scale smart grid design and shape design of full-parameter blended-wing-body underwater gliders in the next stage.

NOTE

1 Based on "Surrogate-assisted Grey wolf optimization for high-dimensional, computationally expensive black-box problems," published in [Swarm and Evolutionary Computation], [2020]. Permission obtained from [Elsevier].

REFERENCES

Dong, H., Song, B., Dong, Z., & Wang, P. (2016). Multi-Start Space Reduction (MSSR) Surrogate-Based Global Optimization Method. *Structural and Multidisciplinary Optimization, 54*, 907–926.

Dong, H., Song, B., Wang, P., & Dong, Z. (2018a). Hybrid Surrogate-Based Optimization Using Space Reduction (HSOSR) for Expensive Black-Box Functions. *Applied Soft Computing, 64*, 641–655.

Dong, H., Song, B., Wang, P., & Dong, Z. (2018b). Surrogate-Based Optimization with Clustering-Based Space Exploration for Expensive Multimodal Problems. *Structural and Multidisciplinary Optimization, 57*, 1553–1577.

Forrester, A. I. J., & Keane, A. J. (2009). Recent Advances in Surrogate-Based Optimization. *Progress in Aerospace Sciences, 45*(1–3), 50–79.

Goel, T., Haftka, R. T., Shyy, W., & Queipo, N. V. (2007). Ensemble of Surrogates. *Structural and Multidisciplinary Optimization, 33*, 199–216.

Guo, Z., Song, L., Park, C., Li, J., & Haftka, R. T. (2018). Analysis of Dataset Selection for Multi-Fidelity Surrogates for a Turbine Problem. *Structural and Multidisciplinary Optimization, 57*, 2127–2142.

Haftka, R. T., Villanueva, D., & Chaudhuri, A. (2016). Parallel Surrogate-Assisted Global Optimization with Expensive Functions–A Survey. *Structural and Multidisciplinary Optimization, 54*, 3–13.

Hajikolaei, K. H., & Gary Wang, G. (2014). High Dimensional Model Representation with Principal Component Analysis. *Journal of Mechanical Design, 136*(1), 011003.

Lim, D., Jin, Y., Ong, Y.-S., & Sendhoff, B. (2009). Generalizing Surrogate-Assisted Evolutionary Computation. *IEEE Transactions on Evolutionary Computation, 14*(3), 329–355.

Liu, B., Zhang, Q., & Gielen, G. G. E. (2013). A Gaussian Process Surrogate Model Assisted Evolutionary Algorithm for Medium Scale Expensive Optimization Problems. *IEEE Transactions on Evolutionary Computation, 18*(2), 180–192.

Liu, J., Song, W. P., Han, Z. H., & Zhang, Y. (2017). Efficient Aerodynamic Shape Optimization of Transonic Wings Using a Parallel Infilling Strategy and Surrogate Models. *Structural and Multidisciplinary Optimization, 55*, 925–943.

Long, T., Wu, D., Guo, X., Wang, G. G., & Liu, L. (2015). Efficient Adaptive Response Surface Method Using Intelligent Space Exploration Strategy. *Structural and Multidisciplinary Optimization, 51*, 1335–1362.

Majumder, P., & Eldho, T. I. (2020). Artificial Neural Network and Grey Wolf Optimizer Based Surrogate Simulation-Optimization Model for Groundwater Remediation. *Water Resources Management, 34*(2), 763–783.

Mirjalili, S., Mirjalili, S. M., & Lewis, A. (2014). Grey Wolf Optimizer. *Advances in Engineering Software, 69*, 46–61.

Müller, J., Shoemaker, C. A., & Piché, R. (2014). SO-I: A Surrogate Model Algorithm for Expensive Nonlinear Integer Programming Problems Including Global Optimization Applications. *Journal of Global Optimization, 59*, 865–889.

Regis, R. G. (2014). Particle Swarm with Radial Basis Function Surrogates for Expensive Black-Box Optimization. *Journal of Computational Science, 5*(1), 12–23.

Regis, R. G., & Shoemaker, C. A. (2013). A Quasi-Multistart Framework for Global Optimization of Expensive Functions Using Response Surface Models. *Journal of Global Optimization, 56*, 1719–1753.

Rodríguez, L., Castillo, O., Soria, J., Melin, P., Valdez, F., Gonzalez, C. I., Martinez, G. E., & Soto, J. (2017). A Fuzzy Hierarchical Operator in the Grey Wolf Optimizer Algorithm. *Applied Soft Computing, 57*, 315–328.

Sánchez, D., Melin, P., & Castillo, O. (2017). A Grey Wolf Optimizer for Modular Granular Neural Networks for Human Recognition. Computational Intelligence and Neuroscience, 2017(1), 4180510.

Shan, S., & Wang, G. G. (2010). Survey of Modeling and Optimization Strategies to Solve High-Dimensional Design Problems with Computationally-Expensive Black-Box Functions. *Structural and Multidisciplinary Optimization, 41*, 219–241.

Sun, C., Jin, Y., Cheng, R., Ding, J., & Zeng, J. (2017). Surrogate-Assisted Cooperative Swarm Optimization of High-Dimensional Expensive Problems. *IEEE Transactions on Evolutionary Computation, 21*(4), 644–660.

Wang, X., Wang, G. G., Song, B., Wang, P., & Wang, Y. (2019). A Novel Evolutionary Sampling Assisted Optimization Method for High-Dimensional Expensive Problems. *IEEE Transactions on Evolutionary Computation, 23*(5), 815–827.

Yu, H., Tan, Y., Zeng, J., Sun, C., & Jin, Y. (2018). Surrogate-Assisted Hierarchical Particle Swarm Optimization. *Information Sciences, 454*, 59–72.

Index